W9-DEV-191

Retrofitting of Buildings for Energy Conservation

Second Edition

Retrofitting of Buildings for Energy Conservation

Second Edition

Edited by
Milton Meckler, P.E.

Published by
THE FAIRMONT PRESS, INC.
700 Indian Trail
Lilburn, GA 30247

Library of Congress Cataloging-in-Publication Data

Meckler, Milton
Retrofitting of buildings for energy conservation / edited by Milton Meckler. -- 2nd ed.
p. cm.
Rev. ed. of: Retrofitting of commercial, institutional, and industrial buildings for energy conservation. c1984.
Includes index.
ISBN 0-88173-183-8
1. Buildings--Energy conservation. I. Meckler, Milton. II. Retrofitting of commercial, institutional, and industrial buildings for energy conservation.
TJ163.5.B84R47 1994 696--dc20 94-4801
CIP

Retrofitting of Buildings for Energy Conservation, Second Edition, Edited by Milton Meckler.

Published by The Fairmont Press, Inc.
700 Indian Trail
Lilburn, GA 30247

Printed in the United States of America

10 9 8 7 6 5 4 3 2 1

ISBN 0-88173-183-8 FP

ISBN 0-13-148370-6 PH

Distributed by PTR Prentice Hall
Prentice-Hall, Inc.
A Paramount Communications Company
Englewood Cliffs, NJ 07632

Prentice-Hall International (UK) Limited, London
Prentice-Hall of Australia Pty. Limited, Sydney
Prentice-Hall Canada Inc., Toronto
Prentice-Hall Hispanoamericana, S.A., Mexico
Prentice-Hall of India Private Limited, New Delhi
Prentice-Hall of Japan, Inc., Tokyo
Simon & Schuster Asia Pte. Ltd., Singapore
Editora Prentice-Hall do Brasil, Ltda., Rio de Janeiro

Table of Contents

CONTRIBUTORS

James M. Archibald, Vice-President, Smith Environmental Corporation, Chapter 23.

Morris Backer, P.E., Sr. Vice-President, Bovay Engineers, Inc., Chapter 18.

Larry W. Bickle, Ph.D., P.E., The Bickle Group, Chapter 11.

Kenneth M. Clark, P.E., Principal, Burns & McDonnell, Chapter 2.

Craig DiLouie, Communications Director, Creative Marketing Alliance, Inc., Chapter 14.

William S. Fleming, Chief Executive Officer, The Fleming Group, Chapter 4.

Francis M. Gallo, Manager, Union Carbide Corporation, Building & Site Operations, Chapter 15.

Michael S. Goodman, The Fleming Group, Chapter 4.

John J. Halas, P.E., Halas Associates, PC, Chapter 16.

Kermit S. Harmon, Jr., P.E., Internal Reserve Board, Chapter 5.

John H. Hirt, Hirt Combustion Engineers, Chapter 22.

Robert C. LeMay, R.C. LeMay Associates, Inc., Chapter 21.

John J. McGowan, CEM, Honeywell, Inc., Chapter 12.

Milton Meckler, P.E., President, The Meckler Group, Chapters 3, 7, 8, 9, 10 and 17.

Paul W. O'Callagham, Garfield Institute of Technology, Chapter 20.

J. Bob Roberson, Southern California Edison Company, Chapter 1.

Donna L. Rybiski, Tenneco, Inc., Chapter 17.

Gideon Shavit, Ph.D., Chief Fellow, Honeywell, Inc., Home and Building Control, Chapter 6.

Fritz A. Traugott, Chairman Emeritus, Robson & Woese, Inc., Chapter 19.

J. Phillip Upton, P.E., Manager, Bovay Engineers, Inc., Chapter 18.

James P. Walts, P.E., CEM, President, Energy Resource Associates, Inc., Chapter 13.

FOREWORD

This second edition of *Retrofitting of Buildings for Energy Conservation* concentrates on the modification, rearrangement, removal or addition to existing energy consuming systems, components, and controls so that the energy consumption is substantially reduced without sacrificing comfort or productivity within the maintained environment. It has been compiled as a comprehensive reference text intended to be used by building owners, managers and operators, investors, architects and design engineers, contractors and plant engineers. Hands-on and operating experience in surveying, auditing, planning, design, construction, and the maintenance of energy-conserving equipment, systems, and processes which have been employed and retrofitted within existing operations, are emphasized. In addition to updating of the previously written chapters, several new chapters in a broad range of topics have been incorporated into the second edition. These chapters cover commercial, institutional and industrial buildings, and other engineered facilities.

While this second edition places special emphasis on practical measures that can be easily implemented, it also deals with innovative techniques and challenging new approaches. The primary intent in preparing this second edition was to bring together as many experts as possible in their respective fields in a collaborative effort to assist architects, engineers, contractors, building owners, managers and operators in reducing energy consumption through state-of-the-art, well-planned and cost-effective retrofitting programs.

With an ever-increasing energy cost, conservation of existing heating, ventilating and air-conditioning (HVAC) systems, elimination of energy waste, improved insulation, and installation of building automation systems are some of the primary options one must definitely employ to provide a practical approach and methodology for the design profession and construction industry. This book addresses a broad range of topics such as energy analysis, historical data, projection of results of proposed new energy-conserving measures, including modeling and computer simulations, evaluation of utility rates, life-cycle and economic analyses and operation and maintenance procedures.

The material presented in this second edition is in four sections. Section 1 (Chapters 1-4) reviews the utility rates, energy audits and economic analysis methods. Section 2 (Chapters 5-10) covers comprehensive design and installation strategies. Section 3 (Chapters 11-14) explores the latest energy management and lighting strategies. This newly revised section includes additional topics such as energy-management techniques for the '90s, demand-side management, and a comprehensive management approach to lighting system upgrade. Finally, Section 4 (Chapters 15-23) concentrates on case studies of retrofit projects for commercial, institutional and industrial facilities including industrial combustion.

Sincere thanks are due to our new and earlier contributors who shared their valuable time and immense knowledge, to make this book possible. Special thanks are also due to Refik A. Sar, Director of Communications at The Meckler Group, who provided invaluable assistance to me in the preparation of this material.

Milton Meckler, P.E., AIC
President
The Meckler Group

Section I

Utility Rates, Energy Audits and Economic Analysis

CHAPTER I

UTILITY RATES AND RETROFIT STRATEGIES

J. Bob Roberson

I. INTRODUCTION

Owners and operators of facilities have adopted the term "retrofit" from the jargon of space technology, where it is used to refer to the solution of unforeseen high-technology problems through the use of new, often innovative technical "fixes," or retrofits. This borrowing of a jargon expression is most appropriate, since more conventional language would have been inadequate under the circumstances.

Most of today's existing inventory of buildings—homes, stores, offices, factories—were built prior to the recognition of what we now call the energy crisis. They were designed and built during a time when energy utilization was seldom even on a list of design constraints. Energy was plentiful, and, more significantly, it was very inexpensive relative to the other costs associated with designing, building, or operating a facility. It is well proved that those halcyon days are past, yet their legacy of an inherently inefficient building inventory remains.

Thus, buildings and processes that were originally designed with very little thought to making them even reasonably efficient users of energy, today in operation yield energy operating costs beyond anyone's definition of reasonable. To be competitive, perhaps to survive, the owner/operator of such facilities must correct the deficiencies of the various energy-using systems—he must retrofit. The reward for energy efficiency, discounting patriotic or moral motivations, is lower operating costs resulting from lowered energy consumption. Consumption and any savings in consumption are measured in therms of gas, or kilowatthours of electricity, or tons of coal, or some other measure whose meaning is equally obscure to the owner/operator. The common denominator, of course, is money—cash. And the translator of such difficult-to-comprehend measures of energy into the universally understood term, money, is the utility rate structure. Utility rates are neither simple nor straightforward, nor do they yield to simple analysis. In spite of this, it is incumbent on the owner/operator to come to grips with these mysterious utility rates and become comfortably familiar with exercising those rates to his benefit. Failure to take this step will likely result in retrofit decisions that may be technical successes, but do not translate through utility rates to the desired common denominator, money saved.

It is the purpose of this chapter, then, to familiarize the reader with the form and types of utility rates along with their rationale, not to dwell on rate specifics. Because, as you will see, utility rates are very sensitive to local climate, geography, utility system configuration, and local/regional politics, and thus are utility-specific.

A. Historical Background

In order to understand and thus deal with utility rate evolution as it is likely to impact the retrofitter of the 1990s, it is helpful to understand the generic concepts that underlie utility rates. We do not need to re-enroll in economic studies in order to deal with retrofit economics. However, the better our understanding of historical and fundamental concepts is, the more easily can we assimilate the implications of innovative, special-purpose rates. Federal impetus to such developments makes them a near certainty for every part of the United States, and less certain but still likely elsewhere. One of the five pieces of legislation making up the National Energy Act, the Public Utility Regulatory Policies Act (PURPA) of 1978, had as one of its principal purposes the conservation of energy by consumers of electricity. To that end, both utility rate structures and load control schemes must be studied in depth.

B. Today's Reality

It is a reality in today's world that a large segment of the population is distrustful of and unsatisfied with its established institutions, "the establishment." This phenomenon extends beyond church and state to "big business," which is exemplified by the utility industry. This unrest, which seems the norm today, is a near perfect backdrop for giving widespread support to critics of utility rates. The common thread of these criticisms is that conventional, historical, cost-based rates are "unfair," or "encourage waste," or "give the wrong price signals." The solution proposed to the "problem" thus identified takes on a wide range of forms, some with wide support, some mere curiosities.

Because traditional cost-based rates are the only approach to utility rate-making for which there is any significant body of experience, all of the many new, innovative, nontraditional, non-cost-based approaches to utility ratemaking are theories—theories begging to be tested. Several such theories have found jurisdictions willing to experiment, and more will follow. Those meeting with success will find quick acceptance in less daring jurisdictions.

Along with nontraditional rate structures, there has come an alternate philosophy to conventional cost spreading. Conventionally and traditionally average costs have been used as the underlying data base. This means that all customers have shared equitably in the average cost of utility plants from the least expensive, turn-of-the-century hydroelectric facility to the most expensive, inflation/energy-crisis-priced nuclear facility. The marginal costing concept has been articulated in such a wide variety of dissimilar specifics that it remains unclear which, if any, of the specific approaches will prevail.

II. BUILDING OPERATIONS

It has often been said (by designers) that inept building operators can defeat the best, most energy-efficient building design. It has also long been recognized that conscientious, intelligent building operators are frustrated by buildings that are incapable of being operated efficiently because of inherently inefficient designs. To solve this latter problem and to ensure that buildings capable of being operated efficiently would become the design norm, a standard to establish minimums of building energy efficiency was established and has been adopted either in the original form or with locally appropriate modifications in many states.

It is less well understood, however, that building operations, efficient or otherwise, have a measurable effect on the utility system, and that this impact translates back to the operator through rates. How, when, and how much building operations impact the utility system is to a great extent dependent on the particular utility system. An even more obscure but nonetheless real relationship has to do with the impact of changes in utility rates on the operators' freedom of action.

A. Impact of the Utility System

Although utility systems are generically quite similar, several utility-specific characteristics are very sensitive to the utility-building interface. Perhaps the most significant and most sensitive of these relationships revolves around the utility generation mix, which is the mix of resources used in the generation of electricity. Attempts to document this important characteristic have met with only limited success. This is so because the scale has been much too large; regional data were used instead of more appropriate and useful utility-specific data.

The capacity of the utility system, its ability to meet the maximum coincident load demanded of it, and the ability of its transmission and distribution circuits to deliver power when and where required, are major constraints that guide the structuring of the rates. It should be fairly obvious, upon reflection, that the above-named constraints are almost always time-sensitive. They are also sometimes seasonal, sometimes influenced by the day of the week, and most frequently affected by the time of day. In order to treat this coincidence time factor, demand rates including the time-of-use and interruptible varieties have been developed. In some utility systems these variables are complicated even further by changes in generation mix on a scheduled or at least on a time-predictable basis.

B. Impact of Utility Rate Changes

For the retrofitter, change in utility rate is perhaps the most ominous of the utility actions that are certain to occur. The change may be as simple as an adjustment to match the inflation-driven rise of fuel costs, or it may be as complex as a basic structure change, such as from an energy-only rate to a demand rate, or from a simple demand rate to a time-of-use (TOU) rate. The point here is that change is inevitable, and that retrofit decision analysis must include an assessment of likely future rate structure changes in the area, their most likely timing, and their most likely magnitude. Adjustments in building operations may mitigate against such changes if done in a timely fashion, and may be reasonable provided the retrofitter took due note of likely forthcoming rate changes at the time of retrofit, and made appropriate provisions for rate change eventualities in the retrofit design. Such provisions, unfortunately, are often easier to describe in the abstract than to evaluate and specify meaningfully in a functional design.

III. TYPES OF UTILITY RATES

Utility rates, which are a useful vehicle for accomplishing a number of unrelated ends, are administered by a wide variety of regulatory jurisdictions with but one thing in common: they are politically controlled. That control may be direct and overt in some jurisdictions, or it may be indirect, subtle, almost invisible in others. But it is always there. Utility rates are too convenient an economic/political tool to be long ignored by most political action groups.

The following discussions will address only a select few of the many types of utility rates that exist today, in the belief that others, and those yet to be proposed, are but variations on, or combinations of, these fundamental types.

A. Commodity Rates

Commodity rates are the simplest and easiest to understand. This is the rate form most familiar to most people. The market advertises melons at 50¢ each, three for $1.00. This is a pure commodity rate with a volume discount. Utilities use such rates more consistently than any other form, and usually with the convention of what are called "descending blocks." For example, a simple two-step gas rate might be $2X per 100 ft^3 for the first 300 ft^3, and $X per 100 ft^3 for everything over the first 300 ft^3. Similarly, because the first block includes the fixed charges associated with such constants as meter reading, bill processing, and so forth, such charges are often spelled out separately as a "customer" charge. In the gas example above, if X were assumed to be $1.00, such a rate might be expressed as $1.50 customer charge, plus $1.50 per 100 ft^3 for the first 300 ft^3, plus $1.00 per 100 ft^3 for everything over the first 300 ft^3. This is a more generally preferred form because the fixed charges are captured independently of the variable charges. A customer utilizing 300 ft^3 or more in a billing period would receive the same billing under either scheme. On the other hand, a customer utilizing only 50 ft^3 during a billing period would receive a bill for only $1.00 from the scheme where customer charges are "buried" in commodity charges, but would receive a bill for $2.25 when called upon to pay a separate customer charge. In electric rates, the same scheme prevails, but the "commodity" is a kilowatt-hour of electricity instead of a cubic foot of gas. This sort of rate is generally applicable only to residential and small commercial applications.

B. Socially Motivated Rates

Occasionally, social objectives are directly addressed through the utility rate structure. Because of the irrationality of the relationship between such goals and energy use, forecasting the impact of such rates, or, even more difficult, forecasting the probability of the future creation of such new rates, is particularly hazardous. The example that most clearly demonstrates this phenomenon is the residential "lifeline" rate. In the lifeline example, the announced but phantom target for "rate relief" was the poor, the elderly, the unemployed, and the disadvantaged of all sorts who were being adversely impacted by rapidly escalating electricity costs. A worthy cause, a documentable problem, a clear need—but the "solution" turned out to be nothing more than a transfer tax. The purpose of the rate, as originally espoused and defended, was almost completely lost in the actual rate. For the retrofitter, such anomalies are probably beyond prediction and seem exempt from rational preventive planning.

C. Demand Rates

Demand rates are perhaps the most misunderstood, or least understood, of the common rate forms. In their most usual form, simple demand rates segregate their charges into three categories: a customer or minimum charge, a commodity charge, and a demand charge. The last charge is the source of the regularly occurring misunderstanding, although it represents a relatively straight-forward concept. Demand may be established in a variety of ways. A typical approach establishes the maximum demand in any month as the measured maximum average kilowatt input during any 15-minute metered interval in the month. Other intervals may be used, of course, such as a 5-minute interval where the demand is intermittent or subject to violent fluctuations. In this approach to the demand rate, the customer charge (minimum charge) is the monthly demand charge, which is $860 for the first 200 kW or less of billing demand, and $4.30/kW for all excess kilowatts of billing demand. That is, if, for example, the monthly billing demand were 300 kW, the demand charge would then be $860, plus (300 – 200) × $4.30 = $430, for a total of $1,290. To extend this example, assume the energy consumption is 135,000 kWh. This particular rate calls for 0.730¢/kWh for the first 150 kWh/kW of billing demand, 0.530¢/kWh for the next 150 kWh/kW of billing demand, and 0.330¢/kWh for the kilowatt-hours over 300 kWh/kW of billing demand. Thus, (150 kWh/kW × 300 kW × 0.730¢) + (150 kWh/kW × 300 kW × 0.530¢) + (150 kWh/kW × 300 kW × 0.330¢) = $328.50 + 238.50 + 148.50 = $715.50. Therefore, for the full example, the total charge would be the sum of the demand charge and the energy charge, $1,290 + $715.50 = $2,005.50.

The application of such a rate in an actual case would be complicated by a number of additions or deletions, such as voltage discount, power factor adjustment, energy cost adjustment, tax change adjustment, and conservation load management adjustment.

D. Time-of-Use Rates

The utility's system peak is built upon the coincidence of many customers' individual demands. One such rate sets a monthly customer charge of $1,075, a demand charge to be added to the customer charge of $5.05/kW for all kilowatts of on-peak billing demand, plus $0.65/kW for all kilowatts of mid-peak billing demand and no charge for all kilowatts of off-peak billing demand, and an energy charge to be added to the demand charge of 0.530¢/kWh for all on-peak kilowatt-hours, plus 0.380¢/kWh for all mid-peak kilowatt-hours, plus 0.230¢/kWh for all off-peak kilowatt-hours. For this example rate, the daily time periods are defined as follows:

On-peak: 12:00 noon to 6:00 P.M. summer weekdays except holidays. 5:00 P.M. to 10:00 P.M. winter weekdays except holidays.

Mid-peak: 8:00 A.M. to 12:00 noon and 6:00 P.M. to 10:00 P.M. summer weekdays except holidays.

Off-peak: 8:00 A.M. to 5:00 P.M. winter weekdays except holidays. All other hours. Off-peak holidays are New Year's Day, Washington's Birthday, Memorial Day, Independence Day, Labor Day, Veteran's Day, Thanksgiving Day, and Christmas.

While the definition of on-peak, mid-peak, and off-peak will be dictated by the characteristics of the local utility system load curves, and the appropriate holidays will vary according to local custom, it can be seen that considerable incentive is provided by such rates to shift the use of energy away from the utility's peak and toward off-peak periods.

E. Interruptible Rates

Some utilities offer a contract rate to the large industrial and/or commercial customer in exchange for the customer's allowing service interruptions during on-peak and mid-peak periods with appropriate notice to the customer prior to the beginning of such an interruption. In the example cited in the reference, not less than a 10-minute notice would be given for the company-controlled interruptible load, and not less than

30 minutes' notice would be given for the customer-controlled interruptible load. This rate takes the form of reductions from the charges established under a standard TOU rate.

IV. TECHNOLOGY—RATE INTERFACE

Retrofit technology covers the full spectrum of the construction disciplines. All of the energy-related retrofits can be equated to utility rates and by that means can express energy savings in the common denominator of money. Some of the techniques are so closely tied to utility rate structures that they deserve special notice here.

A. Demand Control

An important industry has grown up around the technology of controlling systems in a fashion intended to optimize energy utilization. The evolution of miniature, solid state electronics and the digital computer have been the driving forces in this control-industry growth. Such controls may be segregated into user-controlled and utility-controlled categories.

1. User-Controlled Loads. The most common, the most widely and actively promoted innovation in control systems has occurred in the area of user-controlled systems to load and unload in patterns intended to minimize the charges resulting from utility demand rates. This same technology has also been employed to address TOU rates. These products are now firmly established in the marketplace in competition with other products and subject to normal market forces. This development is referred to as the "sustained commercial activity" step in the diffusion of new innovations.

The trick in these control schemes is always to set up an objective and then design systems and select hardware to meet just that objective. Failure to follow this sequence foredooms the control system to almost certain "overkill," which tends to remove, or at least reduce, the economic benefits to be derived through use of the system. Unfortunately, many applications where the unstated objective could have been realized by a simple time-clock installation were rendered ultimately uneconomic by the purchase of elaborate and expensive hardware capable of control far beyond the needs of the basic objective. However, this is not to imply that there are not many economically justified applications for advanced technology control systems. To the contrary, the optimization of control strategies using computer technology is receiving increasing attention.

Also, potential retrofitters should be aware that some electronic systems devote their innovative efforts to "tricking" the utility rate, or more precisely, the utility meter, rather than attempting real load control. Such efforts are ultimately self-defeating as they are met with utility electronics and with reduced and/or floating demand intervals.

2. Utility-Controlled Loads. This technology has been marked by the development and widespread testing of systems for utility control of customer-owned energy utilization hardware. These systems have generally been employed to interrupt or cycle the operation of customer devices such as air conditioners, water heaters, pool circulating pumps, and space heaters on the command of a utility-initiated signal. Such signals may be delivered via carrier, radio, or some hybrid variation and are scheduled on the basis of the needs of the utility system. In theory, and in practice on the basis of as yet limited testing, significant savings in generation capacity requirement may be realized through these systems without any undue inconvenience or discomfort to the controlled customers.

B. Storage

An old, long neglected technology is being recycled in response to the introduction of time-of-use rates. Heat storage utilizing off-peak electricity has had widespread application in some winter-peaking American utility systems, and is a way of life in a number of areas outside the United States, notably Australia and parts of Europe. Yet little has been done with cool storage since the ice bank application for theaters and churches went out of fashion 25 or more years ago. Today, in TOU rate areas, ice, chilled-water, or brine storage systems are making a strong comeback, without a great deal of fanfare. This is proven technology utilizing proven off-the-shelf hardware; only the right price signal was needed to get it dusted off and in use again. It seems clear that as word of these new, economically successful applications spreads, more and more system designers will join the parade. To further that end, the electric utility industry has sponsored research to quantify the interface effectiveness between utility systems and customer cool storage systems.

V. DECISION-MAKING PROCESS

In the case of retrofit decisions, the original opinions are often clouded by overly enthusiastic salesmanship, wishful thinking, inaccurate technical assessment, or, perhaps worst of all, a preconceived decision in search of justification. The scope of this chapter does not extend to a discussion of cost-benefit analysis, life-cycle costing, or engineering economics, but must include some guidance for the use of utility rate information as input into those analysis techniques. Therefore, the following checklist is offered in the hope that its use will assist the potential retrofitter in translating energy savings into dollars:

- Contact the local utility serving the project to become familiar with its rate schedules and with its forecasts of probable rate changes over the expected life-cycle of the retrofit system.

- Either exercise your retrofit alternatives through all reasonable combinations of present rates and probable future rates, or have a professional consultant do so for you. Be sure to assign probability corrections to the various future rate alternatives using the best available opinion.

- Do as much analysis as you can, playing the "what if" game in your analysis. Continue this exercise until the decision resultant of the analysis is self-evident, until you feel comfortable in the decision.

- As much as possible, keep your options open. Allow enough flexibility in your selected systems and hardware to accommodate future rate changes that are not presently thought to be most likely, but which, if they did come to pass, would materially alter the economic consequences of your retrofit system.

- And finally, to reiterate an earlier caution, before alternative systems are considered, before utility rates are studied, before economic analyses are begun, identify clearly and unequivocally the objective of the proposed retrofit. You must know where you are headed before you have any chance of selecting the proper road to get there.

In summary, the retrofitter is urged to become thoroughly familiar with the utility rates that will apply to the project in question. There is no substitute for knowing the rate impact of potential decisions, and that certain knowledge can come only from doing one's homework.

Chapter 2

Managing the Energy Use Survey and Audit

Kenneth M. Clark. P.E.

I. Presurvey Organization
II. Use of Thermography during the Energy Survey
III. Other Important Considerations in the Energy Audit
IV. Obtaining Data for Computer Use
V. Waste Energy Measurement
VI. Economic Aspects of Energy Audits
VII. Survey Process (by Facility)
VIII. Who Should Perform Energy Audits

I. PRESURVEY ORGANIZATION

A systematic approach to the survey or audit is essential. Before the survey or audit is performed, a complete checklist should be made. The alternative is more time-consuming data gathering and the likely omission of some information. It is difficult to obtain all the data needed without overlooking something.

Organization is the key. A checklist of all items to be investigated must be made. Here is a list of items that should be addressed in making an energy audit:

A. Data Gathering

(1) Utility Bills. Utility bills provide a history of the energy consumption in a building. This historical record can be quite informative in evaluating the use of the facility, the growth of the facility, usage of energy, and the cost of energy. It is important to get as many years' past records as possible. At least three years are necessary to obtain a trend in energy consumption. This avoids the use of data from a typical year. Utility bills should include electricity, gas consumption, fuel oil purchases, propane purchases, and water bills.

(2) Lighting Systems. Lighting has a major impact on energy in any plant or facility. A thorough survey should be made to determine what types of lighting are used (i.e., fluorescent, incandescent, HID, etc.). It is also important to determine the efficiency of the lighting system. This requires answering the following questions: Is lighting in the right location for tasks? Were the ballasts removed when the fluorescent tubes were? Is the lighting intensity reasonable? It is also important to obtain the lighting schedule (when the lights are on and off).

(3) Heating and Air Conditioning Systems. Considerable data must be gathered regarding the heating, ventilating, and air conditioning systems. These data should include the capacity of the system, the condition of the equipment, the operating schedule of the equipment, the operation effectiveness of the outside air dampers, and temperature control diagrams and information.

(4) Ventilation and Exhaust Systems. Much heat can escape up the stack, consuming great quantities of energy. It is important to determine the exhaust quantities of energy. In many cases, the exhaust requirements are necessary and essential. However, in some cases the exhaust is not required all of the time or not in the quantities that are being used. When gathering the exhaust data, get the fan size, the motor horsepower, and the rpm if possible. If these data are not available, the manufacturer's nameplate data, model number, etc., should be obtained so that the design capacity of the fan system can be determined.

Data on ventilation systems are also important. Again, the size of the system, the horsepower of the fan, and the manufacturer nameplate data are important information. From these data, it is possible to ascertain just how much ventilation is being used. Also, it is important to determine the condition of the ventilation and exhaust systems. A system in poor condition may or may not be operating at the design air flow quantities. A check of all ventilation system filters is important to determine their condition. Clogged filters can significantly affect the static pressure of the system and, therefore the horsepower required. The condition of the coils in an air-handling unit should also be investigated. If the filters are clogged, there is a good chance that the coils are also in poor condition. An overall system evaluation is important to determine the efficiency and level of maintenance of the system.

(5) Temperature Controls. Heating and air conditioning control systems can significantly affect the energy consumption in a building. The temperature controls should be investigated from two standpoints: (1) the function of the controls and (2) their condition. Temperature control diagrams should be obtained to determine if the temperature controls were installed according to the plans and specs. If this is the case, then proceed with the analysis to determine if the function of the controls is in the owner's best interest. As far as the condition of the controls is concerned, it is important to determine whether the controls are operating as designed. In many cases, building operators adjust their temperature controls to satisfy some particular building needs. Therefore, the outside air quantity, for example, may be substantially different from the design. Other items to look for include dampers not closing tightly or control functions jumpered out of the circuit for the convenience of the operator.

(6) Process Loads. Process loads can play an important function in the building energy picture. Process loads consume energy that must be accounted for and may be a potential heat source for energy recovery. Exhaust quantities from any process should be determined in order to evaluate the potential of recovering this waste heat.

(7) Potential Heat Recovery. Heat recovery can come from two sources: (1) the process loads mentioned above and (2) building exhaust. If a high outside air ventilation rate is required in a plant, there is a good opportunity for recovering heat from this exhaust by preheating incoming outside air. The potential for heat recovery also exists any time there is a simultaneous heating and cooling requirement. That is, one plant process may require cooling while the other process requires some form of heating. This situation could occur over the full range of the temperature spectrum anywhere from a refrigeration process up through steam heating requirements. Obtain all information possible on waste heat and its potential uses.

(8) The Building Shell. The wall and roof construction must be determined, and the size, quantity, and condition of all windows must be tabulated. Also, size, quantity, and types of all overhead doors should be observed including the frequency of opening and length of time that they are open. An overall observation of the condition of the building shell should be made to determine if there are any large cracks or leaks in the building. Thermal leaks from the building can be costly and are less apparent than water leaks.

(9) Building Schedules. The building schedule greatly affects the total amount of energy consumed in a building. First, the number of shifts of operation should be determined. The payback period of energy conservation modification may be significantly faster in a plant operating on three shifts than in a plant operating on only one shift. Also, the energy consumed may be considerably less in one shift of operation.

Second, the occupancy of the building should be determined, and the schedule of the occupancy obtained; ascertain the number of personnel occupying the building during each of the 24 hours of the day.

Lighting Schedules. It is quite possible that the lighting schedule will not follow the occupancy schedule. Cleanup crews at night may require lighting even though the building is not occupied. Skylights in the building may permit the lighting to be shut off during daylight hours.

HVAC Equipment Scheduling. This information needs to be determined not only on a daily basis but on a weekly and seasonal basis. HVAC equipment may be shut down on the weekends, nights, and holidays. Also, the cooling season may be locked out until a certain time of the year; this needs to be verified with the owner.

Process Equipment Schedules. The actual operating schedule for each piece of equipment in the plant needs to be determined. This schedule is important, but it is also important to determine the schedule required for each piece of equipment. Some pieces may be operating 24 hours a day, but they may be required for only a few hours of that day.

(10) Plumbing Systems. The number and types of plumbing fixtures need to be tabulated. The quantity of domestic hot water required must be determined. These data should be compared with utility bills for water. Great discrepancies between the water utility bill and the calculated consumption may indicate either a leak or wasteful consumption.

(11) Fuel Purchase Records. Usually, utility bills are readily available for electricity and natural gas. However, do not overlook the purchase of propane or fuel oil as a standby fuel. These fuels can have a significant impact on the winter energy consumption in the plant, especially if the plant is on an interruptible gas supply.

B. Preparing a Checklist

We have talked primarily about what should be included in the checklist. Some thought needs to be given to the way the checklist should be set up. The purpose of the checklist is for organization before the survey and to minimize omissions during the survey. Therefore, it is important that the checklist be efficient, organized, and an effective reminder of all the items that should be observed during the audit. Several lists may be required. A list of items to be covered when talking with management is helpful. A list of items to be observed while touring the plant and a list of items to be covered with others, such as utility companies, are desirable. Perhaps you would prefer to break down the list by mechanical items, electrical items, etc. The way checklists are prepared will depend upon the number of people making the audit and their areas of expertise.

C. Equipment Required

The sky is the limit on the amount of equipment that can be used in an energy audit. It is nice to have every tool available that you need. However, from the practical side, carrying a suitcase of equipment around can be exhausting. The equipment required will depend somewhat on the type of audit being made. As a minimum requirement, the following equipment is recommended:

(1) Thermometers. A thermometer is needed to measure space temperature as well as air temperature leaving air handler hot and cold decks. Also, thermometers for measuring hot and chilled water temperatures are needed. These instruments may range from a pocket mercury bulb thermometer to an electronic sensing device. The electronic sensing device is handy for measuring air temperatures, plus hot or chilled water temperatures. A sling psychrometer is also valuable to determine the wet bulb for humidity conditions in the air.

(2) Light Meter. Light level readings need to be taken in all occupied spaces. These data are among the easiest to gather, since most readings are taken at eye or desk level.

(3) Tape Measure/Folding Rule. Both these measuring devices are extremely handy during the audit procedure. The tape measure should be a minimum of 50 ft long, and preferably the 100-foot length. A folding rule is essential for measuring shorter distances; it can be handled by one person.

(4) Flashlight. Invariably, the data needed will be on a nameplate in a deep, dark, barely accessible corner. A flashlight is thus an essential tool. It is also important to carry a pocket knife. Invariably, the nameplate data illuminated by the flashlight will have been painted over by maintenance personnel, and the paint will have to be scratched off to get to the data.

(5) Hand Tools. A pair of pliers and a screwdriver can save valuable time. When you are the greatest distance from assistance or tools, the data needed may be behind an access panel mounted with screws or "stuck" wing nuts.

The above list is the minimum equipment required for an effective audit. Several other very helpful items are:

(6) Camera. Dozens of times after returning to the office to begin the analysis, you will kick yourself for not photographing certain items that you thought would be indelibly etched on your mind. Take as many pictures as plant personnel will allow. Be sure to clear all photography with the plant operations people first. Many people are very sensitive about the confidentiality of their processes, etc. Also, avoid taking pictures of plant personnel. This tends to be disruptive to plant operations and, therefore, tends to displease the client.

(7) Walkie-Talkie or Two-Way Radio. This is a handy item if there is more than one person making the survey in a large facility. A two-way radio can also be a valuable tool for observing a piece of equipment that requires someone at a remote location to start and stop it.

(8) Air Flow Meters. Some audited facilities will be in older buildings, where the air flow has probably changed substantially from its original design. Taking data from a set of building plans may give you a false idea of the real energy consumption or the real problems. Therefore, it is valuable to make spot checks of air flow for air measurement readings.

The easiest device to carry in an energy audit is the rotating vane anemometer. This is a small instrument approximately six inches in diameter that measures the air flow over a period of time. This instrument will allow you to take air flow readings across louvers, through doorways, and across filter banks, and also give you a rough idea of the air coming from diffusers. It is an all-around tool, lightweight, compact, and easy to carry.

(9) Volt-Ammeter. This instrument is almost an essential tool. It is easy to take nameplate data, and this may be sufficient for the level of the survey, but, in many cases, the nameplate horsepower on the motor may be significantly higher than the actual operating horsepower. I have reviewed energy audits in which it was assumed that all equipment was running at 100% of nameplate horsepower. A grossly inaccurate picture of the actual energy consumption can be presented in such cases. Voltammeters are handy tools in determining electrical loads.

D. Security Clearance

One of the first things to consider before auditing a facility is the security clearance. An initial meeting with the client should clarify whether special badges are required, or special keys for access to certain areas, or if plant personnel must accompany you at all times. The plant security force should be alerted to the fact that audit personnel are on the premises; they should be told when the audit will take place. It may be necessary to do part of the audit at night, to have special access to the facilities.

On military bases, clearance to certain facilities may be more complicated; it may be necessary to carry a letter of identification at all times. If the energy audit takes place in a multi-building facility, it is a good practice to alert the appropriate people that audit personnel are coming, instead of walking in cold, expecting them to be at your service. Proper identification is essential while making the audit; this may consist of a plant identification badge, a letter of clearance from the client, or your own company identification card. It is essential that proper security arrangements be agreed upon with the client and then adhered to.

E. Meeting with Client

Before the survey is started, meetings should be held with the client to gain as much knowledge about the facility and operation as possible. The following items should be discussed during this initial meeting. If the audit work is being performed entirely in-house, the same information needs to be clarified with management.

(1) Owner's Future Plans. Obtain the owner's projections for future plans at this facility (plans for expansion, production changes, etc.). This information needs to be obtained for near-term and long-term planning. Any contemplated change in plant schedules or number of shifts should be obtained. Records of the plant's increase in production should be obtained, including future projections. Also, gather data on any new equipment or processes the owner anticipates installing or deleting.

(2) Specific Problem Areas. Often the owner will have a particular problem area in his building or systems that should be addressed in the energy review. These problem areas may also suggest some potential energy saving areas.

(3) Scope of Study. It is important that the client and auditor both understand exactly what is to be accomplished, what the energy audit survey is to cover, and what it is not to cover. Also clarify what the report will include or exclude and what details the report will address (i.e., computerized analysis, life-cycle costing, etc.)

(4) Occupancy-Usage Profiles. Obtain schedules from the client on building occupancy and number of personnel by areas. In addition to daily schedules also obtain weekly, monthly, and annual ones. Determine if there are any unusual shutdowns, such as an annual vacation schedule, when the entire plant is shut down at once.

(5) Historical Information. Obtain all the information available about the building or facility: when it was constructed, when additions were added, when modifications were made, what problems have been experienced by the owner, etc.

F. Utility Information

During the energy survey, all the utilities serving the client should be contacted. Some of the information that should be obtained directly from each utility is: (1) the availability of future service to the client, (2) a projection of how long the utility will be able to serve the client, and (3) how much additional capacity can be provided, or, if a reduction in service is anticipated, the expected future curtailments. Obtain price projections in the future, both short- and long-term. Also investigate specific items relating to the client, such as additional transformer capacity, etc.

It is valuable to obtain fuel information for fuel sources not currently being used by the client. For example, standby fuels such as propane, heavy or light oil, etc., might be considered. Coal may be a viable alternate fuel in the future. Information should be obtained on each of these potential fuels, including delivered price, anticipated availability, and problems relating to storage.

G. Time of Survey

When the survey is conducted can have an impact on the data gathered. Obviously, operating conditions are going to be different between the summer and winter months. Some things will be more apparent during winter, such as building air leaks. It is important to remember the seasonal considerations when making the survey. During a summer audit, try to anticipate the conditions that may be different during other seasons of the year. Consult with building operating personnel about variations and operations during other seasons. The audit should include surveys when the facilities are unoccupied as well as occupied. A night-time survey can provide additional information regarding building system operation, lighting levels, etc.

H. Weather Data

Design weather data can be obtained in the ASHRAE handbooks; however, more detailed information may be required, depending on the scope of the study. Check with the client; at the location of his facility, weather conditions may vary from the normal or official weather data for that area.

I. Maintenance Operation

An important consideration in the evaluation of building energy systems is the maintenance factor. Information should be obtained on the type and frequency of maintenance performed. It is important to make a general observation of the condition of equipment and the apparent level of maintenance throughout the facility. Maintenance procedures followed can have a significant impact on energy usage. Some maintenance procedures will be obvious. However, other information will have to be requested, such as chiller tube cleaning, boiler tube cleaning, control systems checked, etc. Information regarding frequency of filter changes is useful.

II. USE OF THERMOGRAPHY DURING THE ENERGY SURVEY

A. General

The use of thermography, or infrared scanning, is becoming more and more popular in the evaluations of building energy systems. Infrared scanning equipment is available at prices ranging from several hundred dollars up to many thousands of dollars. There are many advantages to the use of infrared scanning equipment. One is expressed by the old saying that a picture is worth a thousand words. If the owner can see visually where great quantities of heat energy are lost, that may mean more to him than many pages of data and calculations. Also, infrared scanning can detect heat leaks and problems not visible to the naked eye, such as lack of insulation in walls and roofs.

B. Building Shell

An evaluation of the building shell is probably the most common use for infrared scanning equipment. Whenever a significant temperature difference exists between the ambient air and the building inside temperature, an infrared scan can detect hot spots through the surface. Scanning equipment is more accurate the colder the outside temperature. An area where the insulation has failed will show up as a significant heat difference compared with the adjacent surfaces. The same thing holds true for windows: if the windows are single pane or if they have many leaks around them, these leaks will show up on the infrared scanning equipment. The scanning equipment will also point out heat losses from louvers and other building openings, which may not have been apparent in the past.

C. Mechanical Systems

Infrared scanning equipment can be effectively used in the evaluation of mechanical systems. An infrared camera is a valuable tool in checking the operation of steam traps, especially when it would be difficult to determine if a steam trap were stuck in the open position without disassembling the trap. An infrared scanner can determine this instantly.

A scan of piping insulation is informative. Any failure of the piping insulation can be determined quickly with an infrared camera. A visual inspection of piping insulation systems may not reveal any problems with the insulation, but if it is wet or if the insulation has broken down under the vapor barrier, the thermal effectiveness of the insulation may be nil. The same applies to equipment insulation and boilers, etc. In many equipment and piping systems, it is not uncommon to have aluminum jacketing over the insulation. A failure in this insulation cannot be detected easily without the use of an infrared camera. Inspection of bearings is another good use for the infrared scanning equipment. This is true not just from an energy perspective but from a maintenance standpoint as well. Overheated bearings will show up on the infrared scanner.

D. Electrical Systems

The use of infrared scanning is not limited to mechanical systems. Electrical system evaluation can also be made. The infrared scanning of electrical systems may be done primarily for maintenance reasons rather than for energy conservation. Infrared scanning is a valuable tool for finding "hot" spots in electrical gear and wiring. Such hot spots will appear in gear and wiring on the verge of failure or drawing excessive current. Motors should be checked. Hot spots in motor windings or bearings will reveal potential problems.

III. OTHER IMPORTANT CONSIDERATIONS IN THE ENERGY AUDIT

A. Overview

It is important to get an accurate picture of the energy usage in the building. To do this, you cannot rely solely on opinions of the owner or management personnel or, for that matter, an operating engineer. All their information is very valuable, but you must make your own observations to verify the actual operating conditions. Your observations may verify what the owner's personnel have told you. If that is the case, fine. Otherwise, further evaluation of the problem might be required.

B. Rapport with Client

Another important consideration in the energy audit is the way in which you relate to the client's employees, including both management and plant personnel. At the very outset, use of the word "audit" can have negative connotations to management personnel. This word typically relates to financial investigations where the company's books are evaluated and the financial management ability of the staff is scrutinized. For this reason, it is sometimes better to use the words study or survey rather than audit. It is always good to reassure management personnel that you are not there to pass judgment on the building operation but merely to help them conserve energy and save money, and to lend your expertise to that end.

The same holds true for plant personnel. Plant personnel can become very defensive and cautious with a person walking around with a clipboard, making observations and taking notes. It is not unusual for them to feel that they are being evaluated personally. It is a good policy to let the plant people know what you are doing if management has not already informed them. Let them know that you are friendly, and not the enemy. Valuable data and insight can be obtained from plant personnel on the operation of the facility. However, it is important not to spend much time with them. This takes time away from their duties and can generate a negative reaction from management. It is best merely to state what you are doing and move on. If they have some energy-related information, they will be eager to tell you. You will not have to prod them for information.

IV. OBTAINING DATA FOR COMPUTER USE

Many energy surveys and studies require the use of a computer to evaluate a building or facility, and computer modeling is becoming more and more important in building evaluation. If a computer model is being used, learn all the input data requirements for that computer program before the survey is made. In generating a computer model, some data are required for the input that would not normally be gathered during an energy audit. One of these factors is the building configuration, the physical shape of the building, rather than just the overall wall area. How the structure might shade other portions of the structure is important. Shading from adjacent buildings or trees must be determined on all sides of the building for proper input into a computer program's solar routine. Another important data input requirement is infiltration. Depending on the computer program, there may be several different ways to enter the infiltration data, through the crack method, the air change method, etc. You will need a weather tape for the computer program; therefore, it might be necessary while you are in the area of the facility to obtain information for weather input.

V. WASTE ENERGY MEASUREMENT

A. Exhaust Systems

Substantial quantities of heat will escape through the building's exhaust system, and during an energy audit determining how much heat is escaping through the ventilation and exhaust systems can be a problem. Exhaust through gravity ventilators can be estimated by measuring the throat area, the building height, and the wind velocity and then referring to catalog data. A rotating vane anemometer can also be used in the throat of a gravity ventilator to determine the air flow.

Powered exhaust systems can present a greater problem. In most cases, fan nameplate data must be relied upon to determine air flow quantities. It is usually impractical to attempt air flow measurement on the fan discharge. However, it might be possible to determine air flow quantities if the exhaust fan is connected to a hood or paint spray booth, etc., where you can take air flow measurement readings across the face area of the hood opening or the area associated with the exhaust system intake. Measuring the exhaust gases can be a problem, because of high temperature and contamination. The exhaust may be too hot or dirty for delicate sensing equipment. Look for temperature gauges already installed on equipment or exhaust stacks that can provide the data needed. Usually plant operators will have a good idea of the exhaust temperatures. In an effort to determine the energy utilization in a process system, it may be necessary to rely on plant operators, maintenance people, or engineering personnel for information. Shop drawings and engineering drawings can also be a useful data source.

B. Solid Waste

Solid waste is being increasingly recognized as a potential fuel source. It behooves the energy auditor to make an analysis of the potential use of solid waste. This requires a determination of the types of trash available and the quantities involved. Solid wastes include not only the trash from the plant but also by-products from plant production or processes. Plant personnel can probably provide a good estimate of the quantity of trash hauled away from the site, or the amount incinerated at the plant. If this information is not available, some visual observations can be made of the amount of trash or by-products of production available. Get the help of plant management; they can require that the trash be weighed before dumping, for a better estimate of the quantities involved.

VI. ECONOMIC ASPECTS OF ENERGY AUDITS

Preparing an estimate of the cost of an energy audit involves many factors. The key to minimizing the cost is organization. It is important that you be well organized before making an audit trip to a plant or facility. If the facility is out of town, make arrangements to ensure that all necessary personnel will be available during the audit. If meetings with representatives of the utilities are required, make sure they will be available also. Even if the facility is in town, it is most economical to try to make the survey in one visit. This is not always possible; it will depend upon the complexity of the survey and the size of the facility. If several people are involved in the audit, organize the work into teams before the survey begins in order to make most efficient use of personnel. Before committing your firm to a fee for an energy audit, be aware of travel costs as well as the cost of special equipment that might be required.

Computer modeling will have an impact on job expenses. This can be a significant expense in a large, sophisticated program. If a new, unfamiliar program is used, be careful of the costs. As with any computer program, budget several reruns for errors and modifications. Obviously, the cost of an audit and analysis can vary drastically, depending on the scope of the work. We have found a cost based upon annual energy consumption to be a reasonable gauge. As an average, a fee of approximately 6% of the client's annual energy bill is realistic. Travel expenses and other expenses peculiar to the particular project should be added to that fee. The percentage will vary, depending on the size of the audit; it is usually somewhat higher for a small study and lower for a very large study.

VII. SURVEY PROCESS (BY FACILITY)

A. General

The following is a general outline for an energy audit. Different types of facilities are addressed, after the general outline, and particular requirements for each type of facility are given.

I. Building Envelope Data
  A. Physical survey of the facility
    1. Type of construction
    2. Dimensions
    3. Orientations
    4. Construction materials
    5. Fenestration systems
    6. Shading techniques
  B. Building area
  C. Inspection of building plans

II. Building Systems Data
  A. Type of mechanical systems
    1. Subsystems
    2. Components
    3. Age of equipment
    4. Primary conversion systems
    5. Energy distribution systems
    6. Control systems
  B. Review of shop drawings
    1. Manufacturer's machinery capacities
    2. Input energy requirements
  C. Lighting systems
    1. Lighting fixture locations
    2. Types
    3. Sizes
    4. Switching circuits
    5. Illumination levels
    6. Functions served
    7. Schedules

III. Occupancy-Usage Profile
  A. Population and occupancy schedules
  B. Building utilization data (hours/day, days/week, weeks/year)
  C. Special conditions of usage
  D. Facility use patterns

IV. Operating Practices-Procedures
  A. Identification of intermittent and sequential loads
  B. Equipment operation and schedules
    1. HVAC systems
    2. Setback control systems
    3. Winter heating mode
    4. Summer cooling mode
    5. Maintenance practices
    6. Process loads

V. Equipment
  A. HVAC-plumbing equipment
    1. Equipment inspection
    2. Controls
  B. Electrical equipment
    1. Electrical wiring
    2. Distribution systems
    3. Lighting conditions for each area

VI. Utility Data
  A. Historical bills or invoices
    1. Billing period
    2. Consumption
    3. Demands
    4. Cost
  B. Energy flow study
    1. Type
    2. Quantities
    3. Periods when energy is entering a facility

VII. Weather Data

B. Schools

Some important items to look for in schools are:

(1) Individual classroom controls

(2) Night setback
(3) Air handling systems for the auditorium, gymnasium, and cafeterias (controls and schedules)
(4) Fresh air shut off at night, etc.

C. Hospitals

Hospitals are notorious energy hogs. The energy consumption in a hospital may be several times that of other types of offices and industrial facilities on a per-square-foot basis. This is partly because of the high fresh air requirements. These air requirements, combined with high ventilation rates, restrictions on cross contamination, and humidity requirements, generally mean a high cost for HVAC systems. It is not unusual to find a reheat type system or dual-duct system in a hospital. Investigate the following possibilities for energy savings:

(1) Reduction in 100% outside air requirements for some areas
(2) Eliminating or applying energy conservation controls to reheat systems
(3) Heat recovery (1) Double bundle chillers (2) Exhaust systems
(4) Use of a computerized control system
(5) Building usage schedules (many areas of a hospital are not used at night)

Caution must be exercised in the evaluation of hospitals, because of hospital code requirements.

D. Industrial Plants

Generally, many opportunities for energy conservation exist in industrial plants. Some of the areas for investigation are:

(1) Lighting systems
 (a) Plants with fluorescent or mercury vapor lighting should be considered for conversion to high pressure sodium lighting.
(2) Energy management control systems
(3) Heat recovery systems (from building exhaust or process heat requirements)

E. Office Buildings

Important items to investigate in an office building are:

(1) Scheduling—of people, equipment, HVAC systems, lighting, outside air, etc.

(2) Mechanical equipment rooms
(3) Electrical rooms
(4) Occupied spaces (Determine such things as windows left open, lighting left on in unoccupied areas, or other equipment used in the space.)

F. Military Installations

Many branches of the service are in the process of making extensive energy studies of their facilities, requiring a substantial amount of work. However, there are unusual requirements peculiar only to military installations:

(1) Security (Advance clearance must be obtained before one tries to make an energy audit on a military base.)
(2) Access to buildings (Keys become a real hassle. It is best to prepare a schedule with the facilities engineer before beginning an audit.)
(3) Central plants and utility distribution systems (Type, condition, utilization, age, automation of controls, etc., are assessed.)

G. Residential Buildings

A residence is much smaller, and will require much less time for a survey, compared with commercial or industrial buildings. However, there are still many factors involved in a residential audit:

(a) Insulation (in both the walls and the roof)
(b) Windows (general condition, tightness, and type)
(c) Whether the building is caulked and sealed tightly (The exterior shell is probably the biggest factor in residential energy consumption.)
(d) The furnace and air conditioning units
(e) An observation of types and usage of lights
(f) Overall usage of the residence

VIII. WHO SHOULD PERFORM ENERGY AUDITS

It is important that a person performing an energy audit be well versed in all phases and components of building systems. This requires someone with an engineering background and several years of experience. The individual need not have a professional engineering registration, but should have a great deal of experience and be certified if he or she is not a professional engineer. An energy auditor should understand wall and roof systems, mechanical systems, electrical systems, and control systems, and be able to communicate effectively with clients and the client's employees.

Chapter 3

Economic Analysis of Energy Retrofit Projects

Milton Meckler, P.E.

I. INTRODUCTION

In planning new buildings, there are many ways to conserve energy that are not capital-intensive, but existing buildings often require unique measures that call for additional investment. Since every modernization job is different, a careful study of a mix of system modifications as well as operating or occupancy schedule changes is required to achieve meaningful energy savings.

Determination of the most productive use of such investments, particularly cost effectiveness in generating energy savings, is facilitated by using one or more of the life cycle costing methods proposed here. The uncertainties we face in these changing times clearly expose the weakness of employing traditional construction economics.

II. RETHINKING CONSTRUCTION ECONOMICS

We are entering an age of limits and uncertainties as to growth, the use and probable future costs of energy, and, above all, our expectations. Our past expectations in the building sector have generally been based on available energy and financial resources, provided a project was well planned and located, and a favorable demand in the near term could be demonstrated. Operating costs were predictable and could be expected to follow historical trends, with normal departures for inflation. Future availability of fuels and electricity was taken for granted, following receipt of a commitment letter from the serving utilities.

Economic analysis for project viability also followed one or more of the traditional depreciation cost-to-benefit schemes, adjusted for tax impacts and with primary emphasis on initial cost to maximize owner/developer leverage.

Against this background, a new scenario emerges. It is especially important that building owners look to life-cycle cost (LCC) for better answers when evaluating building design and operations decisions, especially competing energy conservation and solar energy investment opportunities.

An LCC can be thought of as a procurement or decision-making technique that considers operating, maintenance, and other ownership costs over the full life cycle, as well as acquisition or initial costs. An LCC approach provides a more systematic treatment of the full set of relevant costs when considering the development, production, and lifetime ownership of a building. It represents a valid method for evaluating alternative investment opportunities, often in terms of an uncertain or soft approximation of future energy cost, system efficiencies, annualized maintenance or operating cost, and the actual or opportunity cost of capital.

The life-cycle cost analysis (LCCA) is based on relating all costs to present value—simply another way of saying that a dollar in hand today is more valuable than a dollar that may be in hand in the future. Some of the more difficult tasks for the architect and design engineer include selection of realistic criteria, organization of input cost data, and estimating performance of proposed but untried systems.

III. PAYBACK ANALYSIS METHODS

Various investment proposals can be ranked in one of several ways. Use of the simple payback, for example, involves computation of the number of years required to recover the initial investment. Payback analysis can also be improved slightly by modifying the approach to include the effects of debt service. For example, if we find that an initial investment of $30,000 results in an annual operating and maintenance savings of $6,000, to conclude that such a proposal has a simple payback period of five years is to ignore the fact that either interest must be paid on a loan or that there is some opportunity cost for this money because it could be more productively utilized on alternate investments. To solve this deficiency, one may employ the following formula:

$$n = \frac{\log \dfrac{S/rC}{S/rC-1}}{\log(1+r)}$$

where

C = capital cost
S = annual operating and maintenance savings
r = interest rate
n = number of years to achieve payback.

IV. DISCOUNTED CASH FLOW ANALYSIS METHODS

Payback analysis corrected for the cost of monies borrowed still suffers from one serious drawback. Since this method does not account for the time value of money, for example, the effects of discounted cash flow, it is subject to major errors when applied to long-term investment decisions, subject to crossover.

One of the ways to overcome this problem is to employ the internal rate of return (IRR) method, which is based on defining an interest rate (r) that sets the present value of the anticipated future receipts equal to the cost of the initial investment (C) as follows:

$$C = \sum_{t=1}^{N} \frac{R_t}{(1+r)^t}$$

where

N = project's expected life (years)
R_t = net cash flow in the year t
k = cost of capital
C = initial investment cost
r = internal rate of return.

The sum of the discounted receipts is thus made equal to the initial cost of the proposed investment (C) at some interest rate, and this value of r is by definition also the value of IRR.

One can also employ the net present value (NPV) method, which is used to establish the present value of the expected cash flow, for example, annualized energy savings (R_t) from a proposed C, discounted at some estimated cost of capital, for instance, cost of borrowing. This value is computed by substituting k for r in the equation for computing C, given above. To establish NPV, we merely subtract this value from C, the initial investment, required by the proposed system modifications. It must be understood that for it to be considered valid, one cannot obtain a negative value for NPV. When comparing a number of competing alternate investments of equal C, the one yielding the highest value of NPV is selected, and the others can then be ranked in descending order. The same basic relationships are employed in computing IRR and NPV. The principal difference is that the IRR method allows one to evaluate the proposed initial investment in terms of the cost of capital available for such investment. If the value of IRR exceeds the interest cost of the indebtedness, the proposal would be considered profitable. In the event the value of IRR is less than the cost of the indebtedness, the investment would result in losses.

V. BENEFIT COST ANALYSIS

Another approach involves the use of benefit/cost analysis, which can be utilized in two ways: to determine whether it is worthwhile to replace an existing system with a new system, and to determine which of two or more alternative new systems provides the highest cost benefit ratio. For example, to determine whether it would be more advisable to retain existing equipment or obtain new, more efficient equipment, the following factors must be considered:

- Maintenance savings to be obtained by installing the new equipment.
- Operating and energy cost savings to be obtained by installing the new equipment.
- Salvage value of old equipment.
- Capital cost of new equipment, including legal fees, design and professional fees, installation charges, cost of equipment.

- Costs of financing.
- Changes in property taxes as a result of installing new equipment.
- Income tax factors.
- Changes in rental income as a result of installing new equipment.

Assume it can be shown that, on a strict monetary basis, installation of new equipment will result in some savings. Obviously, such measures should be strongly considered. However, one must also take into account the ease of maintenance and possibly the ability to eliminate necessary manpower, lowered number of tenant complaints, or greater reliability.

The so-called benefit/cost analysis (B/C) ratio or profitability index is computed by dividing the present value of future receipts (also discounted at the cost of capital) by the initial investment cost:

$$B/C = \sum_{t=1}^{N} \frac{R_t}{(1+k)^t} \div C$$

Since the numerator is directly equal to the present value of the expected cash flow, all proposals yielding a B/C greater than 1.0 are acceptable. In evaluating several candidate proposals, the one yielding the largest B/C ratio should be selected and the others ranked in descending order. While it is probable that most situations employing either IRR, NPV, or B/C methodology will yield identical rankings, there are some notable exceptions:

- When investment requirements vary significantly among the various alternate investments compared.
- When the direction of cash flow generation varies among the several alternate investments compared.
- When there is wide divergence in the time frame—for example, variations in project life associated with the various alternate investments compared.

VI. EMPLOYING LCCA IN ENERGY DECISIONS

We can now employ the above criteria in escalating energy conservation modifications in buildings, such as energy management or automation programs and systems controlling one or more buildings, and to evaluate economic performance of a proposed solar energy system.

Where energy cost savings are all positive with each of the earlier-described discounted flow methods, the difference is that cash flow yields should not now be considered important. Where possible, project lifetimes for various competing alternate energy conservation investment opportunities should be held the same.

It may be helpful to review briefly some of the strengths and weaknesses of these various discounted cash flow methods. Recall that the IRR method involves an iterative trial and error process, whereas use of NPV or B/C methods is often more straightforward. However, where the cost of capital is believed to be a major factor, the IRR method is often more useful than either the B/C or NPV methods. B/C methods are recommended for analysis in situations where capital is constrained and one is faced with a number of cost-effective energy conservation measures over the same project life. Under these circumstances, it would be possible to rank each of the proposed energy conservation alternatives, starting with the highest-ranked and cumulatively adding them down the list until all available capital had been committed. The NPV method is probably the simplest and most flexible of the recommended methods discussed, although most viable energy conservation proposals will show a positive NPV under the assumed life of the building, and thus make long-term financial sense.

Often, timing may prove poor for the building owner. Under such circumstances, it is usually desirable to establish an acceptable payback period and then compute the NPV for a given set of proposed energy conservation measures for that time period. If the NPV is found to be positive, the building owner's criterion is satisfied, and the proposed energy conservation measures should be dropped and a new NPV computed. This process is usually repeated until a positive NPV is obtained.

VII. APPLYING THE ALGORITHM

Another possible approach is to develop an LCC algorithm. In a recent project, it became necessary to evaluate the NPV benefit resulting from installation of an automated energy management control system (EMCS) system for a major university campus in California. It was determined that this system could be characterized by three main elements of cash flow:

- Initial investment
- Annual maintenance cost attributable to the EMCS system
- Annual energy savings resulting from operation of the EMCS system

If we assume an EMCS system salvage value at the end of useful life to be negligible, there will be no appreciable savings realized in the first year operation due to normal start-up problems and familiarization needs. Since maintenance costs for the first year operation are included as part of the initial EMCS system investment, one may write the following cost algorithm:

$$NPV = S\left[\left(\frac{1+f}{1+i}\right)^2 + \ldots + \left(\frac{1+f}{1+i}\right)^n\right] - C$$

$$\text{minus } m\left[\left(\frac{1+h}{1+i}\right)^2 + \ldots + \left(\frac{1+h}{1+i}\right)^n\right]$$

where

C = EMCS system initial investment ($)
S = annual fuel savings at present cost basis ($/year)
f = annual fuel cost escalation rate (%)
n = expected serviceable lifetime of EMCS system (years)
m = estimated annual add-on maintenance cost, such as outside contract for EMCS system ($/year)
h = annual inflation rate for EMCS system maintenance cost (%)
i = investment discount rate (%)
NPV = net present value benefit for EMCS system.

An Illustrative Example

Employing the NPV algorithm developed above, it was determined that for an EMCS system capital cost of approximately $950,000, an estimated annual fuel cost escalation, annual inflation rate for add-on maintenance, discount rates respectively of 15%, 8%, and 4%, and an estimated EMCS system useful life of 10 years requiring an initial annual maintenance add-on cost of $60,000, that the EMCS system had an effective payback of just under seven years and an NPV of approximately $925,250. This is calculated as follows:

n	$\frac{1+f^n}{1+i}$	$\frac{1+h^n}{1+f}$
1	0	0
2	1.23	1.08
3	1.37	1.13
4	1.52	1.17
5	1.69	1.22
6	1.87	1.27
7	2.08	1.32
8	2.31	1.37
9	2.56	1.42
10	2.84	1.48
Total	17.47	11.46

The effective payback period of seven years was also computed by trial-and-error iteration using the above tabulation and searching for a value of n that makes the NPV equal to zero.

VIII. A BOILER PLANT RETROFIT

When a southern California aerospace company approved the installation of a heat recovery system in its plant, it had every reason to expect a rapid payback. Even though fuel prices dropped precipitously in the period between the original approval and completion of the project, its optimism was rewarded.

The thermal recovery unit employed in the installation is a heat-pipe unit. This counterflow, air-to-air heat exchanger reflects the outward appearance of a conventional plate-fin chilled water or steam coil. It differs, however, in one major aspect. Instead of each tube being connected to another by a return bend or header, each individual tube is a heat-pipe. These heat-pipes operate independently, acting as superconductors of heat.

A heat-pipe is basically a tube that is fabricated with a capillary wick structure. It is evacuated, filled with refrigerant and permanently sealed. Thermal energy applied to either end of the heat-pipes causes the refrigerant at that end to vaporize. The vapor then travels to the other end of the heat-pipe where the thermal energy is removed. As that happens, the vapor condenses back to liquid, thereby giving up the latent heat of condensation. The liquid flows back to the other end of the pipe to be reused, thus completing the cycle.

In this boiler plant retrofit, hot air enters the counterflow air-to-air heat-pipes at a flow rate of 20,290 cfm and 480°F. It leaves with the same flow rate, but its temperature drops to 290°F. On the cold air side, air enters at a flow rate of 18,500 cfm and 105°F. When the air exits on its way to the combustion chamber, it is heated to 313°F.

Actually, boiler exhaust temperatures vary between 450°F and 510°F. Higher steam loads at winter peak demand result in the highest stack temperatures. Inlet air temperature from a stratified layer at the top of the boiler house averages 105°F. This was somewhat higher than the 80°F temperature at ground level originally assumed.

The new forced-draft fans employ a high-efficiency, backwardly curved wheel in a radial-tip fan housing. They are dual-drive equipped with an electric motor. The fans were installed with inlet boxes. Existing motors and inlet damper positioners were retained for this installation (refer to Figures 3-1 and 3-2).

This project involved the installation of a heat recovery system on the separately metered boiler plant that provides process steam. Feasibility studies revealed that the plant's gas consumption averaged 2,768,416 therms per year at an operating cost of approximately $1,635,000. Recovering boiler heat would result in a conservatively estimated 5% annual gas savings. The cost of natural gas was approximately $0.59 per therm. As

Figure 3-1. View of Dual Recuperators Installed in Boiler Room.

Figure 3-2. View of Dual High-Temperature Combustion Air Fans Retrofitted with Electric Motors and Steam Turbine Drivers in Tandem.

the construction started, natural gas prices dropped further to approximately $0.35 per therm. Preheating the combustion air results in substantial fuel savings of about $42,700. A simple payback period of 2.73 years was achieved. Additionally, this southern California aerospace company received a rebate of $23,500 from Southern California Gas Company for its heat recovery program.

CHAPTER 4

RECONCILING ACTUAL RETROFIT PERFORMANCE WITH PROJECTIONS

William S. Fleming
and
Michael S. Goodman

I. INTRODUCTION

Our nation is experiencing an energy crisis that we must live with. Many of our buildings waste energy needlessly and incorporate few, if any, conservation or reclamation features. As engineering consultants, we owe it to our clients and the nation to incorporate energy efficiency into our evaluations for all buildings. We also must be able to present to our clients reasonable projections for potential savings that can be incurred through retrofit construction and operational procedures. Life cycle costing and comparative analysis can be used in evaluations and in selecting retrofit items, and to minimize the costs of energy consumption and ownership of buildings. Marketplace leverage will ensure that manufacturers and vendors of building systems provide to the designer products that are continually improved to maximize efficiency and ease of maintenance. This is beginning to happen already, and better products are being introduced almost daily.

II. TYPES OF ENERGY AUDITS

To project the performance of an item being considered for retrofit, data must be gathered that will allow the engineer to understand how the building operates. The gathering of data is called an energy audit. There are three types of audits, labeled as Class A, B, and C audits.

Class A energy audits consist of:

1. An on-site visit by the energy auditor. This visit may or may not include instrument readings of the building to provide a detailed analysis of the energy consumption, on a system-by-system basis.

2. Evaluation by the auditor of building energy consumption and energy systems. This includes analysis of energy and economic savings likely to result from selected modifications, and information necessary to calculate actual savings resulting from the modifications. This information is usually furnished to the building owner/operator in a report,manual, or handbook.

Class B energy audits consist of:

1. A questionnaire filled out by the building owner/operator.
2. An analysis, by the energy auditor, similar to that of a Class A audit, but based on information supplied by the building owner/operator, and known factors, such as climate conditions and fuel costs.

Class C energy audits consist of:

1. A workbook that enables a building owner/operator to evaluate specific modifications by himself.
2. Information necessary to perform or contract for the modifications that prove economically justifiable.

The degree of complexity for gathering data depends upon the accuracy desired for estimating the potential of retrofit recommendations for saving energy. The Class A audit will provide the best information, but it is the most expensive type to perform. Both the Class A and B audits can be analyzed with the aid of a computer program or manual calculations. The analysis of a building for energy conservation requires a professional who is familiar with energy conservation techniques. These techniques evaluate both the building envelope and the building systems. There are four basic methods of reducing energy consumption in a building:

1. Eliminate unnecessary systems or functions.
2. Reduce consumption of over energized systems.
3. Improve the efficiency of systems and the building envelope.
4. Reduce the consumption of necessary systems.

Reducing energy consumption should be done systematically. An energy audit should be performed before modifications. Items having little or no effect on working conditions should be performed first, and the first energy audit need not necessarily be the last.

III. RETROFIT PERFORMANCE EVALUATION

Many building modifications that may be implemented as a result of recommendations made after the completion of an initial energy audit usually can be performed at little or no cost, and result in energy savings. These modifications do not require detailed energy and economic analysis and are generally the first to be implemented. Of course, any modification that requires large capital expenditures should have energy and economic analyses performed to indicate that the modification will be cost effective.

The energy conservation program should not end with the implementation of operational or design modifications. The operating characteristics of a building change with time. Building remodeling and additions, replacement of large pieces of equipment, energy conservation modifications, replacement of systems, and changes in building personnel all contribute to this change. It is desirable for the building owner to have energy audits performed every few years, especially if a large sum of money has been spent on retrofit construction for energy conservation. It is advantageous to the building owner/operator to keep monthly energy records to evaluate the success of the building changes. In addition, a sudden jump in energy usage can be noticed immediately, and the situation can be corrected. Intensified maintenance programs and other human factors should be checked periodically, or lapses will occur. Such checks enable the owner to determine if the investment is performing as predicted, and also serve as indicators to determine if further action is needed.

The evaluation of a retrofit item does not begin after it has been implemented, but before. In order accurately to evaluate a retrofit item, data representative of the operation of the building before retrofit should be readily accessible. An energy audit is the easiest way to obtain these data. Some items that may be addressed both before and after the retrofit are:

1. Contacting the local utility company for a demand survey.
2. Instrumenting and observing the building over a period of several days.

Energy bills can only provide mean daily energy consumption. A great deal can be learned from an hourly energy consumption profile for occupied and unoccupied days. This is especially true when energy systems with highly variable energy consumption rates are present. To conduct this type of survey, universal recording watt meters are connected to the major electrical systems. In addition, electric, gas, and water meters are read hourly. Building occupancy by the hour is also noted. This type of survey can reveal information not normally obtained, such as:

1. Time of peak demand.
2. Building systems that operate inefficiently. (This may occur only in certain situations.)
3. If a building is used for more than one function, the one that is the most energy intensive.
4. If energy reclamation systems are under consideration, whether storage will be required, and if so, how much.
5. Length, in time, of morning pull-up or pull-down load.
6. Operation of systems at times when they are not required.

The biggest mistake an energy auditor can make is to assume that a building operates as designed. Even if this is true on the day the building opens, it has been found that buildings tend to settle into an operating condition somewhat different from the design condition. Equipment performance can change with age, and the quality and frequency of maintenance. Air dampers leak, insulation or equipment covers may not be properly replaced after repairs have been made, lights become dirty, and weather stripping cracks. Air pollution changes the air flows around a building. Control systems may or may not control the building, even when in perfect working order. Addition or removal of interior partitions can have a considerable effect on building energy consumption. Redecoration of offices can change the amount of light absorbed or reflected by interior surfaces. Desks or filing cabinets may block heat registers.

Since these factors, and innumerable others, can affect building performance, it is wise to investigate as much as possible and assume as little as possible. If performance is estimated, the estimation may be based on the assumption of an efficient operation, and this may not be the case.

A study, completed for the Department of Energy ("Energy Conservation in Office Buildings," Phase 1, DOE contract #EY-76-C-02-2799.000), addressing energy conservation in existing office buildings, reinforces the notion that operational data both before and after the implementation of energy conservation programs should be available. This study indicated that:

> Most building owners or their representatives do not have ready access to or knowledge of the information required for an adequate evaluation of their building's energy consumption. Generally they do not know how much their energy consumption changes from year to year, the quantitative benefits from changes they have implemented or will introduce, and how they compare with similar buildings. For the most part, their decisions with respect to energy conservation appear to be intuitive or based on qualitative assessment.

The study found through questioning various owners that only 10% had monitored their building's energy consumption.

IV. CASE STUDIES

The firm of W.S. Fleming and Associates, Inc., contacted two of its clients who had had detailed and instrumented Class A energy audits performed within the last few years. The clients were asked if they had quantitatively evaluated the operational, envelope, and HVAC modifications that the firm recommended. Both clients had kept records of the performance of their buildings after recommendations were implemented. The following cases explain the data gathering process and the results of implementation.

Case 1. A detailed energy audit was performed by our firm on the Agway Corporate Headquarters building in DeWitt, New York. A variety of methods and resources were utilized to evaluate many alternatives and to make recommendations for retrofit. Hourly and daily energy use profiles were obtained by measurement and observation. Mechanical system operational familiarity came through interviews and discussion with

Agway's plant engineering staff, as well as through observation. Long-term energy use data needed to place results in historical perspective were obtained from Agway personnel, who had kept remarkably complete records for many years. All these data were sifted through, plotted, analyzed, and used to modify and validate a computer simulation model. With this valuable tool, alternative mechanical systems and structure modifications were evaluated, both in yearly energy use and in economic terms.

Of the many ways to reduce energy consumption in the Agway building, only a precious few were practical from the standpoint of cost, ROI, installation, and human comfort. Any energy conservation option must meet the above criteria if it is to be a realistic option. Input for option selection was obtained from the Agway personnel most closely associated with building operation and maintenance. In addition, a mechanical contracting firm was brought to the site to aid in the selection process as well as to provide realistic material and installation prices for energy conservation equipment.

To enhance the accuracy of the audit, a computer simulation was performed. The utility of simulating a building depends upon the end-use requirements of a particular client. A design/build contractor, for instance, may only be interested in the relative performance of building materials or mechanical systems in a hypothetical building. In this case, the program's user is not overly concerned with the accuracy of input data other than for the thermal/mechanical systems he wishes to study.

A building owner's needs are different. The computer simulation is desired to provide management with a decision-making tool to use on a real building. In this case the building was a 15-year-old corporate headquarters building. In our application of a computer simulation, the accuracy of input data was extremely important. Each building is unique in its daily operation, mechanical systems, and equipment quirks. It is the responsibility of those performing the simulation to identify what is unique and different, and accurately input this information into the program. The ultimate purpose of these computer simulations was to measure the impact of alternatives to the current double-duct variable volume HVAC system in terms of energy consumption and economics. Our strategy to save money on operating costs was:

1. Perform a detailed energy audit to determine accurately the current state of total energy use within the existing building.

2. Correlate the results of the energy audit with long-term data that were available to gauge the consistency of energy use on a yearly basis.

3. Input the data into our program to generate a validation of the energy studies. This validation process is the key to studying alternate systems and the economics thereof.

4. Decide what alternate operational procedures, envelope characteristics, or systems are practical from both a retrofit and a new construction standpoint. Input for this step comes from (a) operating personnel, (b) the firm's past experience and research, and (c) consulting mechanical contractors. Because we are an independent firm, we were not tied to any particular manufacturers, utility companies, etc. This allowed us to explore freely any and all system alternatives, including those that required no purchase of equipment, or architectural/engineering design changes.

5. Assemble all data pertaining to systems performance and economics and compare the results of the simulation versus the real case validation run. This was done in confidence because of (a) accuracy of the energy audit, (b) accuracy of the validation run, and (c) correlation of long-term data to the above.

The results of the energy audit, and analysis of long-term data, resulted in the definition of an "average" year for many parameters. After examining five years' worth of energy use at the building, an average year was constructed for total heating gas, total electric consumption, kilowatt demand, and hot water consumption. The results of the validation are indicated in the following table.

Validation Comparison

Parameter	Computer Program	Real Data	Accuracy
Total MBtu for space heating	10053.4	10842.4	92%
Total gas (CCF) for hot water	14490	15130	96%
Total kWh	4696145	4987328	94%
Total kWh/ft^2	28.13	30.57	92%

After the computer model had been validated against the actual operating data, various alternatives were evaluated. In making a choice of retrofit alternatives, several basic factors were recognized and considered:

1. Building's purpose
   (a) Life expectancy of building
   (b) Type of tenant(s) and activities
   (c) Anticipated change in use

2. Geographical location
   (a) Type of climate—heating, cooling, or both required

3. Tenant comfort
   (a) Air temperature control (minimum variation)
   (b) Humidity control
   (c) Air movement control—air changes, ventilation, exhaust
   (d) Sound level

4. Aesthetics—appearance of equipment, inside and outside of building

5. Space limitations for equipment (piping takes up less space than ducting)

6. Costs
   (a) Installed costs
   (b) Operating and maintenance costs

7. Reliability
   (a) Frequency of required service
   (b) Inconvenience caused during service

A variety of alternatives were considered and many recommendations made. Those recommendations that have been implemented since the completion of the energy audit consist of:

1. Modification to the operating procedures of the building plant and systems.
2. Modification of the control system of the variable volume air-handling unit.
3. Installation of an energy management control system.

The energy saving that could be realized from the implementation of the above recommendations was estimated to be approximately 18%, with a payback period of five years. The building manager has kept records of the building's energy consumption since the implementation program was completed. Three full years have elapsed since the program was completed, the manager verifies that the annual energy savings have been approximately 20% or $52,000. Given the professional cost of the energy study and the capital expenditure for the purchase of products, the payback period to Agway, Inc., in present value terms, shall be less than five years.

Case 2. The Marriott Corporation, which had detailed instrumented energy audits performed by our firm for their restaurant division on three of their restaurants (Hot Shoppe, Joshua Tree, and Roy Rogers), has kept data that indicate the differential in energy consumption after certain energy-conserving recommendations had been implemented.

The energy study was divided into two main categories: (1) electric demand and (2) natural gas consumption. Under electrical consumption, subcategories were established and instrumented for:

1. Electrical consumption
2. Refrigeration
3. Kitchen equipment
4. Exhaust fans

Under natural gas consumption, the subcategories investigated were:

1. Domestic hot water
2. Food preparation

Each of the above subcategories was subjected to a series of measurements to allow for an energy consumption by hour analysis. This, in turn, was related to customer count and energy costs.

The data-collecting procedures involved both observation/recording and instrumentation/recording. For example, the domestic hot water (DHW) cycles were measured with a running time meter to obtain the duty cycle (average time on per hour). The quantity of gas consumed was observed and recorded by the hour and for selected one-minute intervals. The electrical load was recorded using a series of six probes for continuous recording plus running time meters and reading the electric meter every hour. Air flow was recorded by an air flow meter for exhaust air volume and infiltration, and water consumption was taken directly from the water meter. These and other measurements/observations were then correlated by time to provide the electric, gas, and water budgets.

After the initial data collection was complete, an energy and economic analysis computer program was utilized to determine the effect various retrofit alternatives had on energy consumption and the economic feasibility of each alternative. Input to the computer program consisted of building envelope description, lighting and equipment energy consumption, operating schedules, weather data, energy costs and escalation rates, estimated equipment costs, etc. The computer simulation was first performed to model the building as it actually existed and validated to within ± 10% of the building's actual consumption. Once the actual model was complete, each of the alternatives was simulated. After the energy consumption of each alternative was determined, the economic analysis was performed to identify what the payback period was, if any.

The following items were recommended for retrofit, based on a low payback period and high return on investment:

1. Ceiling insulation—payback period of three years, return on investment 42%.
2. Set back-set up thermostat system— payback period of two years, return on investment 72%.
3. Electrical demand limiter—payback period of four years, return on investment 33%.

There were other alternatives that were recommended; however, the above alternatives were the first to be implemented in the restaurants. The estimated savings in energy consumption for the above items was 8%. Also implemented in the restaurants, but not included in the retrofit analysis, were recommendations determined from operational observances:

1. Promote a preventive maintenance schedule for all energy-consuming equipment.
2. Promote an energy conservation program for employees.
3. Eliminate unnecessary lighting.
4. Prepare a time on-time off schedule for equipment.

The total estimated savings for the both retrofit and operational recommendations was 15% to 20% of annual energy consumption. The Marriott Corporation had collected data for the year prior to that when the recommendations were implemented and the year when the recommendations were implemented. The range in savings ran from a low of 10% to a high of 28%. The Marriott Corporation is very pleased with the results achieved so far and plans to implement these recommendations in as many restaurants as their yearly budget will allow.

V. CONCLUSION

In summary, the retrofit evaluation program should not end when all the options have been implemented. It must become a permanent part of building operations. Comprehensive maintenance programs and compliance with guidelines by building personnel must be periodically checked. In addition, changes in building occupancy, remodeling, and/or repartitioning may require changes in an energy reduction plan.

Energy consumption and cost should be monitored monthly. This will enable building operators to keep track of the success of their programs. Since energy consumption is weather-dependent, monthly degree-days should be charted as well. After several years, energy consumption variations with weather conditions will become predictable. It is also useful to monitor the energy consumption of the large building systems.

The application of energy conservation options will alter the way a building operates. The energy audit should be repeated periodically to determine whether new methods of operation are amenable to the other retrofit options. New types of energy-conserving equipment are constantly being developed. These should be screened for possible building retrofit. This is especially true when existing building systems wear out and must be replaced. Because of the large number of products entering the market, cooperation between building operators and manufacturers will be required.

Section 2
Design and Installation Strategies

Chapter 5

HVAC System Analysis

Kermit S. Harmon, Jr. P.E.

I. Energy Accounting
II. General Analysis
III. Energy Audit Review
IV. Preliminary Retrofit Opportunities List
V. Temperature Control Systems
VI. Equipment Efficiency Check

I. ENERGY ACCOUNTING

A too frequently committed error in energy analysis stems from neglecting the universal law of energy conservation that states that energy can neither be created nor destroyed. The total amount of energy in the universe remains constant. By accepting this law, we can astutely analyze how a given quantity or constant flow of energy is utilized. A helpful analogy is to think of energy as a bank account: you cannot draw out more cash than the account possesses. The closely related and nearly synonymous terms "work" and "energy" should not be confused. The "work" concept may be considered as forming a crucial link between early Newtonian physics and the modern term "energy."

The nineteenth-century term "energy" has been defined as the capacity to do work, and, as such, "energy" may be present in many different forms: we may have electrical, mechanical, thermal, magnetic, chemical, or nuclear energy. The closely related term "power" is the rate of doing work, or the rate at which energy is transformed from one form to another. During the process of energy analysis, many forms of energy transformation and transmission may be realized, yet the energy will not cease to exist. Only the energy's form will change. Refer to Figure 5-1 for a simple hypothetical example containing the following elements of a typical HVAC system:

1. 10-kW sealed hermetic compressor
2. 3-hp fan motor (2.75-bhp load)
3. Two 1-hp condenser fan motors (1.85 bhp total load)
4. Two 50-VA control transformers (35-VA load each)

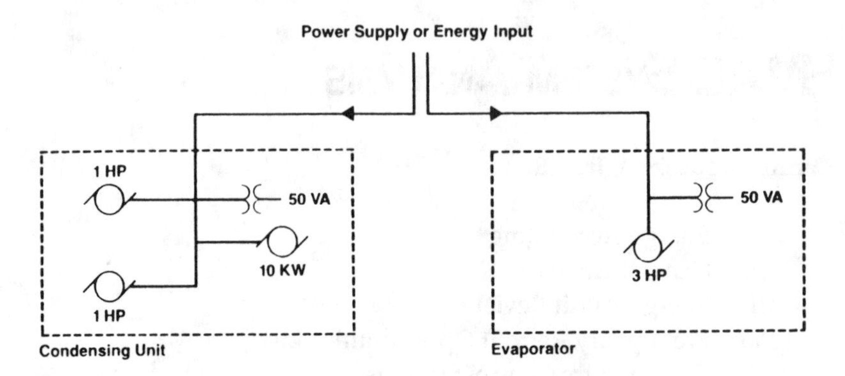

Figure 5-1. One-Line Electrical Diagram.

Assume that motor efficiency and drive efficiency are 0.85 and 0.95 respectively, for the evaporator fan and the condenser fan (items 2 and 3). The loads in kilowatts of this equipment may be expressed as follows:

1. Compressor = 10.00

2. Evaporator fan $\dfrac{2.75\text{ hp}}{0.85 \times 0.95} \times \dfrac{0.746\text{ kW}}{\text{hp}} = 2.54$

3. Condenser fan $\dfrac{1.85}{0.85 \times 0.95} \times \dfrac{0.746\text{ kW}}{\text{hp}} = 1.71$

$$4.2 \times 35\text{ VA} \times 0.90\text{ PF} \times \frac{\text{kW}}{1{,}000\text{ W}} = 0.06$$

Total power input = 14.31 kW

If this system provides ten tons of cooling, the system energy efficiency ratio (EER) can be obtained by dividing 10 tons by 14.31 kW, as follows:

$$EER = \frac{10\ \text{tons}}{14.31\ \text{kW}} \times \frac{12{,}000\ \text{Btu}}{\text{ton–hr}} \times \frac{10\ \text{tons}}{1{,}000\ \text{watt}} = 8.39\ \text{Btu/watt–hr}$$

The 10 or 120,000 Btu (126.7×10^6 ton-hr joules) of cooling is not energy from the system, but rather the absence of energy within the building. Heat energy was removed from the building by pumping the heat to the outdoors with mechanical refrigeration. So what happened to the energy purchased from the power company if it was not the 120,000 Btu of cooling? The 14.31 kWh of purchased electrical power is equivalent to 48,840 Btu (51.5×10^6 J) of thermal energy (3413 Btu/kWh). This conversion rate is for boundary energy, not raw source energy. Raw source energy would be approximately three times as great, owing to generation, transmission, and distribution system losses. What actually happens is that the 120,000 Btu (126.7×10^6 J) of heat pumped from inside the building and the 48,840 Btu (51.5×10^6 J) of converted electrical energy, are transferred to the outdoor air by the condensing unit coil. Figure 5-2 illustrates how almost all of the 168,840 Btu (178.2×10^6 J) total is transferred to the outdoor air by the condenser coil. A portion of the heat from the sealed, hermetic, compressor casing and hot gas piping and all of the heat from the condenser fan motors are transferred directly to the outdoor air by conduction, convection, and radiation.

The important point is that all energy considered should be accounted for. The fact that the system receives energy from more than one source [48,840 Btu (51.5×10^6 J) from the power company and 120,000 Btu (126.7×10^6 J) from room air] should not confuse the basic issue. The system eventually transfers the 168,840 Btu (178.2×10^6 J) total to the

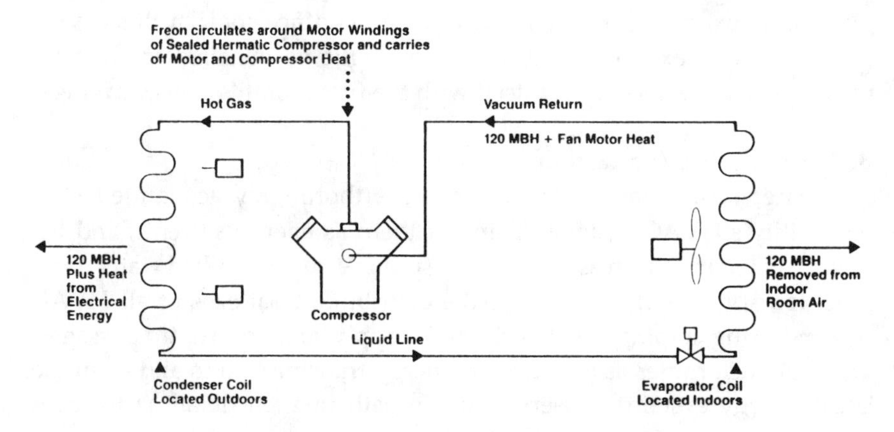

Figure 5-2. Refrigerant Piping Diagram.

outdoor air.

Of what value is such accurate energy accounting? Such energy analysis has caused the discovery of many energy-conserving opportunities, heat reclamation systems being probably the most significant discovery. The 168,840 Btu (178.2 × 10^6 J) of otherwise lost heat, if entirely reclaimed, can heat 579 gallons (2.19 m^3) of domestic water from 70°F (21°C) to 105°F (41°C). If the cooling load is for a space that requires winter cooling, such as interior computer rooms, the reclaimed heat can be used to heat the building perimeter spaces.

Please note that reclaimed heat is free, and is replacing an electric resistant heat equivalent of 50 kWh of boundary energy or 150 kWh of raw source energy. If purchased at 5¢ per kWh, 50 kW of resistant heat would cost $2.50 per hour. Reclaimed heat use, therefore, could easily amount to substantial energy cost savings over a year's operation.

II. GENERAL ANALYSIS

A. Facility Familiarization

When analyzing any HVAC system, first become familiar with the facility that the system serves. Review exterior elevations, floor plans, the roof, and each type of space within the facility. Find out how the facility is used, the number of employees usually present, employee tasks, and normal working hours.

Ask sufficient questions to identify the better retrofitting opportunities. For example, through questioning you may find that one of the vice presidents works ten hours every Sunday and operates the 500-ton central plant at a cost of $25.00 per hour to keep his office cool. In this case, a separate direct expansion (DX) system might be installed in the vice president's office and pay for itself with the first month's energy savings.

B. System Familiarization

After facility familiarization, become thoroughly acquainted with the facility's HVAC systems. Think of them as energy systems, and list them in categories such as unitary DX, split DX, central chilled water, two-pipe, four-pipe, and so on. Note the distribution patterns of all HVAC power wiring, piping, ductwork, and supply and return air passages. Think of such routes as avenues for energy transmission to and from the listed energy systems. Observe the air path in such detail as to learn exactly how air enters and leaves each environmental space.

Become knowledgeable about the HVAC system controls. The two basic categories of controls are manual and automatic. Manual controls require an operator to make observations, to make a decision based on these observations, and to manually adjust the controls to create the desired environmental results. Many exhaust systems, pumping systems, air-handling units, and similar types of equipment are manually controlled. All automatic controls have a form of master manual control; this may be only the circuit breaker or fusible switch that feeds power to the control system or the HVAC system.

Automatic controls, when energized automatically, respond to such devices as thermostats, pressure sensors, or flow switches. Most automatic temperature control devices are reset manually like typical room thermostats. However, some control devices are reset from a sensor signal, a combination of sensor signals such as a discriminator control circuit, or a totalizer control circuit. Time clock or computer controls may further automate the control system, thus reducing or eliminating manual tasks. Obviously, computer-made decisions must be better than manually made decisions if energy is to be saved.

With this information in mind, review system installation drawings, temperature control drawings, and the actual systems. Note control locations and how they really work. In particular, note room thermostat locations, since these thermostats are the final points of temperature control. Also, note what device, such as damper, valve, or electric reheat coil, the room thermostat triggers to achieve final room temperature.

Barring major complications, the foregoing effort should be completed within a day or two for a 100,000-sq-ft facility. Avoid being delayed by unimportant details as you become acquainted with the overall facility and its HVAC systems. Later in the analysis, you will want to focus in greater detail on areas where opportunities appear to exist.

C. System/Facility Interaction Analysis

After becoming familiar with the facility and its HVAC systems, analyze how well the HVAC systems respond to facility requirements. Interview operators, take detailed notes of their daily activities, review their daily log sheets, and review occupant complaint logs. Complaint call logs are very important; complaints may frequently be from energy-abusive areas that are overheated or overcooled.

The key question is: Does the system interact with the facility so that the system fulfills occupant comfort needs without being forced into an energy-abusive mode of operation? Related questions are:

- Do some occupants use portable heaters?

- Can computer rooms or other areas used 24 hours per day be operated without operating the entire central plant?
- Do any occupants open windows?
- Is extensive makeup air required by a starved exhaust hood that keeps the building under a vacuum?
- Must the heating boiler be operated all summer to provide a small amount of domestic hot water?
- Are fans computer-cycled for demand control instead of being slowed down and fitted with smaller motors with better power factors to continuously save energy?
- Has lighting been reduced to save energy without corresponding reductions in fan horsepower?

Such questions are numerous and will continue to be refined as we gain experience in analyzing HVAC systems for retrofit opportunities.

III. ENERGY AUDIT REVIEW

A. Review Energy Consumption for Suspected Energy-Abusing System

Although most energy audits do not investigate HVAC systems in as much detail as would most design consultants, audit data can be most informative. For certain types of buildings, air-conditioning energy usage can be estimated by its absence from energy bills during winter months, and heating energy usage can be estimated by its absence from energy bills during summer months. Cooling and heating energy usage can be even more accurately broken down when a fossil fuel is used for heating while electrical energy is used for cooling. It is easy to establish fairly accurate energy consumption estimates for lighting, fan motors, and pumps; this can be done by verifying their loads and multiplying them by known operating hours. The numerous notes contained in some energy audits may also provide ideas for retrofitting energy-consuming areas.

B. Perform Efficiency Tests on Suspect Equipment and System Functions

Completely thorough efficiency tests on equipment and systems are costly, and may not be necessary in some cases. In most cases, the fossil fuel boiler is not operating efficiently. The primary reason for testing the boiler would be to prove to the owner that efficiency upgrading could pay for itself within two years. Chiller efficiencies can, in some cases, be calculated from logged data obtained from chiller instrumentation and water balance and service reports. Chiller efficiency is very low if most of the operating hours are at low load, particularly if load is often added to keep the chillers on line. Fan efficiencies should be checked by calculating the air transport factor as defined in ASHRAE Standard 90-75. Deteriorated cooling towers offer visual evidence of water chilling plant efficiency loss.

C. Establish Levels for All Energy-Consuming Categories

Although considerable data have been published on energy budgets, the breakdowns have often been in percentages. Unfortunately, percentages are practically meaningless if only one category of energy consumption changes significantly.

Computer-modeling programs probably provide the best reference for a breakdown of energy use by categories. However, there is still a need to establish the levels of accuracy of some programs by modeling buildings that are equipped with meters to break down the energy.

It is proposed that levels be established for the energy use categories against what experience has shown can be achieved through conscientious design effort. For example, internal loads can be reduced through high-efficiency luminaires and efficient office or processing equipment. Use generous-sized ductwork and piping to minimize fan and pump horsepower. The known achievable loads of these items, when calculated against operating hours, will provide realistic energy quantities against which to compare estimates of the existing systems.

IV. PRELIMINARY RETROFIT OPPORTUNITIES LIST

A. Order of Cost/Benefit Ratio

At this point, sufficient familiarity with the facility and HVAC systems has been gained. Now, an idea-generating exercise can prove beneficial. This can be an excellent group exercise, and it may include individuals familiar with the facility as well as individuals who do not know the facility but are known to have good retrofit ideas. To gain maximum benefit from the exercise, each participant must feel free to suggest all types of ideas, even those that may seem impractical. A seemingly impractical idea suggested by one participant may be complementary to an idea from another participant.

After completing the retrofitting list, review the list and delete items that are too impractical. This might be accomplished by a group vote on each item. Arrange the remaining items in descending order of cost/benefit ratio, with those items that have the best cost/benefit ratio listed first. It is not necessary to determine accurate cost/benefits at this time; the cost/benefit of ideas can be estimated by group matrix evaluation. For example, each group member can give each idea a cost/benefit value on a scale of one to ten, and the total number of points received for each item would establish that item's appraisal value for priority listing.

Select the highest 5, 10, or 20 items, or as many as appropriate, for pursuing an accurate evaluation. Conceptual design documents for each of the selected retrofit ideas will be required for pricing. The conceptual design documents may be freehand schematic drawings or written descriptions for some items, but they must be comprehensive enough that estimates may be obtained from a qualified contractor.

Both energy savings and resulting cost savings should be calculated for each retrofit item to determine the actual cost/benefit ratio. These calculations will require careful and astute judgment. Be practical; use shortcuts and approximations for calculating savings from items that represent low capital cost but could cost a lot of engineering hours if highly accurate calculations were produced. For example, it may be possible to install an adjustable time delay for $150.00 on outside air dampers so that the dampers remain closed during startup. It could easily cost over $200.00 in engineering time to take the weather data for each day of a typical year and manually calculate the savings or model the operation on a computer. However, a reasonably accurate estimate could be calculated in about 15 minutes by selecting a few typical weather conditions throughout the year. Energy-saving calculations for retrofit items that require considerable capital investment should be investigated selectively. Do not hesitate to invest greater engineering effort and computer time if necessary for accuracy.

The cost/benefit ratio may be expressed as simple payback period according to the following formula:

$$\text{Simple payback period} = \frac{\text{Capital cost of retrofit item}}{\text{First–year savings in energy cost}}$$

If energy cost is escalating at a greater rate than interest on investments, the actual payback period will be shorter than the period as calculated above.

A simple payback period of five years represents a 20% annual rate of return on the first year of the investment. The rate of return on investment will increase as energy rates increase; so such investments are obviously excellent for those owners who want an investment that exceeds inflation rates.

V. TEMPERATURE CONTROL SYSTEMS

A. Of all subsystems, elements, and components comprising an HVAC system, the temperature control systems generally offer more energy cost-reduction opportunities than are found anywhere else. Conversely, temperature control system malfunction is frequently the greatest energy abuser. For this reason, thoroughly inspect the system to determine if the controls are operating as intended. This inspection will also better acquaint you with the system and should reveal control system retrofit opportunities. The number of items that have been wrongly installed, or that have been tampered with, and remain undiscovered for lengthy periods, would shock some of the "theoretical" geniuses who engineered them. Tampering is often the result of an unsuccessful attempt to remedy a malfunctioning control. Obviously, the retrofit program should include correcting all defective controls.

B. After thoroughly inspecting and testing the temperature control system, develop a list of possible control enhancements. Some enhancements may have already been included in the previously discussed preliminary list. Evaluate the list and process the items in a manner similar to that used for the preliminary items retrofit list.

The following typical list is by no means comprehensive, but it should help you get started:

1. Outside Air Closed on Startup. Outside air dampers on HVAC systems with automatic temperature controls have routinely been programmed to open upon evaporator fan startup and to close upon fan shutoff. Imagine the energy abuse during the fifties and sixties when it was common to operate large central systems 24 hours per day. Now it is common to operate systems in office buildings for 12 hours per work day, although the occupants may use the building for no more than 9 hours per day. Early system startup helps bring building temperature under control before office hours. The outside air load during warmup or pulldown is wasted energy. Startup might be delayed by half an hour if outside air dampers remain closed, reducing the load, until office occupancy hours. Some operators wastefully operate cooling or heating systems in the more climatically severe afternoons or on holidays for a few hours, without occupants in the building, to more easily bring the building under temperature control the next morning, again with outside air dampers open. Time delay devices, time clocks, or remote manual overrides can be installed to control outside air flow on an "as needed basis."

2. Temperature Scheduling. If HVAC system fluids temperatures can be limited to minimum deviation from final control point objectives, considerable energy can be saved. In other words, keep all temperatures as neutral as possible and yet deviate enough from neutral to permit the systems to do their job. The less the magnitude of the temperature differentials is, the less the rate of heat transfer. Savings are increased by diminishing undesirable gains and losses such as those through insulated walls of buildings, pipes, ducts, or storage tanks. Also, mechanical refrigeration efficiencies improve at lower heat pressures.

Supply Air. Supply air temperatures can be reset by outdoor air temperature, or by zone thermostats that have the greatest demand for cooling. Control circuits that reset from a thermostat having the greatest demand are commonly referred to as discriminator control circuits. Each air system must be analyzed individually to see how supply air temperature resets can be achieved to save energy. For example, on a hot and cold deck multizone air unit, it may be practical to control the cold deck temperature reset with a discriminator control circuit and to control the hot deck temperature reset by outdoor air temperature. In this case, heating water might not flow to the warm deck coil when the outdoor air temperature is 60°F (16°C), and yet it might begin flowing to the coil as the outdoor air temperature drops below 60°F. The graduation could be such that the hot deck temperature uniformly rises from 80°F (27°C) to 120°F (49°C) as the outdoor air temperature decreases from 60°F to 10°F (–12°C). The savings thereby achieved on the multizone air unit are primarily due to minimizing the mixing of cooled air and heated air for temperature control. Savings continue to be realized even for zones in full-heat or full-cool mode, because of the inevitable air mixing that occurs from normal damper leakage. An average double duct system can be viewed as a multizone unit. The hot deck and cold deck are extended by the hot duct and cold duct. The zone mixing dampers are located in double duct mixing boxes.

In variable-air-volume (VAV) systems, normally constant-temperature supply air is dumped into the return air plenum by terminal units as room thermostats become satisfied. Return air can therefore become even colder than room temperature, thus causing energy abuse by excessive temperature gain from the plenum's warm walls and roof. Supply air temperature can be reset by either outdoor air temperature or return air temperature. Additional benefit is gained by better air movement at low loads within the conditioned rooms. These types of VAV systems do not throttle the fan; therefore, fan horsepower savings are not lost from supply air temperature reset.

In VAV systems, resetting supply air temperature is detrimental to the fan horsepower savings achieved through throttling the supply air. However, savings in reheat energy may be much greater than would otherwise have been saved in fan horsepower. Furthermore, it may be necessary to increase supply air temperature because of lighting load reduction. Air movement within conditioned rooms can become too low as lighting energy is reduced to approximately 2 watts per square foot. Reset control for this type of VAV system can be accomplished with the discriminator control circuit. Obviously, supply air temperature reset and the discriminator control circuit would be effective on systems that are zoned by reheat.

Heated and Chilled Water. Generally, heated and chilled water temperatures can be effectively reset on a schedule based on outdoor air temperature. The reset schedule should be field-adjustable so that optimum values can be determined by trial and error. For maximum energy conservation, chilled water temperatures should never be lower than necessary to provide air temperatures that will meet cooling demands. Before concluding that increased chilled water is the best option, determine what additional savings might be achieved in fan horsepower by slowing down fans and installing smaller fan motors. Where lighting loads have been greatly reduced, it is possible to increase chilled water temperature in addition to reducing supply air flow. If automatic temperature reset based on outdoor air dry-bulb, wet-bulb, or enthalpy is deferred because of cost, implementation of seasonal manual reset can also save energy.

Condenser Water. In past years, most systems were designed to control condenser water at 85°F (29°C). Although systems are designed for 85°F (29°C), in most cases it is unnecessary to operate at this temperature. The chiller manufacturer should be consulted to determine the recommended low limit for condenser water. Setting condenser water at the lower limit can expend more energy in cooling tower fan horsepower, but the savings in reduced chiller power will generally exceed the penalty.

In areas where very low condenser water temperatures are achievable for several months of the year, and the chiller must operate at very light loads, the "free-cooling" cycle should be considered. This cycle is a chiller modification that enables a chiller to operate with the compressor shutdown.

3. Additional Time Clocks. Adding time clocks to automatically turn "on" and "off" various items of equipment or entire systems can be an effective and inexpensive means of energy-conserving automation. A variety of reliable time clocks have such features as 24-hour carryover for power failures and skip-a-day for automatic shutdown on weekends. Clocks with 24-hour dials can generally be programmed much more accurately than clocks with seven-day dials. Minicomputers now becoming available at low cost can perform all time clock functions, gather data, and monitor equipment. Careful analysis is required to ensure that time clocks are applied in the most effective manner possible. Connecting time clocks in control circuits to prevent operation of manually started equipment during unauthorized periods is a good application. Where time clocks are easily accessible, the manual override switch should be key-operated. The computerized-type switch uses a password to accomplish this same function.

4. Dead-Band Thermostats. Dead-band thermostats which can be directly substituted for existing thermostats are also available. The dead-band thermostat, as the name implies, permits the building's temperature to drift within a preselected dead-band range without calling for heating or cooling. For example, a thermostat with a 5°F (3°C) deadband could be set to call for cooling at 76°F (24°C) and heating at 71°F (22°C). The throttling range, if completely outside the dead-band, permits greater overshoot; i.e., temperature would drop to 69.5°F (21°C) and rise to 77.5°F (25°C) with 1.5°F (1°C) overshoot.

An alternative to the dead-band thermostat is a thermostat that throttles within the semidead-band. A thermostat with this characteristic would have only a 1°F or less true deadband, and would modulate (throttle) heating or cooling in proportion to room temperature deviation from setpoint. If the room temperature ever reached the limits, 71°F (22°C) or 76°F (24°C), the thermostat would (at that point) be calling for full heating or cooling. This type of thermostat will save energy and, at the same time, provide greater human comfort.

5. Reheat Coils. Reheat coils are often used for zone splitting to provide individual room control on multizone, VAV, or even double duct systems. The main zone thermostat is located in the room that has the predominant cooling requirement, and the reheat coil thermostats are located in the rooms that would tend to overcool. (Refer to Figure 5-3.)

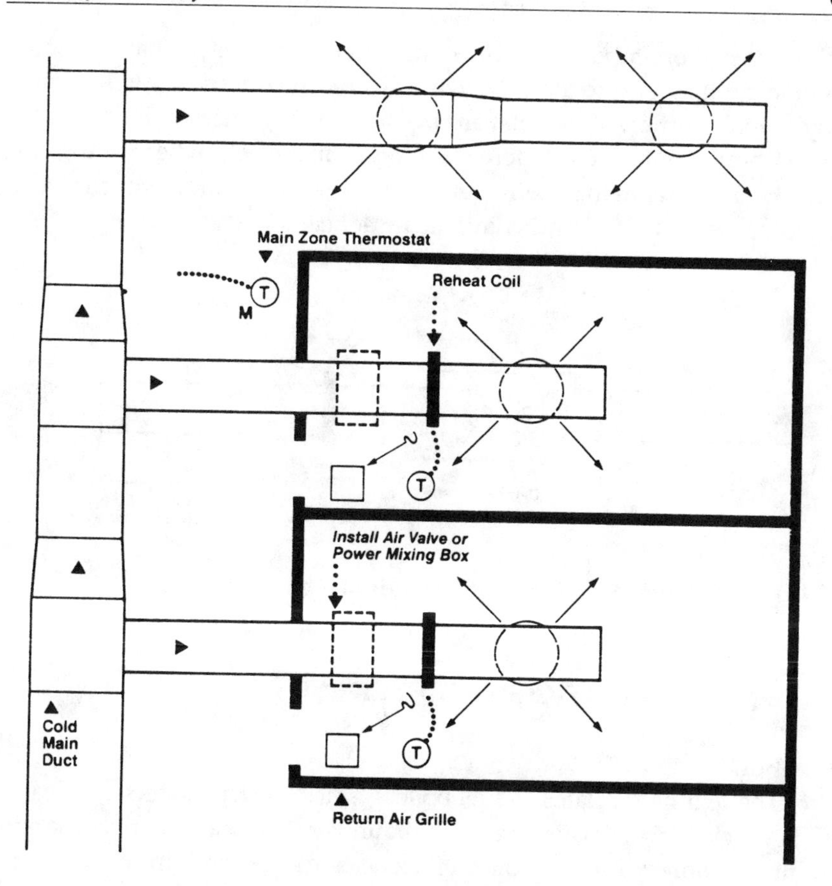

Figure 5-3. Partial Air Conditioning Plan.

In other cases, all thermostats on the zone would control reheat coils, and the zone duct temperature would be reset through a discriminator circuit from the room that has the greatest demand for cooling. In either case, energy-conserving modifications can be made to the systems.

The first step is to explore whether or not the heating coils can be deactivated or eliminated completely by rebalancing the air, relocating the main zone thermostat, and similar measures. The next step is to determine whether the coil can be replaced with an air valve to provide VAV control. If neither of the foregoing remedies can be successfully employed, install a powered mixing box ahead of the heating coil and sequence the mixing box damper with the heating coil valve as shown in Figure 5-4. The damper fully closes to the cold deck, and warm air from the heated return air is reclaimed before any new energy is used. This arrangement can also improve a room that is at times starved for cool air. Consider the extent of volume variation imposed on the central fan which does not have variable volume control. Where air valves or powered mixing boxes will control only a small portion of the total air, the system may tolerate the modification without a noticeable difference in performance elsewhere in the system. The diversity in load being reflected in the air requirements may also help other areas that at times have an air deficiency.

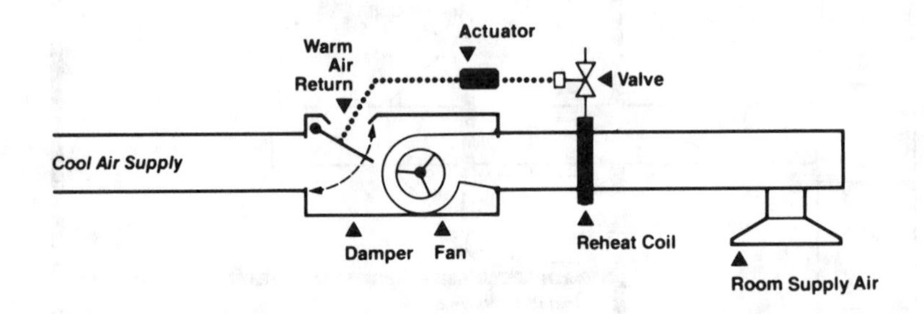

Figure 5-4. Schematic Detail of Reheat Retrofit.

VI. EQUIPMENT EFFICIENCY CHECK

A. Boilers

The text here relates to fuel boilers, since electric boilers are nearly 100% efficient in transferring heat (boundary, not raw source energy) within the boiler. Precise boiler efficiencies can be determined only by laboratory test under fixed conditions. However, approximate combustion efficiency can be determined by flue gas analysis. By recording flue gas temperature and percentage of carbon dioxide (CO_2) and referring to a chart or table for the fuel in use, approximate combustion efficiency can be found. The formula is as follows:

$$\text{Combustion efficiency} = \frac{\text{Input} - \text{Flue loss}}{\text{Input}} \times 100\%$$

Refer to the operating log to identify an increase in flue loss by a rise in flue temperature. This can result from scale buildup on heat transfer surfaces. The objective is to approach complete (stoichiometric) combustion by minimizing excess air (thereby minimizing oxygen, O_2, in the flue gas) without an increase of carbon monoxide (CO) in the flue gas.

Overall efficiency is always less than combustion efficiency owing to radiation loss from the boiler outside surface. The formula is as follows:

$$\text{Overall efficiency} = \frac{\text{Input} - \text{Flue loss} - \text{Radiation loss}}{\text{Input}} \times 100\%$$

Overall efficiency may also be expressed as follows:

$$\text{Overall efficiency} = \frac{\text{Gross output}}{\text{Input}} \times 100\%$$

Determine an approximate value for overall efficiency by measuring input and output over a specific time period under a stable load condition. Measure input from the gas or oil flowmeter after turning off other loads served by the flowmeter. Water flow rate and temperature difference in supply and return are obtainable. For steam with 100% condensate return, measure the condensate return flow rate and temperature.

If the boiler is not already equipped with a modern draft burner, probably over 40% of the purchased fuel is being wasted. When retrofitting a boiler, explore the use of equipment that will reduce excess air, preheat incoming air, increase flame temperature, increase firebox turbulence, and maintain boiler cleanliness. Only 1/8 inch (3 mm) of soot buildup on heat transfer surfaces can decrease heat transfer by up to 25%. Also, explore flue gas heat recovery by means of waste heat boilers. Claims are being made that less than one-year payback is common for boiler retrofitting. Fire tube boilers can be retrofitted with devices, installed within the tubes, to increase turbulence and uniformly distribute fire flow through the tubes. Consult boiler manufacturers before making this alteration.

B. Chillers

Conduct all tests of reciprocating liquid chillers for rating or verification of rating in accordance with ASHRAE Standard 30-77. Derive or verify centrifugal or screw-type liquid chiller ratings tests in accordance with ARI Standard 550-77. Such precision testing is costly, and such a degree of accuracy is not necessary to determine if an opportunity exists for chiller retrofit or replacement. Condensing temperature or temperature lift, load, and fouling affect chiller efficiency, most often referred to as kilowatts per ton or kilowatt-hours per ton-hour. These units can be converted to coefficient of performance (COP) or energy efficiency ratio (EER) as follows:

First, invert the expression of kW/ton to tons/kW. Then multiply the results by units of conversion as follows:

$$\text{COP} = \frac{\text{tons}}{\text{kW}} \times \frac{12{,}000\ \text{Btu}}{\text{ton–hr}} \times \frac{\text{kWh}}{3{,}413\ \text{Btu}} = \frac{\text{Btu cooling}}{\text{Btu input energy}}$$

$$\text{EER} = \frac{\text{tons}}{\text{kW}} \times \frac{12{,}000\ \text{Btu}}{\text{ton–hr}} \times \frac{\text{kW}}{1{,}000\ \text{watts}} = \frac{\text{Btu cooling}}{\text{watt–hr}}$$

Note that EER = COP × 3.413, and that input energy is boundary energy. Additional credit should be given to absorption chillers that utilize raw source energy.

Review the operating log sheets to determine which chiller load exists for the greatest number of operating hours. Perform efficiency tests during the time when this predominant load is relatively stable. Use either a kilowatt strip chart recorder or a watt-hour meter on the line side of the chiller starter. An adjustment in readings will be necessary if some operating chiller accessories are not fed through the metered feeder. Verify flowmeter and thermometer accuracy on the chilled water circuit, and then record readings throughout the test duration. The following readings will be used for a sample calculation:

Test period = 2 hr
Energy consumed = 300 kWh (1.08×10^9 J)
Flow rate = 450 gpm (0.0284 m^3/s)
Entering water temperature = 55°F (13°C)
Leaving water temperature = 45°F (7°C)

$$\text{Ton–hours} = \frac{450\ \text{gal}}{\text{min.}} \times \frac{8.33\ \text{lb}}{\text{gal}} \times \frac{60\ \text{min.}}{\text{hr}} \times \frac{\text{Btu}}{\text{lb°F}}$$

$$\times \frac{(55-45)\text{°F} \times 2\ \text{hr}}{12{,}000\ \text{Btu/ton}} = 375\ \text{ton–hr}$$

$$\text{Performance} = \frac{300\ \text{kWh}}{375\ \text{ton–hr}} = \frac{0.80\ \text{kW}}{\text{ton}}$$

Check this number against the manufacturer's published performance curves. If the number reasonably agrees with the curves, you probably have a clean condenser and the equipment is performing satisfactorily. If the chiller is old and manufacturer performance data are unavailable, a comparison should be made against a likely replacement chiller. A payback calculation may reveal that a new, more efficient chiller will pay for itself within five years.

C. Heat Exchangers

Air-to-air, water-to-water, water-to-air, refrigerant-to-air, and refrigerant-to-water heat exchangers lose operating efficiency by becoming clogged and from film buildup on heat transfer surfaces. Unfortunately, cleaning is not always easy, and a thorough visual inspection may be difficult. For this reason, filters, strainers, and water treatment systems have a very important role. Cooling coils can clog completely, even when filtered, if there is enough dirt buildup to support mold and bacteria growth. It is not unusual to find leaves and debris clogging up to one-half the condenser tubes in a wooded apartment complex where someone has removed the strainer basket.

Cleaning a heat exchanger is much easier than performing detailed heat transfer calculations to predict performance compared to that for which the heat exchanger was designed. However, you can check actual performance against published manufacturer data for similar, if not identical, equipment. If entering and leaving temperatures for the heat exchanger low side and high side have been logged periodically, deterioration in performance will be apparent.

Heat exchangers can be retrofitted with better instrumentation to monitor performance, better water treatment systems, filters, and even an automatic brush cleaning system, which may be a necessity when brackish water is used for condensing purposes. Heat exchangers can often be supplemented by additional heat exchangers to upgrade operational efficiency.

D. Pumps

When analyzing pump efficiency, first analyze the water circuits to determine if pumping requirements can be reduced. On systems where three-way valves have been used extensively, replacing such valves with two-way control valves will cause water volume variation according to block load demand. This often means more than a 20% reduction in peak water flow. If the three-way valves are relatively new, explore throttling down or completely shutting off the bypass circuit balancing valves.

After evaluating the water circuits and determining existing flow and heat requirements, compare pump, drive, and motor in-service efficiency against replacement candidates. Be certain to calculate energy *input* difference when determining payback, rather than the difference in brake horsepower energy required. (Overall efficiency = pump efficiency × drive efficiency × motor efficiency.) Always consider a change of impeller, motor, or both, as a potentially viable alternative for improved efficiency.

Explore variable speed pump drives, since the region of best efficiency, like the system curve, follows a parabolic curve (head versus flow) to zero as pump speed decreases. Inverters (SCR variable frequency) now available can efficiently vary squirrel-cage induction motor speed. Impeller diameter reduction is commonly employed to cause a pump to operate at a specified point, although pump efficiency usually decreases with appreciable reduction of impeller diameter. A false head (overpressure, as with balancing valves) is also used to cause a pump to operate at a specified flow rate. This practice can often appear to yield a higher efficiency on the pump performance curve than that of the trimmed impeller, but the apparent benefit can be deceiving. The efficiency indicated on the curve may be higher, but since the pump does more work to overcome the increased heat, more energy is used in this case than a trimmed impeller would use. Commonly used alternatives to the variable speed drive for limiting overpressure are:

1. Multiple pumps operating in parallel
2. Multiple pumps operating in series
3. Multispeed pumps

E. Fans

Much of the philosophy that applies to pumps also applies to fans. Explore all opportunities for reducing head and flow requirements (static pressure and cubic feet per minute). The energy required may be reduced, and some nuisance noise problems may disappear.

Excessive static pressure, like pump overpressure, wastes energy. Review the complete air circuits to determine where constrictions exist; these may be modified to relieve excessive static pressure. Return air passage area may be inadequate. Sound attenuators may be misused so as to add more noise than they attenuate, through increased static pressure.

Excessive use of balancing dampers will create unnecessarily high system static pressure. For a well-designed air supply system, only minimal adjustments are required by balancing dampers. However, system loads always change, and the existing system may be in serious need of rebalancing. But rebalancing may be inadequate without redesign of the ductwork. Load changes in various zones can be drastic in many buildings that have had lighting reduced to conserve energy and satisfy thermal lighting codes. Obviously, rebalancing must include reducing air quantities to areas having reduced loads, and slowing down the central fan as required. Sealing leaking ductwork should not be overlooked; it can save considerable fan horsepower in some projects.

A number of items must be considered to achieve optimum fan efficiency. (Overall efficiency = fan efficiency × drive efficiency × motor efficiency.) If the fan blades and scroll are clean, fan efficiency should be as reflected on manufacturer performance curves. Check the belts for wear, alignment, and tension. Check the bearings to ensure a minimum of drive loss. With the fan cleaned and slowed down and the drive assembly serviced, the motor may be oversized for the required duty. Most motors operating under light load lose efficiency and have a low power factor. Low overall power factors create additional charges on most utility rate structures. Replacement with a smaller motor may be a good investment.

F. Motors

Thoroughly analyze and minimize all system elements, including drive losses that contribute to motor loads, before seriously analyzing the motor. Additionally, determine and quantify the resulting specific load requirements in brake horsepower (BHP).

Where loads are constant, simply select a motor that can carry the given load with the least quantity of input energy. It is very important to think of selecting the most efficient motor in these precise terms. For example, by installing a new, high-efficiency, T-frame, squirrel-cage, polyphase, induction motor that is oversized for the application, you can end up with a less efficient installation than one using a less efficient motor correctly sized for the load.

Where the load will vary, as with VAV systems or variable water flow systems, the minimum load must also be determined and quantified in brake horsepower. However, even this load may be insufficient for complete evaluation if the efficiency is very low at minimum load, but the motor has only a few operating hours at that point. It is often necessary to determine the number of operating hours at various loads. Computer modeling is ideally suited for developing such load profiles.

Figure 5-5 illustrates how efficiency and power factor of a typical induction motor vary with load. Use the performance curves developed from testing the specific motor under consideration, since there can be a wide variation in characteristics for different sizes, types, and manufacturers. Use the performance curves in combination with the load profile to calculate difference in energy consumption, operating cost, and payback.

The power factor is important not only because of increased 12R energy loss in electrical distribution systems, but also because of energy

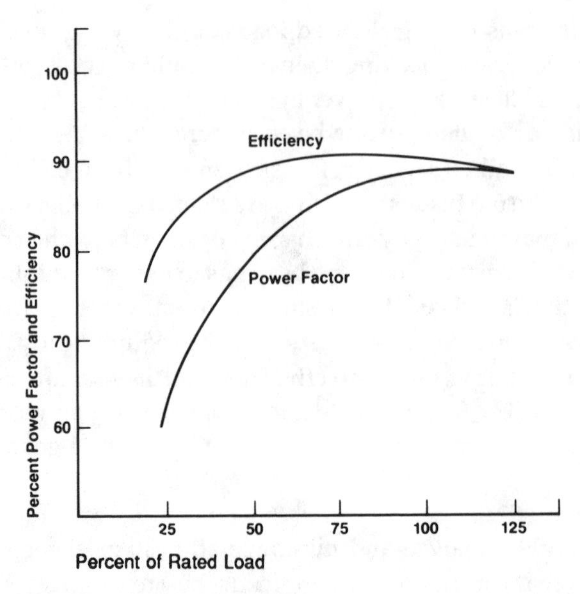

Figure 5-5. Variation of Efficiency and Power Factor Against Load for a Typical Induction Motor.

cost penalties levied in most utility company rate structures. These penalties should be included in the payback analysis mentioned in the preceding paragraph. Remember also that the power factor can be effectively corrected with capacitors for the existing motor. Furthermore, do not overlook an opportunity to relocate or exchange the function of existing motors within a large facility. Motors oversized for their load may be ideally suited for replacing larger motors that are also oversized.

The preceding discussion on motor selection has been intentionally limited to energy considerations. It assumes that other considerations such as torque, speed, duty, and current characteristics will also be taken into account. It is important to note that unusual service conditions may result in the consumption of additional energy. The usual NEMA service conditions for polyphase induction and synchronous motors are:

1. Ambient temperature, 0°C to 40°C
2. Altitude, under 3,300 feet
3. Installation on rigid mounting surface
4. Installation in locations that do not interfere with ventilation
5. Voltage tolerance, ± 10%
6. Frequency tolerance, ± 5%

Be cautious about replacing old U-frame motors with the newer, more common, T-frame motors. Studies funded by the now defunct Federal Energy Administration revealed that 1955/56 motors below ten horsepower were considerably more efficient than 1975 standard commercial motors. The U-frame motors are still available from some manufacturers, and they should be considered, since, in some applications, they are more efficient than many of the new "high efficiency" motors. U-frame motors are rugged and easily repairable.

You may also find that totally enclosed, fan-cooled (TEFC) motors generally have 2% to 3% better efficiency than open drip-proof type motors. It is also important to specify a cast iron main frame and end bells for efficient replacement motors. Higher-speed induction motors are inherently more efficient than most other types. Motors for pumps, fans, and compressors that start unloaded can accept lower locked rotor torque than other types can, which also allows for higher designed-in efficiency.

For small fractional horsepower motors, such as for direct drive fans, consider permanent split-capacitor motors; they are inherently more efficient than shaded-pole fan motors. The capacitor-start/capacitor-run motor is inherently more efficient than either capacitor-start or split-phase motors. Ball bearings with rolling friction are also inherently more efficient than sleeve bearings with sliding friction.

G. Cooling Towers

Cooling towers, can be classified into two basic categories. The first category involves direct contact between heated water and atmosphere; this category is normally called cooling tower. The second category involves indirect contact between heated fluid and atmosphere; this category is called closed-circuit fluid cooler, evaporative condenser, or closed-circuit cooling tower.

The discussion here will be limited to cooling towers, since closed-circuit fluid coolers are less efficient in cooling range and, therefore, should be used only where other benefits may be realized. For example, the closed-circuit fluid cooler is an alternative to the double bundle chiller application in a heat reclaim system.

The direct contact category of cooling towers can be further subclassified into gravity atmospheric towers, atmospheric spray towers, ejector towers, chimney towers (hyperbolic), ponds, spray ponds, spray module ponds and channels, induced-draft counterflow (single or double), and forced-draft counterflow. Gravity atmospheric towers or ponds might be considered the most efficient, since they require the least amount of operating energy. However, since either would present a space problem on almost any modern project, we must limit tower size considerations to available space. Using available space as a constant, we may then analyze first cost against owning and operating costs. Consider the possibility of locating the new cooling tower in a location other than the space occupied by the existing cooling tower. Additional space will provide the opportunity to consider a more efficient tower. Two related axioms are well worth repeating:

1. The greater the effective volume, the better the performance, other things being equal.

2. The greater the wetted surface, the greater the cooling, other things being equal.

In calculating owning and operating costs, be sure to include water consumption and maintenance, since these items can vary widely between cooling tower solutions. The energy consumption difference between induced-draft and forced-draft types can also be considerable. Two-speed motors used for condenser water temperature control can also save energy.

Do not overlook the possibility of refurbishing the existing cooling tower as a viable alternative if the basin and a major portion of the tower structure are in good condition. Many existing induced-draft cooling towers can be refurbished by patching, painting, and adding new spray heads. Such cooling towers can be packed with new high-efficiency fill to realize a 20% increase in original tower cooling capacity.

CHAPTER 6

CENTRALIZED VERSUS DISTRIBUTED FAN SYSTEMS IN HIGH-RISE BUILDINGS

Gideon Shavit, Ph.D.

I. INTRODUCTION

Traditionally, in commercial buildings conditioning of air is done by a fan system centrally located in the building. This approach usually has the least initial cost. The fan systems in a high-rise building are very often located in a mechanical room. This approach is used to minimize noise as well as to facilitate maintenance.

The development of the Life Safety Code for high-rise buildings in the early 1970s led to the creation of safety havens in buildings. This requirement is important because, in the event of a fire, the occupants cannot all leave the building in just a few minutes. In order to allow the occupants to get to a safety haven, it is necessary to contain the propagation of smoke in the building. This requirement is implemented by pressurization of the floors above and below the fire floor and in stairwells. The centralized fan system now has an additional function of providing life safety. The pressurization process is achieved through dampening.

An alternative to pressurization by a central fan system is the compartmentalization approach, in which each floor has an independent fan system. This approach provides a minimum number of openings, such as elevator shafts and stairwells, between floors. The shafts have fans dedicated to pressurization. The main difficulty in this approach is that the floor above and below the fire cannot be pressurized if smoke starts to enter the floor.

Energy consumption has drawn increased attention ever since the energy crisis of 1973, and has become as important a factor in design of buildings as has life safety. Some designers continue with their traditional approach of designing centralized fan systems, resorting to a perimeter system that has heating and cooling, such as the four-pipe system. Redundant heating and cooling are eliminated through improved zoning and implementation of energy control strategies. For example, a fan system for each of the perimeter zones and one for the interior zone may provide a very good alternative.

The concept of a fan per floor looks attractive, since the system requires a relatively small amount of energy to transport the air around the floor. However, in this approach, it is difficult to justify the HVAC system along the perimeter. With a constant volume system, the fan has to satisfy the load of all the zones at all exposures, which maintains a continuous high demand on the system. The tempering of the air requires some level of redundant heating and cooling. This problem is reduced by the installation of a variable-air-volume system (VAV) per floor and reheat along the perimeter. The system now matches the load of the zones at a lower penalty. However, during the heating mode in a given zone, redundant cooling and heating take place.

The purpose of this chapter is to study building performance when several fan system architectures (centralized, semicentralized, and distributed) and types of fan systems (reheat, double duct, and VAV) are subject to various energy conservation strategies. The results are used to determine promising configurations of fan systems that will minimize energy consumption.

The same problem of selecting optimum configuration of fan systems in new buildings exists with a somewhat different emphasis in the retrofitting of existing HVAC systems. Most of the centralized HVAC systems designed prior to the energy crisis in 1973 were reheat and double-duct fan systems. These systems are now energy wasteful, and it becomes necessary to upgrade their performance. Presently there is continuing activity in retrofitting double-duct fan systems to VAV systems. However, there is no activity in retrofitting reheat fan systems to VAV systems. The improved performance of the few systems that have implemented integrated zero energy band control (enthalpy control, reset from the zone of highest demand, and dead band thermostat) shows that this strategy can be a viable alternative for retrofit. The purpose of this chapter also is to provide information on fan system retrofitting alternatives and their corresponding energy savings.

II. DEFINITION OF PROBLEM

The problem in designing a mechanical system for the life cycle of a building is to provide comfort with the least amount of energy consumption. The designer has to analyze the optimum configuration of the system and its operation, using the following parameters:

1. Centralized fan system versus distributed fan system
2. Type of mechanical system
3. Energy conservation strategies

The same parameters are involved when the designer retrofits a building. The magnitude of retrofit effort may include the following: establish new zones, converting from one mechanical system to another, and determining the energy conservation strategies that are important.

The issue of a centralized fan system versus a distributed fan system provides the designer with three major potential possibilities:

1. One fan system for the entire building
2. One fan system per floor
3. One fan system per exposure for the perimeter zones with or without fan coil systems and one for the interior zone (i.e., five fan systems)

Other possibilities in this category are rejected as not being advantageous. For example, it is possible to have a self-contained (heating and cooling) fan coil system in each office in the perimeter zone. This falls into the category of a fragmented approach, which creates great difficulty from a maintenance point of view.

Every fan system architecture encompasses the following possibilities:

1. A reheat fan system
2. A reheat fan system in the interior and a four-pipe fan coil in the perimeter zones
3. Double duct fan system

4. VAV with reheat in the perimeter
5. VAV with a four-pipe fan coil in the perimeter zones

Each of the above fan systems can be operated in different modes of control for energy conservation. The concept of zero energy band with its requirement of a wide comfort range (68-78°F, 20-26°C) is very promising. The following strategies are considered for control:

1. Dry-bulb economizer only
2. Enthalpy control and reset of the discharge temperature from the fan system on the zone of highest demand
3. The integrated concept of zero energy band (enthalpy, load reset, and deadband thermostat) [Strategy #3 includes a wide comfort range of 68-78°F (2026°C) in the zone.]

The designer faces the following problem in designing a new building as well as in retrofitting an existing one: What is the best configuration to minimize energy consumption?

III. ANALYSIS

The purpose of the analysis is to determine the configuration that will provide minimum energy consumption during the life cycle of a building.

A. Description of the Building and Systems

A typical, but fictitious, office building in Washington, D.C. was selected for this analysis. It has 19 stories, and the northern exposure is directed to the north (no offset). The cross-section of the building is 150 ft × 100 ft (46 m × 305 m) with a total area of 285,000 ft^2 (26,477 m^2). The core of the building, 40ft × 75 ft (12.2 m × 23 m), is considered to be unconditioned space. The net area for conditioning is 228,000 ft^2 (21,181 m^2). The exterior wall structure is made of 6-inch (15-cm) concrete with 35% window area. The windows have two panes of standard glass with 1/4-inch (0.64 cm) air space. White venetian blinds are installed on the windows (shading coefficient of 0.55). The floor thickness is 6 inches (15 cm).

Internal loads in the building are mainly artificial lighting and people. The lighting level maintained during daily operation is 2 watts/ft^2 (22 watts/m^2). One person occupies 100 ft^2 (9 m^2) of the net floor area. The mechanical system provides 0.2 cfm/ft^2 (0.001 m^3/s/m^2) for ventilation all year long. The building HVAC system and lighting are turned on at 7 A.M., and occupancy time is 8 A.M.

The support systems of the building are turned off at 6 P.M. A night setback temperature of 60°F (16°C) is maintained during the heating season. In summer, no air conditioning is provided during the night. The setting of the HVAC control system and the gain of the coils are given in Table 6-1. The space temperature is always held at + 1.5°F (2.7°C) throttling range. In the conventional mode of operation, the space setpoint is maintained at 75 + 1.5°F (23.9 + 2.7°C), whereas with deadband it is maintained at 68 + 1.5°F (20 + 2.7°C) and 78— 1.5°F (25.5—2.7°C).

The central plant consists of a boiler that has a capacity of 6.87×10^6 Btuh (7.29×10^6 kJ) and a single centrifugal chiller that has a capacity of 928 tons (3.26 MW). The chiller has a low limit cutoff at 100 tons (0.35 MW). The fan power requirement was determined for each individual configuration. An inlet vortex damper regulates the air flow in the VAV fan system.

B. Simulation Program

Analysis of this type of problem is complex. Yet ability to analyze the performance of the system is necessary, since control strategies are an important alternative. The full interaction of the HVAC systems with the building subject to the control strategies has to be analyzed in the same manner as it is implemented in a real building. Honeywell's BLDSIM program simulates closed-loop operation of the HVAC system. An energy balance, made every minute, calculates the operating temperatures throughout the building for the mechanical and control system, as well as for the building structure. The operation of the HVAC components is then determined loop by loop, subject to the individual loop setting and control strategies coordinating the operation of the individual loops. A functional description of the program is illustrated in Figure 6-1. The effect of the central system on operation of the fan system is very important. For example, the temperature of the air leaving a cooling coil for two different settings of throttling range is illustrated in Figures 6-2 and 6-3.

A total of 45 individual cases were analyzed. In every case, the base included the type of fan system for a given fan system architecture and dry-bulb economizer control for the intake of the outdoor air. In each case, only one parameter was changed.

IV. RESULTS

The simulation program calculated annual building energy consumption for each of the following categories: artificial lighting, fans, chiller, total electricity, and natural gas (all heating is done with natural

Table 6-1. System Gain, Setpoint, and Throttling Ranges,°F (°C).

	Reheat	Double Duct	VAV	Fan Coil
Mixed air set point	54 (12.2)	54 (12.2)	54 (12.2)	—
Mixed air throttling range	10 (5.6)	10 (5.6)	10 (5.6)	—
Cooling coil setpoint	57 (13.9)	57 (13.9)	57 (13.9)	—
Cooling coil throttling range	10 (5.6)	10 (5.6)	10 (5.6)	—
Cooling coil gain	30 (16.7)	30 (16.7)	30 (16.7)	30 (16.7)
Reheat coil gain: exterior	40 (22.2)	40 (22.2)	30 (16.7)	35 (19.4)
interior	20 (11.1)	—	—	—
Preheat coil gain	—	—	30 (16.7)	—
Economizer damper cutoff	70 (21.1)	70 (21.1)	70 (21.1)	—
Night setback	60 (15.6)	60 (15.6)	60 (15.6)	—

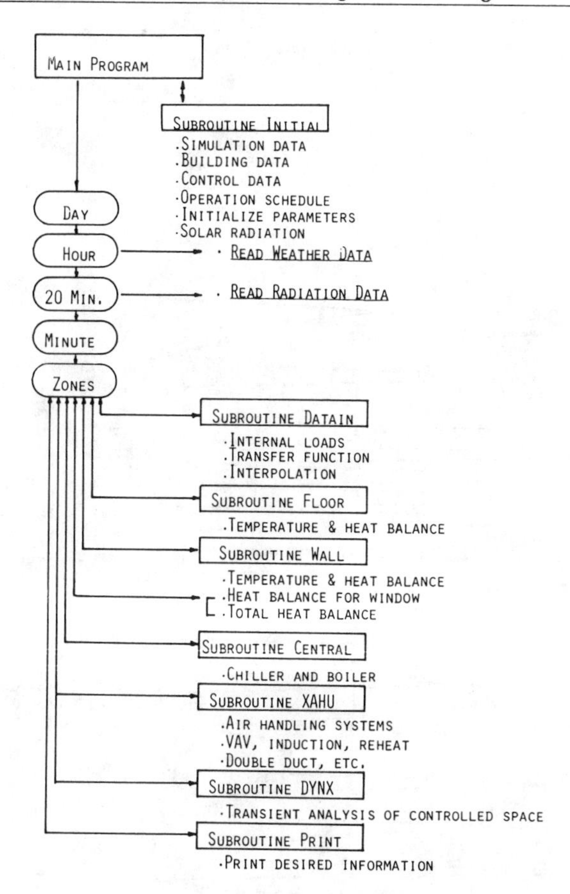

Figure 6-1. Honeywell's BLDSIM Program.

gas). The results were then analyzed to determine the optimum fan system architecture, the energy conservation due to the control strategies, the optimum type of fan system, and retrofit considerations.

A. Fan System Architecture

1. Reheat Fan System. Data in Table 6-2 show energy consumption for a reheat fan system with fan coils along the perimeter, for the fan/building and fan/exposure configurations. The fan/floor in Table 6-2 is for a reheat fan system only, since it is not cost-effective to have a fan system per floor, along with the fan coils around the perimeter. The data are given for two energy control strategies: conventional dry bulb economizer and integrated zero energy band. The systems with economizer control are discussed first. Fan/floor is the most economical for transport-

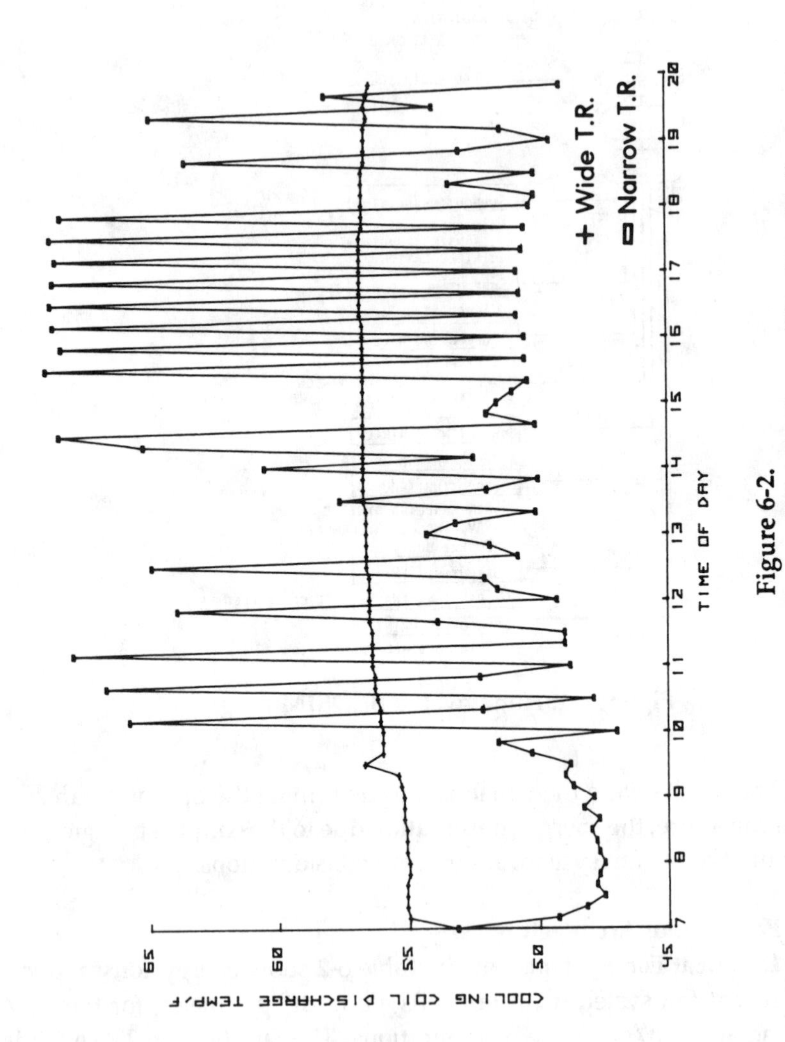

Figure 6-2.
Effect of Different Throttling Range Settings of Cooling Coil
(Summer—North-Facing Exterior Zone)

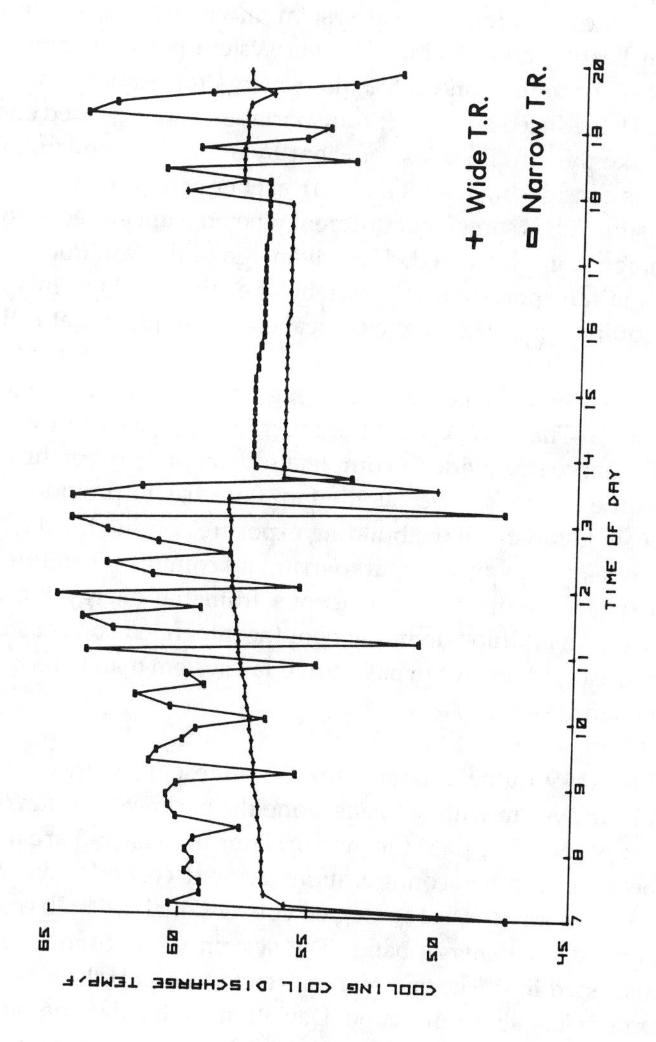

Figure 6-3. Effect of Throttling Range on Discharged Temperature (Winter—South-Facing Exterior Zone)

ing the air. This is expected, since the power for transportation is proportional to the (cfm^3 (m^3/s). The energy requirements for cooling and heating in the fan/floor are much higher than in the other two configuration, fan/building and fan/exposure. The main reason for this is the ability of the system with the fan coil to match the load of the zones with the least possible redundant heating and cooling, whereas the fan/floor performs as a conventional reheat system in which there is considerable redundant heating and cooling. The fan system per floor requires 38% more energy for cooling and 135% more energy for heating than the fan/exposure. The reduced energy for transportation and increased energy for cooling make the fan/floor less efficient by 8.5% than the fan/exposure when one is considering the total electrical energy requirement.

The situation is somewhat different when the integrated zero energy band strategy is implemented. The advantage of the fan/floor in energy required for transportation is offset by the disadvantage in increased cooling requirements. The total electrical energy requirement is the same for fan/floor and fan/exposure. It is expected that the fan/building would not be as efficient because of the higher level of energy required for transportation. The fan/exposure is much more efficient in heating requirements, since it provides a complete separation between heating and cooling in the fan coils, whereas the fan/floor has to provide colder air continuously to satisfy all the building exposures during the day, and the off peak zones still require reheat to maintain comfort. Therefore, the fan system/exposure is most advantageous from the energy requirement point of view. In addition, to implement the integrated control strategy, it is much more cost-effective to have a five-fan system than to repeat it in 19 individual fan systems.

2. Variable-Air-Volume. Data in Table 6-3 represent energy consumption for a VAV fan system with fan coils along the perimeter for fan/building and fan/exposure. The data for the fan/floor in Table 6-3 are for a VAV with reheat, since other configurations are not cost-effective. Data are given for two energy control strategies: conventional dry bulb economizer and integrated zero energy band. The system with economizer control only is discussed first. The energy required to transport the air in all three fan systems is just about the same. Deviations in the data do not appear, because of the rounding off of numbers. The fan/floor requires greater amounts of energy for heating and cooling. The main reason for the difference in cooling is the nature of zoning in the system. In the fan/floor, a single fan supplies the entire floor for all exposures. The VAV terminal box does not shut off completely in order to maintain circulation in the

Table 6-2. Reheat Fan System with Various Fan Systems Architecture and Energy Control Strategies.

	Economizer Only						Integrated ZEB Strategy					
	Fan/Bldg Reheat & FC*		Fan/Floor Reheat Only		Fan/Exp Reheat & FC		Fan/Bldg Reheat & FC		Fan/Floor Reheat Only		Fan/Exp Reheat & FC	
Lighting (kWh/sq ft/yr)[1]	(6.9)	74.5	(6.9)	74.5	(6.9)	74.5	(6.9)	74.5	(6.9)	74.5	(6.9)	74.5
Fan (kWh/sq ft/yr)[1]	(3.5)	37.8	(3.5)	37.8	(3.5)	37.8	(3.5)	37.8	(3.5)	37.8	(3.5)	37.8
Chiller (kWh/sq ft/yr)[1]	(5.5)	59.4	(5.5)	59.4	(5.5)	59.4	(5.5)	59.4	(5.5)	59.4	(5.5)	59.4
Total elec. (kWh/sq ft/yr)[1]	(15.9)	171.7	(15.9)	171.7	(15.9)	171.7	(15.9)	171.7	(15.9)	171.7	(15.9)	171.7
Boiler (cf/sq ft/yr)[2]	(48.5)	14.8	(111.6)	34.0	(47.5)	14.5	(17.7)	5.4	(63.6)	19.4	(13.8)	4.2

*FC: fan coil
1kWh/m2/yr.
2kl/m^2/yr.

zone and to supply air for ventilation. In the configuration with fan coils along the perimeter, the control system can turn off valves, and energy will not be consumed. The higher heating load on the fan/floor is due to the reheating needed to maintain comfort in the perimeter zones. Therefore, fan/exposure or fan/building is more economical than fan/floor.

This situation is somewhat different when the integrated zero energy band strategy is implemented. There is no change in the energy requirement for transporting air. The energy required for cooling is now the same for fan/floor and fan/exposure and is somewhat lower than for fan/building. Therefore, total electrical energy requirements are the same in both cases and slightly higher in the fan/building. However, reheat in the perimeter zone of the VAV fan/floor causes higher energy requirements for heating than in the configurations of the fan/building and fan/exposure. It can be concluded that in the VAV fan system there is a considerable reduction in the energy requirement. There is no advantage in the fan/floor over the fan/exposure when electrical energy is considered, and when it is operated with the integrated zero energy band strategy. However, the fan/exposure and fan/building are considerably more economical than the fan/floor in regard to the energy requirements for heating.

B. Retrofitting the Fan System

There is still a trend to retrofit double-duct fan systems into VAV systems. The main motive for the retrofit is to improve fan system performance and to conserve energy. The double duct fan system is the primary target, since the mixing box can be retrofitted to VAV boxes with relative ease.

Annual energy consumption for the building in question using a single fan system with a reheat or a double duct fan system subjected to the dry bulb economizer and integrated zero energy band control strategies is presented in Table 6-4. The energy requirement for heating and cooling is lower in the double duct fan system than in the reheat fan system with the economizer control strategy. Therefore, the question is asked: Why is there no activity in the country to retrofit a reheat fan system to a VAV? The double duct fan system consumes 34% less heating energy and 11% less cooling energy than the reheat fan system. Additionally, there are more reheat fan systems than double duct fan systems in high-rise buildings.

The addition of integrated zero energy band conservation strategy to the reheat and the double-duct fan systems reduces energy consumption in both fan systems. The impact of integrated zero energy band is greater

Table 6-3. VAV Fan System with Various Fan Systems Architecture and Energy Control Strategies.

	Economizer Only			Integrated ZEB Strategy		
	Fan/Bldg VAV* & FC**	Fan/Floor VAV & RH***	Fan/Exp VAV & FC	Fan/Bldg VAV & FC	Fan/Floor VAV & RH	Fan/Exp VAV& FC
Lighting (kWh/sq ft/yr)[1]	(6.9) 14.5	(6.9) 74.5	(6.9) 74.5	(6.9) 74.5	(6.9) 74.5	(6.9) 74.5
Fan (kWh/sq ft/yr)[1]	(1.4) 15.1	(1.4) 15.1	(1.4) 15.1	(1.4) 15.1	(1.4) 15.1	(1.4) 15.1
Chiller (kWh/sq ft/yr)[1]	(4.4) 47.5	(5.3) 57.2	(4.3) 46.4	(3.8) 41.0	(3.6) 38.9	(3.6) 38.9
Total elec. (kWh/sq ft/yr)[1]	(12.7) 137.2	(13.6) 146.9	(12.6) 136.1	(12.1) 130.7	(11.9) 128.5	(11.9) 128.5
Boiler (cf/sq ft/yr)[2]	(13.3) 4.1	(36.1) 11.0	(13.3) 4.1	(5.5) 1.7	(12.7) 3.9	(4.1) 1.2

*VAV: variable-air-volume
**FC: fan coil
***RH: reheat (along the perimeter).

[1]kW/m^2/yr.
[2]kl/m^2/yr.

on reheat fan systems than on double-duct systems. The net effect is that the double duct fan system is now more efficient than the reheat by 7% and 2% for cooling and heating, respectively (Table 6-4).

Conversion of the reheat fan system to a VAV with reheat along the perimeter zone reduces energy consumption by a significant amount. Table 6-4 also gives annual energy consumption for a VAV with reheat. Energy savings are manifested now in all three categories: fan energy, energy for cooling, and energy for heating. The conversion of a reheat fan system with only economizer control to a VAV with reheat reduces energy consumption by 47%, 30%, and 67% for the fan, cooling, and heating energy, respectively. This is a considerable savings and, therefore, an attractive alternative for retrofit.

The VAV and reheat fan system also provides very attractive options for retrofit when integrated zero energy band strategy is implemented. Table 6-4 illustrates the potential of this alternative. The retrofit of a reheat fan system with zero energy band to VAV with reheat and zero energy band provides savings of 47%, 22%, and 80% on the energy for transportation, cooling, and heating respectively. The maximum possible savings are obtained when a reheat fan system with economizer control only is retrofitted to a VAV fan system with reheat subject to the integrated zero energy band strategy.

C. Retrofitting the Energy Control Strategies

Tables 6-2, 6-3, and 6-4 contain data regarding the retrofitting of the control system to provide energy-efficient operation. The retrofitting of a mechanical system is rather costly and extensive. Conversion of fan system control from the simple dry bulb economizer to the sophisticated integrated zero energy band is less costly and results in significant savings. The integrated zero energy band strategy includes the following: enthalpy control, reset of the discharge temperature from the zone of highest demand, and a wider comfort range with a dead band thermostat.

Table 6-4 demonstrates that energy consumption in the reheat fan system with the zero energy band strategy is 39% and 43% less for cooling and heating, respectively, than it is with economizer control. The primary reasons for this are the minimization of redundant heating and cooling and the wider comfort range. The savings due to enthalpy control are secondary. Table 6-4 also demonstrates the attractiveness of this methodology in the double-duct and VAV fan systems. For example, the reductions in energy consumption in the VAV fan system with zero energy band over the one without it are 32% and 65% for cooling and heating energy, respectively.

Table 6-4. Single Fan System/Building (Lighting Load of (74.52 kW/m^2/year) (6.9 kWh/sq ft/year).

Type Fan System	Economizer Energy		Integrated ZEB Energy		% Saving
Reheat					
Fans (kWh/ft^2/yr)[1]	(6.4)	69.1	(6.4)	69.1	0
Chiller (kWh/ft^2/yr)[1]	(7.6)	82.1	(4.6)	49.7	39
Total elect. (kWh/ft^2/yr)[1]	(20.9)	225.7	(17.9)	193.3	14
Total gas (cf/ft^2/yr)[2]	(111.6)	34.0	(63.3)	19.4	43
Double Duct					
Fans (kWh/ft^2/yr)[1]	(6.4)	69.1	(6.4)	69.1	0
Chiller (kWh/ft^2/yr)[1]	(6.8)	73.4	(4.3)	46.4	37
Total elect. (ft^2/yr)[1]	(20.1)	217.1	(17.6)	190.1	12
Total gas (cf/ft^2/yr)[2]	(73.9)	22.5	(62.7)	190.1	15
VAV & Reheat					
Fans (kWh/ft^2/yr)[1]	(3.4)	36.7	(3.4)	36.7	0
Chiller (kWh/ft^2/yr)[1]	(5.3)	57.2	(3.6)	38.9	32
Total elect. (kWh/ft^2/yr)[1]	(15.6)	168.5	(13.9	150.1	11
Total gas (cf/ft^2/yr)[2]	(36.1)	11.0	(12.7)	3.9	65

V. CONCLUSIONS

1. The architecture of a fan system/exposure is the most efficient, and the fan/building is more efficient than the fan/floor when the conditioning is done with a reheat or VAV fan system and fan coils along the perimeter, subjected to the dry bulb economizer.

2. The architecture of a fan system/exposure is more efficient for heating consideration when the conditioning is done with a reheat or VAV fan system and fan coils along the perimeter, subjected to the integrated zero energy band strategy. There is no difference between the fan/floor and the fan/exposure in terms of total electrical energy usage.

3. The architecture of a fan/building is the least efficient for a reheat or VAV fan system and fan coils along the perimeter when one is considering only total electrical energy subject to the integrated zero energy

band. It is more efficient than the fan/floor when considering heating energy when subjected to the integrated zero energy band.

4. It is not sufficient to consider only the energy savings due to transportation when comparing fan/floor and fan/exposure architecture. The energy conservation strategies for each architecture should also be considered.

5. Retrofitting of a reheat fan system per building to a VAV fan system and reheat along the perimeter provides 67% and 30% reduction in the requirement for heating and cooling, respectively. Compared to an economizer control system, the integrated zero energy band provides a reduction of 47%, 80%, and 22% in the energy requirement for transportation, heating, and cooling, respectively.

6. A VAV fan system is more efficient than a reheat fan system. Significant effort should be directed to retrofitting reheat fan systems to VAV fan systems.

7. The architecture of a VAV fan system per exposure and fan coils along the perimeter is the most efficient configuration.

8. Implementation of integrated zero energy band control provides significant energy savings without retrofitting of the fan systems.

9. The results presented here are true only for the given structure, fan systems arrangement, control setting, dynamics of the given control and HVAC system, and the given location. The results should not be extrapolated, since they may result in erroneous data.

CHAPTER 7

COGENERATION AS A RETROFIT STRATEGY

Milton Meckler, P.E.

I. Introduction
II. Cost Analysis
III. Feasibility Analysis
IV. Prime Movers
V. Induction Generation
VI. Developing Load Profile
VII. Options and Research

I. INTRODUCTION

Cogeneration is defined as the combined production of power and useful heat by the sequential use of energy from one fuel source. It involves the use of rejected heat from one process for energy input into a subsequent process. Cogeneration systems are used at many of today's facilities to produce electricity and thermal energy on a cost-effective basis. Energy savings for both the plant and utility are beneficial byproducts of an efficiently utilized cogeneration plant.

Pioneered in the late 1800s, today's improved cogeneration technology is simple and can be applied in various ways. Equipment used can vary from turbines, diesel engines or fuel cells, depending on fuel choice or the energy input to be used. In any cogeneration application, the goal is to increase overall efficiency of the system. An average electric generating plant converts fuel at an annual efficiency of 25% to 35%. Mechanical and electrical systems integrated by cogeneration attain efficiencies up to 75% to 85%.

In 1930, 30% of total U.S. generation capacity was supplied by onsite electrical generating facilities. The 10-year period from 1965 to 1975 showed an increase of 40% to 50% in 500-MW and greater capacity electric utility plants. Today, a study of the U.S. cogeneration market shows that this capacity could rise to roughly 39,350 MW by the year 2000.

II. COST ANALYSIS

A cost analysis will often determine that operating costs for a cogeneration system are lower than those for a system using purchased energy. The first step in a complete analysis is to compile an annual cash flow list for every year of the projected life of a proposed system (Table 7-1 shows a summary of annual costs for a typical application).

The estimated useful life of cogeneration systems using gas engines or turbines ranges from 20 to 25 years. Systems using steam turbines have an estimated useful life to 30 to 35 years. It has been proven that within limits and appropriate conditions, cogeneration will save money and costly fuel. Yet, due to natural restrictions in conversion of fuel for conventional power use, less than 40% of this energy is converted into power. The remaining "waste heat" is left unused during the generation of electricity or operation of mechanical equipment.

The use of waste heat in processes other than power generation increases the amount of fuel energy available. Still, not every power generation facility is able to put this additional energy source to use. However, many industrial, commercial and institutional facilities (in and outside of the U. S.) utilize this valuable energy saving commodity. Installation of energy consuming equipment that relies on heat-installed power (i.e., absorption cooling rather than mechanical cooling) is common. It represents a dual advantage by simultaneously decreasing overall power requirements and increasing the use of waste heat, while improving control of electrical utility channels. Waste heat recovery can improve cost and energy savings, but these savings are available only under the proper conditions. A cogeneration feasibility analysis is needed to determine these conditions for each facility.

III. FEASIBILITY ANALYSIS

The cogeneration feasibility analysis involves three distinct phases: (a) data collection, (b) analysis, and (c) report documentation. Developing a meaningful analysis is a complex process. It requires a working knowledge of sound engineering and economic principles, orderly procedures

Table 7-1.
Summary of Annual Costs for High-Pressure Boiler with Steam Turbine.

HIGH PRESSURE BOILER WITH STEAM TURBINE

INITIAL INVESTMENT: $576,000

Year	Rate	ANNUAL ENERGY USAGE						
		Purchased Electricity	Fuel Cost @ 3.90/MCF	Water	Maintenance & Operation	Cost Sub-total	Capital Recovery	Total Annual
	$/KWH	$/Yr	$	$/yr	$/yr	$	$	$
1980		1,302,056	692,250	41,800	187,000	2,223,106	63,160	2,286,266
1981		1,540,766	865,312	45,980	205,700	2,657,758	63,160	2,720,918
1982		1,822,878	1,081,640	50,580	226,270	3,181,368	63,160	3,244,528
1983		2,085,459	1,352,050	55,640	248,897	3,742,046	63,160	3,805,206
1984		2,397,952	1,690,063	61,200	273,787	4,423,002	63,160	4,486,162
1985		2,758,188	1,842,169	67,320	301,165	4,968,842	63,160	5,032,002
1986		3,005,579	2,007,964	74,050	331,282	5,418,875	63,160	5,482,035
1987		3,276,840	2,188,681	81,460	364,410	5,911,391	63,160	5,974,551
1988		3,569,803	2,385,662	89,600	400,851	6,445,916	63,160	6,509,076
1989		3,893,147	2,600,372	98,560	440,936	7,033,015	63,160	7,096,175
1990		4,242,532	2,834,405	108,420	485,030	7,670,387	63,160	7,733,547
1991		4,624,468	3,089,502	119,260	533,533	8,366,763	63,160	8,429,923
1992		5,041,126	3,367,557	131,190	586,886	9,126,759	63,160	9,189,919
1993		5,494,675	3,670,637	144,300	645,575	9,955,187	63,160	10,018,347
1994		5,989,457	4,000,994	158,740	710,132	10,859,323	63,160	10,922,483
1995		6,527,640	4,361,084	174,600	781,145	11,844,469	63,160	11,907,629
1996		7,115,735	4,753,582	192,470	859,260	12,921,047	63,160	12,984,207
1997		7,755,913	5,181,404	211,280	945,186	14,093,783	63,160	14,156,943
1998		8,452,513	5,647,730	232,400	1,039,705	15,372,348	63,160	15,435,508
1999		9,214,215	6,156,026	255,640	1,143,675	16,769,556	63,160	16,832,716
2000		10,043,191	6,710,068	281,210	1,258,042	18,292,511	63,160	18,355,671
TOTAL		100,154,133	66,479,152	2,675,700	11,968,467	181,277,452	1,326,360	182,603,812

NOTE: ELECTRICAL ESCALATION AS FOLLOWS: 18% FOR THE YEARS 1980-81, 15% FOR THE YEARS 1982-83 AND 9% THEREAFTER.
NATURAL GAS ESCALATION AS FOLLOWS: 25% FOR THE YEARS 1980-84, 9% THEREAFTER.
CAPITAL RECOVERY FACTOR @ 9%/YEAR FOR 20 YEARS.
ANNUAL FUEL CONSUMPTION: 177.500 X 10^6 CU-FT.

and calculations. A broad perspective, good judgment and objectivity also are required.

Important factors of the overall feasibility analysis include technical and economic analyses. The technical analysis consists of determining energy characteristics and consumption to select suitable equipment types and sizes to meet system requirements. For example, single waste heat recovery is a primary function of cogeneration prime movers, waste heat boilers, and associated recovery equipment that requires special attention. Consideration should be given to alternative methods of supplying energy needs (mechanical vs. absorption cooling, or electric motors vs. prime mover direct drives, for instance). This is necessary to optimize the demand balance between prime movers and waste heat recovery. After a technically feasible system is selected, the equipment performance characteristics can be applied to its energy load requirement. This will determine primary energy consumption such as fuel or electricity, usually for a period of one year. The economic analysis helps to estimate the initial investment and annual operating cost of each alternative system analyzed in the technical analysis. Cost estimates are developed by suitable procedures in order to determine the most economically feasible system.

A. Cost Factor Considerations

Assuming that cogeneration packages become more standardized for commercial applications, most of the high costs associated with today's cogeneration market will be alleviated. Still, research and development work is needed to reduce specific costs such as those related to connecting cogenerators with utility grids. For utilities, a major concern is that a great number of small cogeneration systems connected to the grid may deteriorate the level of safety maintenance on the grid. Their aim is to maintain high power quality, prevent variations in load, current and voltage, and to protect powerline workers from hazards.

Several interconnection guidelines can be used for safe operation. Protective relaying is an important requirement for operating a cogeneration system in parallel with the grid, and it also protects the cogeneration system. There is, however, a lack of published technical data on interactions between the grid and cogeneration hardware, making the policies regarding interconnect practice inconsistent. These uncertainties are causes of unnecessary costs and delays for cogeneration development. Another barrier limiting applications for packaged cogeneration is the high cost of installation. In California, for example, installation cost is often 50% of the total commercial cogeneration system costs. An interesting research program now being conducted, aimed at reducing installation costs, involves the direct integration of cogeneration systems with commercial rooftop units. This work is based on valid foresight which predicts that the future of cogeneration in the retrofit market (i.e., for application to restaurants, hotels and other commercial buildings) depends on inexpensive integration with packaged air-conditioning units generally found in commercial applications.

The concept developed by Gas Research Institute (GRI) of Chicago involves a patented heat transport system utilizing a refrigerant to couple the direct expansion (DX) coil of existing rooftop HVAC equipment and the absorption chiller of a cogeneration package. A direct refrigerant interface takes away the need for water coils, pumps, and other expensive modifications. This concept has been tested successfully with common rooftop units and has positive results for the technical feasibility of retrofitting commercial HVAC systems with cogenerated space cooling and heating.

IV. PRIME MOVERS

Because of the difference between electrical and steam loads, two methods are used to size equipment for a cogeneration plant. These are: (a) selecting equipment best suited to supply electrical demand and supplementary boiler firing when steam demand is higher than generated steam, and (b) choosing equipment to meet steam demand without supplementary boiler firing.

Reciprocating engines are probably the most widely used prime movers due to their comparable low-cost and various ranges of size and speed. They can be fueled by oil, natural gas, liquid propane gas, digester gas, or a combination of these fuels. Dual-fuel engines which can switch fuels while running offer a higher degree of reliability. Complete "total energy" modules consist of an engine generator unit, heat recovery unit, switchgear and controls. These units lower equipment and installation costs, decrease construction time, and place the burden of responsibility on one supplier. Generator ratings normally are based on continuous duty operations at a power factor of 0.8. The most common size and speed engine generators used in cogeneration systems, are shown in Table 7-2.

Factors favoring gas turbines as prime movers include lightweight, minimum foundation requirements, rapid start-up capability, and automatic operation with minimum supervision. Unfavorable factors include a thermal efficiency generally less than reciprocating engines, rapidly falling efficiency at part load for the single-shaft gas turbines, and sub-

Table 7-2. Common Generator Sizes and Speeds.

kW range	Rpm	Cycle	Fuel
150-800	1,200	4	Oil, gas, LPG
450-3,000	720 or 900	2 and 4	Oil, gas or dual fuel
1,650-5,000	514	2	Dual fuel

stantial loss of power output with higher ambient temperatures and high initial power cost.

A. Gas Turbine Categories

Based on shaft arrangement, gas turbines fall into two categories: single-shaft and two-shaft. The single-shaft units are more suitable for constant speed applications such as generators. Their design is mechanically simpler for reducing cost. Also, their ability to handle speed variations is excellent with simple fuel control. Two-shaft gas turbines permit a wide variation of full power speeds. There also is an inherent lag in response between the gas producers and power turbines, though each can be equipped with a speed governor. Only the gas producer turbine requires a starter to accelerate and pressurize the system. Gases leaving the gas turbine flow through the free power turbine forming a fluid coupling. The remaining energy is absorbed by the power turbine and transferred to the output shaft which drives the load of the electric generator.

Turbine speeds vary from 40,000 rpm to 60,000 rpm for 50 hp to 2,000 hp (37 kW to 1490 kW) to as low as 3,600 rpm for 20,000 hp to 30,000 hp (14,920 kW to 22,380 kW). Starter systems include compressed air, gas starter (popular in the U.S.), ac/dc motors, batteries, chargers, motor-generator sets, reciprocating engines, high-pressure impingement, and hydraulic and hand cranks. Some turbines burn two types of fuels, usually gas and oil, incorporating two complete fuel systems that could be switched while in operation.

Gas turbines meet required steam and electrical demands ranging from 21,160 pounds per hour (lb/h) at 4 MW to 25,000 lb/h hour at 4.8 MW. They can be designed to meet demands with or without supplementary boiler firing, depending on the capacity range of operation and the number of individual turbines used. Steam turbines, either condensing or non-condensing, are used for many industrial applications. Mechanical drive steam turbines are ideally suited to drive mechanical equipment such as refrigeration compressors, pumps, fans, and other plant auxiliaries. They are efficient and economical. Monetary considerations tend to group turbines in terms of pressure, temperature, and steam flow, as indicated in the pressure and temperature classes shown in Table 7-3.

Table 7-3. Turbine Types (Pressure and Temperature Classes).

Unit size	Pressure (psig)	Temperature °F
Small	150-400	500-750
Medium	400-600	750-825
Large	600-900	750-950
Large	900-2,000	825-1,050

B. Back-pressure Turbines

Most mechanical drive turbines in process industries are either back-pressure or reducing-valve types. An assured demand for low-pressure steam in a central plant can make back-pressure turbines very attractive to use.

In self-generation, the cost of high pressure boilers often can be justified. This entails using a high-pressure, multistage turbine exhausting into a medium-pressure turbine drive for a centrifugal refrigeration machine and exhausting this turbine into an absorption machine at low pressure. Due to high steam rates, back-pressure turbines must be highly efficient to fit into the heat balance of an existing plant. Therefore, multi-stage turbines are required. A back-pressure turbine is installed with an exhaust pressure regulator to provide constant exhaust pressure.

C. Condensing Turbines

In a condensing turbine, steam expands from throttle conditions to a condensing pressure. Its main advantage is a low steam rate. The higher the pressure at inlet conditions, the lower the steam rate for the same condensing pressure. Throttle conditions are constant and determined by economics, except in some process industries where there is an abundance of waste steam.

A single-stage turbine is economically attractive in small sizes. However, additional equipment is necessary to implement the condensing cycle. This equipment may include a surface condenser, hot well pumps, steam ejectors or vacuum pumps to maintain condenser vacuum, a cooling water system involving cooling towers, and other accessories which increase initial costs.

Condensing turbines are best suited when the demand for process steam is not simultaneously required with power generation or shaft output, since steam would then become a waste product. When available, waste steam cannot be used by the condensing turbine. A bypass, however, can be used to conduct it directly to the surface condenser.

D. Extraction Turbines

Extraction turbines consist of two turbines in series enclosed within the same casing. Extraction steam serves process needs, or feeds to auxiliary equipment at one pressure. The second turbine, if non-condensing, may serve other process needs at another pressure. A second condensing turbine drives an electric generator.

An extraction turbine may be either the condensing or non-condensing type. In an extraction turbine, all of the steam works up to the extraction point. This portion of the work is done at a higher water rate. The amount that proceeds to the condenser is the only part of the steam that has done work at a low steam rate.

In many central plants, it is possible to use non-condensing extraction turbines for power generation with an exhaust steam extraction point for hot water heating, absorption refrigeration, or other needs. With condensing turbines, an extraction point can feed refrigeration machines. When it is not needed, the steam can proceed to the condensing section to generate power. An extraction turbine is not built for less than 500 hp.

E. Single-Stage Turbines

A single-stage turbine is one of the most basic prime movers in industrial plant operations. It is generally considered the "workhorse" of the industry due to its inexpensive drives and design flexibility (depending on first cost for intermediate or low efficiencies).

The relatively high steam rate of a single-stage turbine makes it very suitable for non-condensing service, and its exhaust steam has various uses in process work. Because of its high range (15 hp to 3,500 hp or 11 kW to 2610 kW), it can be used successfully in central plant work for auxiliary pumps, fans, etc., in a centrifugal absorption arrangement. As a condensing turbine, the single-stage turbine is at more of a disadvantage than the multistage turbine due to its higher efficiencies, which can keep steam rates low and steam plant size within reasonable limits.

At certain times, there may be more energy available than needed. The logical solution is to consider selling it back to the utility company. This is a complex activity that must be considered before building or modifying the cogeneration system. Two basic approaches can be used: (a) a cogeneration system running parallel with utility systems, or (b) separate cogeneration systems capable of transferring loads back and forth (i.e., import and export).

If the cogeneration system is separate (that is, it cannot connect in parallel to the utility system), a switching scheme will be necessary (refer to Figure 7-1). A parallel scheme will be easier to operate because the connection between the grid and the cogeneration system is direct (refer to Figure 7-2). With this system, however, the cogeneration scheme must meet utility and government standards.

If cogeneration capability is less than 20 kW, this involves installing controls and devices for suitable connection to the utility. In this case, the devices are required only if the cogeneration system maintains its output when disconnected from the grid.

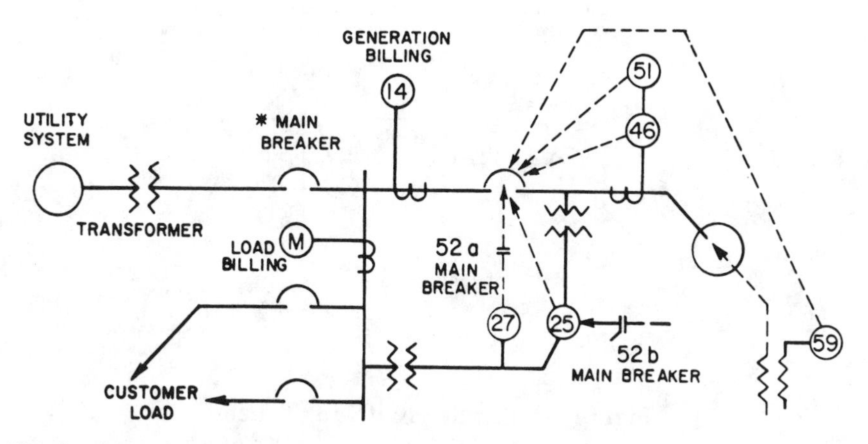

*Synchronizing equipment required if customer desires to close breaker with both sides energized.

Relay Identification

25- Synchronizing
27- Undervoltage
46- Phase-Balance
51- Phase-Overcurrent
52- Main Breaker Auxiliary Switch
59- Neutral Overvoltage
M- Meter

Figure 7-1. Cogeneration Switching Scheme.

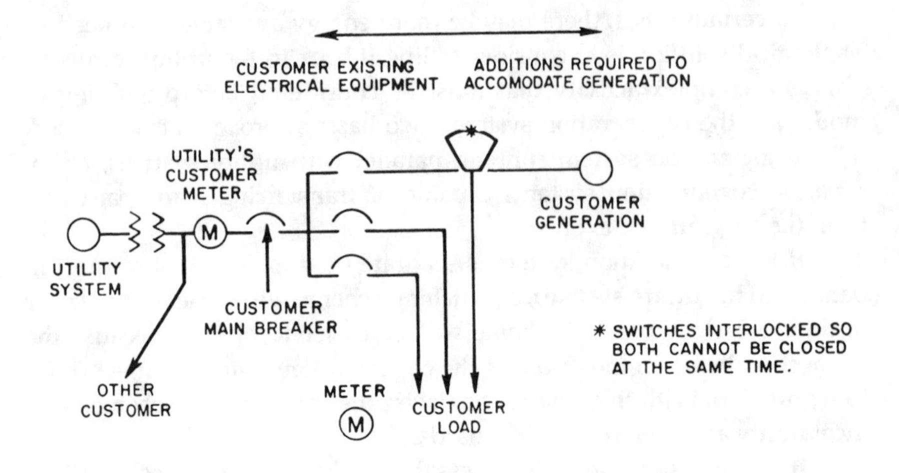

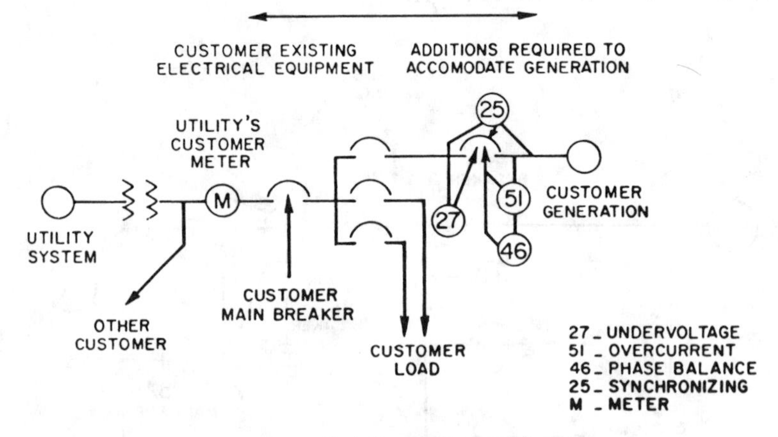

Figure 7-2. Parallel Switching Scheme.

With this scheme, voltage regulation equipment also will be required. Designs for all of these items must be approved by the utility. Low- or high-voltage surges are not allowed, and the cogenerator is responsible for any damage to the utility or other user equipment. Cogeneration systems with a capacity of more than 20 kW will require various trip devices, detectors, special controls and assorted interlocks. In addition, cogeneration systems with outputs high enough to affect utility generation or voltage regulation must have monitoring and control equipment that can be remotely operated by the utility. With systems of this magnitude, the utility will have priority over cogeneration systems including start-up, shutdown and output operations.

The cogeneration system operators also must be aware of operating requirements for these larger systems and must notify the utility before any operation. And, the cogeneration system must comply with utility requests for maintenance work shutdown and/or emergency situations. Weekly, bi-weekly and monthly records should be kept for maintenance reporting procedures. For example, a one-week record on engines will indicate the need for cleaning filters and other engine maintenance. These records also should indicate any overlapping of operating schedules. Ultimately, all complete data (i.e., log sheets) should be systematically filed to maintain continuous records. These records will help in preparing an effective preventive maintenance program which is essential for dynamic equipment use in the overall system.

V. INDUCTION GENERATION

When operated parallel with a utility, induction generators must have a frequency greater than electrical bus frequency (that point in a system where we have nominal voltage and frequency). This will avoid a slip between the generator and bus.

Important differences between a synchronous generator and induction generator are:

(a) A synchronous generator can operate separately from the utility. The induction generator draws its excitation from the bus. When other generators in the utility system fail, the induction generator cannot operate.

(b) A synchronous generator can improve (raise) plant power factor by carrying a reactive load. Since the induction generator takes its excitation from the power line in the form of reactive power, it has a tendency to lower the overall plant power factor, unless corrected. Lower power factor will increase the power bill.

(c) When going on-line, the induction generator and the bus do not have to be as close in frequency and phase as in the case of a synchronous generator. Therefore, paralleling the induction generator with the utility is easier. This is accomplished simply by closing the circuit breaker.

(d) Induction generators require fewer controls - often just a circuit breaker and simplified relay protection. Controls for a synchronous generator include a regulator, reverse-power relays, voltage control, frequency control, differential and loss of excitation protection.

The generator enclosure may be of the open or totally enclosed water/air-cooled (TEWAC) type. For the TEWAC type, air inside the generator is isolated from the ambient air so that the dust and moisture cannot enter. A fan moves the air inside the generator through windings. Heat is removed from the air by a water-cooled heat exchanger. TEWAC generators are used for outdoor installations, dusty locations and applications requiring minimum noise and attractive equipment. TEWAC generators also provide better heat removal than open generators.

Generators can be built with a variety of internal and terminal connections. Some available electrical connections include delta connection, star (or wye) connection, line-to-line, line-to-ground, 3-wire, 4-wire, 6-wire and 7-wire. The type of connection used will not affect the requirements of the generator driver and will have minimum effect on its controls.

Stator temperature detectors (STDs) are devices embedded in the stator windings to detect the temperature windings. Two per phase can be supplied if stator resistor temperature detectors (RTDs) are specified. If one RTD in a phase fails, other RTDs can be used since replacing a stator RTD is a major repair job.

Generator controls are needed to connect the generator to the customer electric power system. Generally, the following items are included in the generator control unit:

(a) **Circuit breaker:** This switch is used to manually connect the generator to the line or take the generator off the line (either manually or automatically). The circuit breakers are sized to trip at circuit demand. They are calibrated 15% above the required current to produce the rated kVa at a rated voltage.

(b) **Shunt trip device:** This causes the circuit breaker to trip in response to an electrical signal. The turbine shuts down in response to a safety device. A switch on the turbine will send an electrical signal to the shunt trip, which causes the generator to go off-line. Some devices that will trip the unit are:

- overcurrent relay,
- generator differential relay,

- thermal overload relay,
- low lube oil pressure,
- transformer sudden pressure relay,
- undervoltage,
- main and bust tie circuit breakers open, and
- control switch.

(c) **Ac ammeter:** This device indicates the output current and the generator input current to the motor.

(d) **Ac voltmeter:** Indicates the bus voltage.

(e) **Wattmeter:** Indicates the real component of the generator's electric power output or input to the motor. This meter is zero centered.

(f) **Recording wattmeter:** Same as wattmeter.

(g) **Current transformers and potential transformers:** Reduce current/voltage from the generator output terminals to proportionally lower values suitable for instrumentation. These are furnished as needed for the 3-phase instruments in the generator control unit.

(h) **Bs voltmeter: Indicates** bus voltage.

(i) **Governor control switch:** Controls the motor-operated speed changer. It also controls the speed setting of the governor, thus controlling the generator load.

(j) **Reverse power relay:** Indicates electric power flowing from (or to) the bus into the generator. It is needed to show whether the generator is in the generating or motoring mode.

(k) **Watthour meter:** Summarizes the total amount of energy delivered to the system.

To illustrate the feasibility, analysis and benefits of a cogeneration system, the following case history of an institutional-industrial application is summarized. High temperature water (HTW) production, quality and reliability were the facility's most important requirements. Because of the critical nature of production losses, continuous electrical and gas services were essential. Cogeneration, fired by natural gas with back-up systems, offered the facility assurance of a reliable source of electrical power and thermal energy.

For each alternative, this analysis required evaluation of the estimated electrical power production, gas and energy consumptions (compared to the existing system). An economic analysis was needed to indicate the most cost-effective scenario and option. The options analyzed were all gas turbine cogeneration scenarios with different applications and modes of operation. A quick review of these options is as follows:

1. **Existing facility:** Three HTW generators, conventionally fired to produce hot water at 400°F. Hot water would be used as the heat transport medium to provide heating for all facilities where required.

2. **Scenario I, Option 1**: Three gas turbines operating at constant full load. The existing three HTW generators would be used as heat-recovery heaters. Excess hot water would produce steam to operate a small steam turbine to generate additional electricity.

3. **Scenario I, Option 2**: One large turbine operating at full load with the turbine exhaust equally ducted to each of the three existing HTW generators. Excess heat would be passed through a heat exchanger to produce steam to operate a small steam turbine to generate additional electricity.

4. **Scenario I, Option 3**: One large gas turbine operating at full load with turbine exhaust passing through a new waste heat boiler sized to match the turbine. The other existing boilers would be used as back-up units. The steam turbine cycle was included, and would generate additional electricity.

5. **Scenario II**: Three gas turbines, each coupled to one of the three existing HTW generators. Each turbine would vary its load and alternately start and stop in accordance with the thermal demand. Therefore, no excess heat would be generated and a steam turbine cycle would not be needed.

6. **Scenario III, Option 1**: One gas turbine sized to meet minimum thermal demand, operating continuously. One of the three existing boilers would be used as a waste heat boiler. Any thermal load in excess of what is provided by the turbine exhaust would be supplied by conventionally firing the other boiler(s).

7. **Scenario III, Option 2**: The basic operation is the same as Scenario III, Option 1. Instead of firing another boiler, however, supplemental firing would be done by directly firing the turbine gases exhausted to the hot water generators.

All scenarios were analyzed to determine gas consumption, generated electrical power and the amount of heat recovery expected from each option (refer to Figures 7-3 through 7-5). New domestic hot water (DHW) systems were calculated for three major areas: the cafeteria, maintenance and the calibration building.

The cafeteria required about 116 gallons/hour maximum demand capability, which required a maximum heating capacity of 96,300 Btuh. Normal demand would probably be reached very sporadically. The maintenance building's maximum demand was calculated to be 770.4 gallons/hour. This also was the storage tank size. To meet demand, a heating capacity of 624,000 Btuh was needed. This system was more expensive and required larger runs of pipe and facility area expansions in addition to the cost and installation of the heaters and concomitant equipment. Maximum DHW demand for the calibration building was calculated as 30 gallons. A 40-gallon storage tank, however, was recommended because of its standard size and only slightly higher cost. The heating demand was calculated to be 4,265 Btuh. Existing HTW boilers operated with a maximum efficiency of 78.3% at a full-load of 25×10^6 Btuh each. Operating at a minimum pressure of 250 psig, the generators were capable of producing hot water at 400°F.

The boilers had dual fuel capability with natural gas as the primary fuel, and #2 fuel oil as the standby fuel source. During boiler plant operation from April to November, only one boiler was operated. Two boilers were operated from November to April. The boilers were instrumented so that the total load was distributed equally between the two boilers. A boiler plant operator was present on a 24-hour basis throughout the year.

HTW was distributed to the facility by a high-pressure piping system. It provided for all domestic hot water and space heating requirements as well as all heating processes required by the building's maintenance facilities. HTW demand was based on actual boiler logs from 1979. Boiler load data were averaged for each month, and the day of the month closest to the average became a typical day. This was done separately for weekdays and weekends. The typical day was based on thermal loading rather than electrical demand.

The electrical system at the facility was serviced by 34.5-kV loop feeders. The 12-kV feeder system was a combined overhead and under-

3 - 800 KW GAS TURBINE GENERATORS EXHAUSTING TO 3 HEAT RECOVERY HOT WATER GENERATOR UTILIZING STEAM GENERATOR AND SUPPLYING DOMESTIC HOT WATER SCHEMATIC. (SCENARIO I, OPTION I)

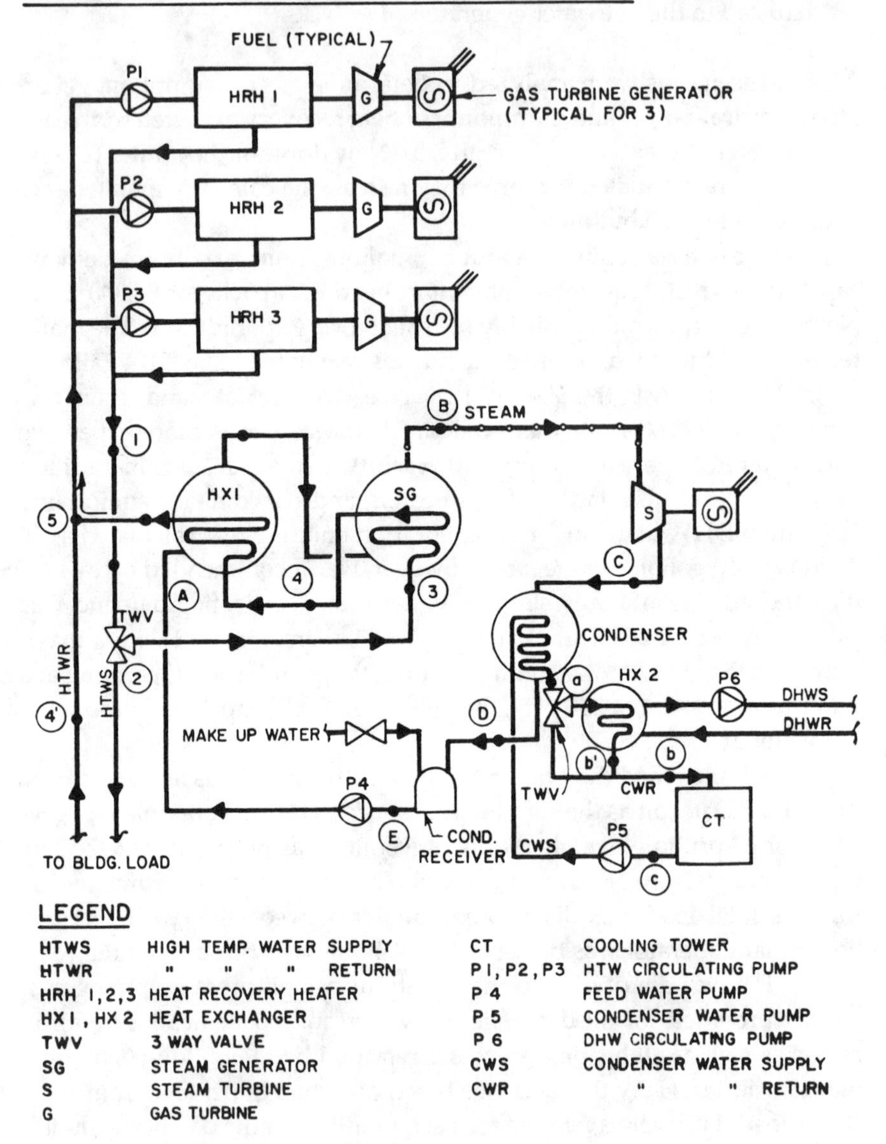

Figure 7-3. Scenario I, Option 1.

1-3200 KW GAS TURBINE GENERATOR EXHAUSTING TO 3 HEAT RECOVERY GENERATORS UTILIZING A STEAM TURBINE GENERATOR AND DOMESTIC HOT WATER HEAT EXCHANGER SCHEMATIC.

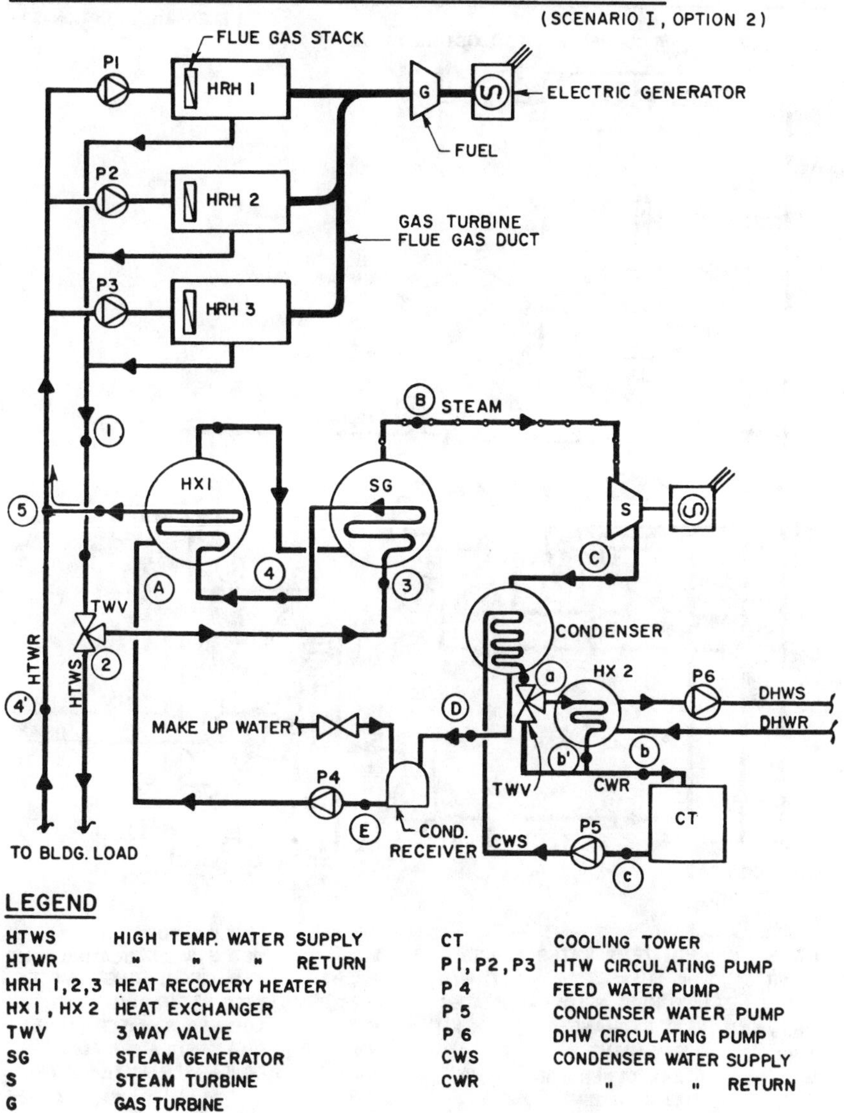

Figure 7-4. Scenario I, Option 2.

I-3200 KW GAS TURBINE GENERATOR EXHAUSTING TO A HEAT RECOVERY HOT WATER GENERATOR UTILIZING A STEAM TURBINE GENERATOR AND DOMESTIC HOT WATER HEAT EXCHANGER SCHEMATIC.

(SCENARIO I , OPTION 3)

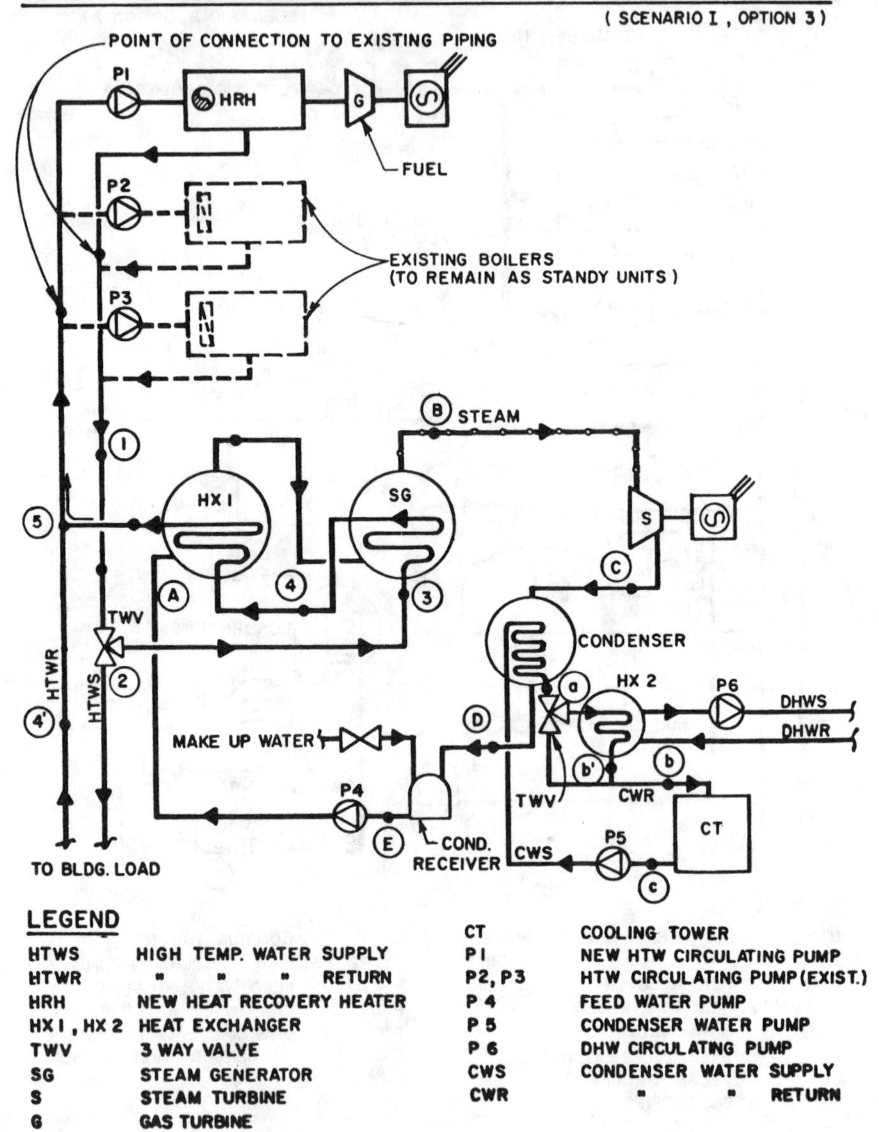

Figure 7-5. Scenario I, Option 3.

ground system. Throughout the base, the 12-kV is transformed to use voltages at various substations. The intent was to interface with the existing 12-kV system at the cogeneration plant site. This would allow base flexibility in operation of the cogeneration plant.

VI. DEVELOPING LOAD PROFILES

Various electrical load profiles were developed from computer print-outs for the previously chosen typical weekday and weekend conditions for each month. Figure 7-6 shows this monthly electrical consumption of the facility for a typical year. Thermal and electrical profiles were determined for each month. The peak thermal demand indicated in December was significant; the cogeneration system must have the capability of meeting the load. Therefore, the thermal load dictated the electrical power generation.

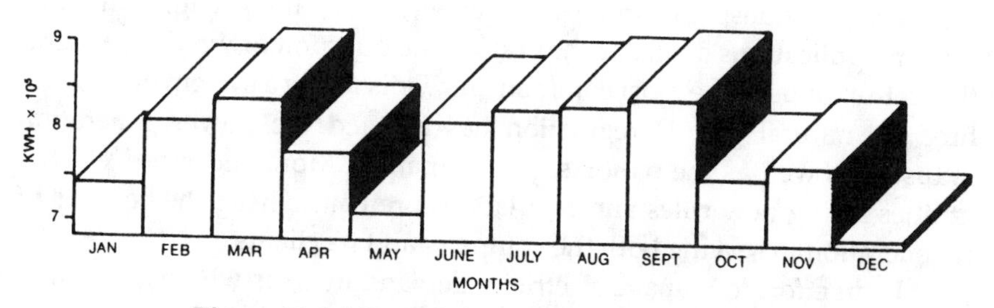

Figure 7-6. Monthly Electrical Consumption

Since the primary concern of the facility was to provide the necessary thermal load, the electrical generator had to be sized coincident with the turbine exhaust system that could best meet the load. The monthly electrical energy use varied from 693,000 kWh/month to 845,100 kWh/month. The peak demand was 4,234 kW.

An analysis of the thermal demand showed wide seasonal variations, unlike the average monthly electrical demand profiles which did not vary significantly. The HTW load showed seasonal fluctuations. Demand profiles for weekday operations were similar for every month. Electrical demand profiles had a low of 400 kW to 600 kW and a high of about 2,400 kW. These occurred at about the same time of the day.

Thermal load analysis was performed using the previous operating year data contained in log books obtained from the facility boiler plant.

The electrical load analysis was based on data obtained from the serving facility and analyzed for demand and consumption. Economic analyses were performed on a 25-year life-cycle basis beginning in 1985. Different scenarios of the total life-cycle costs were compared to the existing facility's total cost, then compared to each other on the basis of savings-to-investment.

Cogeneration provided an effective energy conservation and cost-saving technology for the facility. Furthermore, a cogeneration system, fired by natural gas with the existing system, was used as back-up. It offered additional protection against production losses if utility electrical service is interrupted. Accordingly, Scenario I, Option 3 was determined to be the most cost-effective scenario and chosen for implementation (refer to Figure 7-5).

VII. OPTIONS AND RESEARCH

Recent widespread interest in cogeneration, along with numerous system applications available, has raised the question of which cogeneration set-up to use. The most common answer is to use a system of turbines fired by natural gas. Deregulation has affected the power generation industry as well as the nation's gas companies. More and more, electric utilities facing new rules and regulations are considering the benefits of cogeneration in serving both the company and its clients.

In an effort to manage electricity demand in areas where generation capacity is strained, some utilities are considering gas-fueled cogeneration as a reliable source for the future. The potential for expansion of cogeneration for commercial application (less than 1,000 kW) is being actively developed since large-scale systems have proven reliable. More and more, big-name engine manufacturers (as well as smaller companies) are marketing new cogeneration packages. Some of these companies are now geared toward the down size or "microsystem" trend (away from the multi-megawatt scale) to only 30 kW or less.

Packaged systems respond well to needs of the commercial market, yet are still expensive on a dollars-per-kilowatt basis. Installation costs of commercial systems are disproportionately high and the market is still not mature. Profitable small cogeneration projects are becoming harder to find. The first commercial ventures were established by third-party developers who would contract the work, provide financing for the user, and in return, get a percentage of the savings in energy costs. This practice has been limited due to a saturated market and smaller energy savings. Thus, to succeed, organizations will need financial strength and staying power.

Currently, the cogeneration market is plagued by unexpectedly high installation costs. Primarily, buildings are not designed the way cogeneration systems would want them to be. Yet, in spite of the temporary handicaps, cogeneration systems are gaining a steadfast place in the market. Many chain organizations, hospitals, hotels, motels and restaurants are using cogeneration systems on a trial basis. Should these systems prove successful, cogeneration could become a standard design option for new and remodeled buildings. Surveys have shown that commercial, industrial and institutional building industries using HVAC equipment have the greatest cogeneration potential for the next decade. Future predictions point to an increase in the use of on-site cogeneration systems during the next five years.

Since 1980, natural gas has been the choice of fuel for more than 65% of all U.S. cogeneration projects - usually large-scale system generators. For smaller systems, the trend is greater, showing a gas usage of almost 90%. In the Gulf Coast region of the U.S., natural gas is in plentiful supply. Cogeneration experts agree that gas turbines offer low equipment cost, high electrical generating capacity and a higher exhaust gas temperature. Accordingly, 60% of cogeneration plants built in the U.S. in the past two years are gas-fired. Yet, some of the earliest and most successful cogeneration plants were those built to use waste fuels such as process gas, petroleum, coke or wood.

A number of other fuels can be used in cogeneration - for example, hydrogen-rich off-gases from the chlor-alkali production. The biggest alternative to natural gas is coal. What makes coal so attractive for cogeneration is the rising popularity of fluidized-bed combustors (FBCs).

Sulfur emissions from FBCs can be closely controlled through the injection of lime or limestone into the combustion bed. This may eliminate the need for stack-gas scrubbers. It is an especially important factor in cogeneration planning because air emissions are being monitored.

New cogeneration technology involves the construction of modular cogeneration systems. These systems supply steam or heating water as well as electricity to office buildings, laboratories and other commercial buildings. Cogeneration systems benefit almost all facilities. The strongest development of cogeneration in the U.S. has taken place in light industrial areas such as California and Texas. It also has grown in the heavy industrial mid-Atlantic states and in most of the upper Midwest and New England.

Cogeneration installation capacity in the U.S. is estimated to be

14,000 MW to 16,000 MW at this time. Within the next few years, annual additions should approximate 2,000/3,000 MW (based on current data). Crediting increases in electrical energy costs, the demand for cogeneration in this market is likely to continue. The future of cogeneration is difficult to predict. yet, it is a common belief that the economics of cogeneration can be important to the continued manufacture of products in the U.S.

CHAPTER 8

ESTABLISHING VARIABLE-AIR-VOLUME AIRFLOW RATES ON RETROFIT PROJECTS

Milton Meckler, P.E.

INTRODUCTION

In single-duct systems, control of the dry-bulb temperature in a space requires a balance between the load and volume of the supply air delivered to the space to offset the load. This is accomplished by varying the supply air temperature while maintaining supply air volume constant for constant-volume systems, and by varying the volume of supply air while maintaining supply air temperature constant for variable-air-volume (VAV) systems. VAV systems can be applied to interior or perimeter zones of a building with a common or independent supply air temperature control, common or independent fan systems, and with or without auxiliary heating devices. The concept of VAV may apply to volume variation in the main system total airstream and/or to the control of individual zones.

Fan-powered VAV boxes or individual zone fans may be utilized to maintain minimum or constant supply air to the zone while the system primary air fed to the zone is throttled. The load is satisfied by recirculating return air with the sum of the throttled system air and the recirculated return air kept constant. This is particularly useful for zones with more than a normal variation of internal or external loads and may or may not be combined with terminal reheat. High-pressure primary air may be used with the induction of room or ceiling return air at each zone. The primary air to each zone and the total system air are throttled back, while the induced air is increased by the same amount to maintain substantially constant supply air volume to each zone. This may or may not be combined with terminal reheat. Both of these techniques can be applied to specific areas without using them in all zones.

In a building, indoor contaminant concentration is a major function of the building ventilation rate. ASHRAE Standard 62-1989 (1) now recommends minimum ventilation rates to maintain acceptable indoor air quality (IAQ) for a variety of building types and occupancies. It calls for a minimum ventilation rate of 15 cubic feet per minute (cfm) per person of outside air based on a maximum space concentration of 1,000 ppm for carbon dioxide. The actual distribution of outside air to the occupied zone of a building cannot be assumed to be perfectly mixed and is a function of several variables, including the configuration of the supply and return air distribution circuits, rate of circulation and outside airflow, mode of operation (cooling or heating), and the presence of indoor partitions, equipment and occupants. Recent tests (2) conducted using actual particulate measurements in several representative office buildings with VAV systems to determine ventilation efficiency, and other tests (3)(4)(5) using tracer gases, indicate that the actual delivery of outside air to the occupied zones of an office space may be substantially less than that calculated using building balance reports. The questions arising here are that, if a VAV system is properly set up and balanced in accordance with the minimum recommendations of ASHRAE Standard 62-1989, does this system deliver the recommended minimum ventilation rates to the occupied space at all times and, if not, how much ventilation is necessary to comply with the ASHRAE Standard 62-1989 guidelines?

There is a need to establish adequate documentation to characterize the types of mechanically ventilated air distribution systems so that the heating, ventilating and air-conditioning (HVAC) system designers can properly interpret the published results of building IAQ performance. The main objective of this paper is to develop and install appropriate measurement and control methods capable of determining actual rates of outside air necessary to ensure satisfactory IAQ along with recirculated, filtered and cleaned return air to a typical VAV air distribution system for nonresidential buildings.

The first consideration is the new ASHRAE Standard 62-1989 (1) containing the IAQ Procedure which allows one to compute acceptable indoor contaminant concentrations, proper air distribution in a given space, use of air cleaners to reduce outdoor air requirements and ventilation parameters such as ventilation effectiveness.

II. ASHRAE STANDARD 62-1989

IAQ is a result of a complex relationship between the contamination sources in a building, ventilation rate and the dilution of the indoor air contaminant concentrations with outdoor air. Among these air contaminants are carbon monoxide, formaldehyde, radon, nitrogen oxides, volatile organic compounds, asbestos fibers, respirable particulates, pathogens and allergens. The contemporary tightly sealed modern office buildings may at times contain air contaminants that cause occupants to suffer from, e.g. eye irritation, headaches, respiratory irritation, fatigue and drowsiness, symptoms described collectively as the sick-building syndrome or building-related illness. Table 8-1 shows some probable causes of sick-building syndrome as a result of 346 indoor air investigations conducted by National Institute of Occupational Safety and Health. Of the 346 IAQ investigations, 179 (or 51.7% of the tested) were attributed to inadequate ventilation.

Table 8-1. Probable Causes of Sick-Building Syndrome.

Probable causes	Percent of cases studied	Cases
Fabric contamination	4.0	14
Microbiological contamination	5.5	19
Outside contamination	11.0	38
Inside contamination	16.5	57
Inadequate ventilation	51.7	179
Unknown sources	11.3	39

In response to energy crises of the early 1970s, reducing the amount of outdoor delivered to a building has been considered as one of the easiest ways to conserve energy by most building operators, owners and managers. ASHRAE Standard 62-1973 recommended minimum ventilation requirements for most building occupancies, which were sufficient until the late 1970s. ASHRAE Standard 90-1979 reduced the ventilation air rates from 15 cfm to 20 cfm to 5 cfm per person. As a result of numerous complaints of occupant comfort and increasing IAQ problems, ASHRAE Standard 62-1981 was issued. This standard differentiated between the smoking-permitted and smoking-prohibited areas in a building. Ventilation for office areas where smoking is permitted was increased to 20 cfm, while nonsmoking areas remained at 5 cfm per person. However, the use of this standard for smoking and nonsmoking areas was confusing, and the recommended concentrations of formaldehyde were challenged (1).

ASHRAE has long recognized the need for effective and applicable IAQ standards that keep pace with current technology and emerging public awareness of IAQ-related problems. ASHRAE undertook an early review of ASHRAE Standard 62-1989: Ventilation for Acceptable Indoor Air Quality (1), which embodies contemporary research and technology. This new standard will no doubt affect the way ventilation systems are designed and operated. Table 8-2 provides a comparison of outdoor air ventilation rates of ASHRAE Standard 62-1989 with those of ASHRAE Standards 62-1973 and 62-1981 for various occupancies.

ASHRAE Standard 62-1981, in addition to having the Ventilation Rate (VR) Procedure allowed the HVAC designer to use any amount of outdoor air necessary as long as he could demonstrate the concentrations of indoor air contaminants were below the recommended levels. However, the IAQ Procedure in ASHRAE Standard 62-1981 failed to account for the effect of different types of air distribution systems (constant volume variable temperature vs. constant temperature-variable volume) on the resulting space contamination concentration of any given species. This flaw was corrected in ASHRAE Standard 62-1989 (2).

Carbon dioxide exhaled by occupants in a space must be diluted to achieve desirable comfort as well as to reduce odors and avoid possible health hazards. As the ventilation load increases, the amount of outdoor air needed to dilute occupant-generated carbon dioxide increases. Therefore, carbon dioxide becomes a convenient surrogate for odor. The 5 cfm per person as the minimum outdoor air, as recommended in ASHRAE Standard 62-1981, has been changed to 15 cfm per person to control occupant odors and ensure that the concentration of carbon dioxide will not exceed 1,000 ppm.

ASHRAE Standard 62-1989 now contains a procedure in Appendix E for the use of clean, recirculated air that can treat both constant volume and VAV (with constant or proportional outdoor air). Employing Table E-1 of ASHRAE Standard 62-1989, one can compute the allowable indoor air

Table 8-2. Comparison of Outdoor Air Ventilation Rates of ASHRAE Standard 62-1989 with those of ASHRAE Standards 62-1973 and 62-1981.

Occupancy	Outdoor ventilation air requirement, cfm/person				
	ASHRAE Standard 62-1973		ASHRAE Standard 62-1981		ASHRAE Standard 62-1989
	minimum	recommended	nonsmoking	smoking	VR Procedure
Ballrooms discos	15	20-25	7	35	25
Bars and Cocktail lounges	30	35-40	10	50	30
Beauty shops	25	30-35	20	35	25
Classrooms	10	10-15	5	25	15
Dining rooms	10	15-10	7	35	10
Hospital patient rooms	10	15-20	7	35	25
Hotel conference rooms	20	25-30	7	35	20
Office conference rooms	25	30-40	7	35	20
Office space	15	15-25	5	20	20
Residences	5	7-10	10	10	0.35 ach
Retail stores	7	10-15	5	25	0.02-0.30 cfm/ft^2
Smoking lounges	—	—	—	—	60
Spectator areas	20	25-30	7	35	15
Theater auditoriums	5	5 10	7	35	15
Transporting waiting rooms	15	20-25	7	35	15

contaminant concentration (C_s) in a given occupied space knowing the outdoor airflow rate (VO), ventilation effectiveness (E_v), filter effectiveness (E_f), flow reduction factor (F_r), outdoor contaminant level (C_o), and the respective supply (V_s) and return (V_r) air rates for any of the seven classes. The IAQ Procedure can be used to verify the adequacy of the outdoor air ventilation rates obtained by the VR Procedure as well as to compute more directly the concentrations of indoor air contaminants in an occupied space.

III. AN EXAMPLE

The following example illustrates the computer-assisted (6)(7) use of the IAQ Procedure by using the equations presented in Table 8-3 for a conventional VAV system (Class VI with constant temperature and outdoor air and a filter in location B as shown in Figure 8-1) at 100% full- and 50% part-load conditions. The 50% part-load condition is assumed to be a minimum to provide acceptable indoor air contaminant levels. Also notice that the number of people in a zone and the contaminant generation rate ($\dot{N}$) are reduced to reflect realistically the reduced activities at 50% part-load conditions. The building under study is a nominal 22,000-ft^2 office building in Burbank, California. It contains four occupancies: (a) telecommunication, (b) conference rooms, (c) offices, and (d) shops. The following data are for the communication room only.

At 100% full load condition:

Volume of space = 1,133 m^3 (40,000 ft^3)
Number of people = 300

$\dot{N}$ = 151 µg/min • m^3 (based on ASHRAE Standard 62-1989)
E_v = 0.8 (based on reported test results for air distribution)
V_s = 361 m^3/min (12,740 cfm)
E_f = 0.5
F_r = 1.0
C_o = 5 µg/m^3

At 50% part-load condition:
Volume of space = 1,133 m^3 (40,000 ft^3)
Number of people = 210
$\dot{N}$ = 10.5 µg/min • m^3 (based on ASHRAE Standard 62-1989)

Table 8-3. Required Outdoor Air Space Contaminant Concentration with Recirculation and Filtration (1).

Class	Required recirculation rate				Required outdoor air	Space contaminant concentration	Required recirculation rate
	filter location	flow	temperature	outdoor air			
I	none	VAV	constant	100%	$V_o = \dfrac{\dot{N}}{E_v F_r (C_s - C_o)}$	$C_s = C_o + \dfrac{\dot{N}}{E_v F_r V_o}$	not applicable
II	A	constant	variable	constant	$V_o = \dfrac{\dot{N} - E_v R V_r E_f C_s}{E_v (C_s - C_o)}$	$C_s = \dfrac{\dot{N} + E_v V_o C_o}{E_v (V_o + R V_r E_f)}$	$R V_r = \dfrac{\dot{N} + E_v V_o (C_o - C_s)}{E_v E_f C_s}$
III	A	VAV	constant	constant	$V_o = \dfrac{\dot{N} - E_v F_r R V_r E_f C_s}{E_v (C_s - C_o)}$	$C_s = \dfrac{\dot{N} + E_v V_o C_o}{E_v (V_o + F_r R V_r E_f)}$	$R V_r = \dfrac{\dot{N} + E_v V_o (C_o - C_s)}{E_v F_r E_f C_s}$
IV	A	VAV	constant	proportional	$V_o = \dfrac{\dot{N} - E_v F_r R V_r E_f C_s}{E_v F_r (C_s - C_o)}$	$C_s = \dfrac{\dot{N} + E_v F_r V_o C_o}{F_r E_v (V_o + R V_r E_f)}$	$R V_r = \dfrac{\dot{N} + E_v F_r V_o (C_o - C_s)}{E_v F_r E_f C_s}$
V	B	constant	variable	constant	$V_o = \dfrac{\dot{N} - E_v R V_r E_f C_s}{E_v [C_s - (1 - E_f) C_o]}$	$C_s = \dfrac{\dot{N} + E_v V_o (1 - E_f) C_o}{E_v (V_o + R V_r E_f)}$	$R V_r = \dfrac{\dot{N} + E_v V_o [(1 - E_f) C_o - C_s]}{E_v E_f C_s}$
VI	B	VAV	constant	constant	$V_o = \dfrac{\dot{N} - E_v F_r R V_r E_f C_s}{E_v [C_s - (1 - E_f) C_o]}$	$C_s = \dfrac{\dot{N} + E_v V_o (1 - E_f) C_o}{E_v (V_o + F_r R V_r E_f)}$	$R V_r = \dfrac{\dot{N} + E_v V_o [(1 - E_f) C_o - C_s]}{E_v F_r E_f C_s}$
VII	B	VAV	constant	proportional	$V_o = \dfrac{\dot{N} - E_v F_r R V_r E_f C_s}{E_v F_r [C_s - (1 - E_f) C_o]}$	$C_s = \dfrac{\dot{N} + E_v F_r V_o (1 - E_f) C_o}{E_v F_r (V_o + R V_r E_f)}$	$R V_r = \dfrac{\dot{N} + E_v F_r V_o [(1 - E_f) C_o - C_s]}{E_v F_r E_f C_s}$

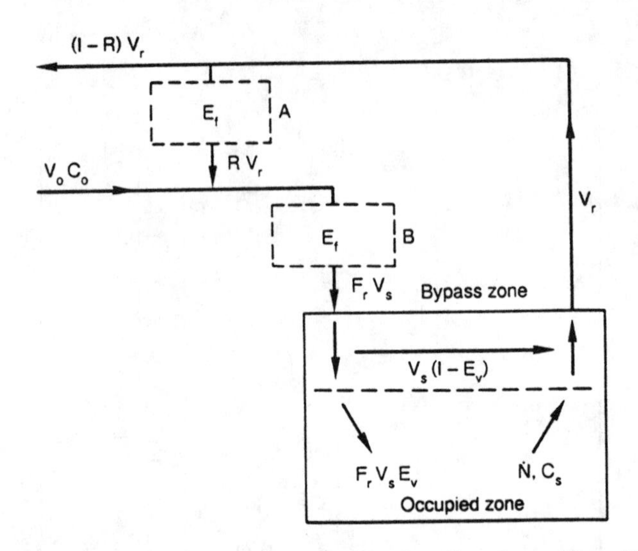

Figure 8-1. Model for Indoor Air Quality Procedure (1).

$E_v = 0.8$ (based on reported test results for air distribution)
$V_s = 361\ m^3/min$ (12,740 cfm)
$E_f = 0.5$
$E_r = 0.5$
$C_o = 5\ \mu g/m^3$

The mass balance equation for V_o in Table 8-3 and the flow balance equation of $V_o = F_rV_s - RV_r$ in Figure 8-1 are simultaneously solved. To satisfy the condition of $C_s < 50\ \mu g/m^3$ and $C_s < 150\ \mu g/m^3$ at long and short terms, respectively, and a minimum outside supply air of 15 cfm/person (as specified by ASHRAE Standard 62-1989), at part- as well as full-loads, V_o is found to be 4,358 cfm with $C_s = 89\ \mu g/m^3$ at 100% full-load and 50% part-load conditions is 15 cfm and 21 cfm, respectively.

IV. AIR DISTRIBUTION IN A SPACE

Air distribution is very important in all HVAC applications, especially for VAV systems. In a VAV system, the selection of grilles and diffusers is a major factor in the proper distribution of air within the conditioned space. The diffusers establish the direction and pattern of the airflow.

In an all-air system, a fast-moving airstream is directed into the conditioned space. As the airstream moves through the room, is aspirates or induces room air into the moving airstream. This aspiration helps circulate air in the room and eliminates stratification. The throw is the distance an airstream travels from the outlet to the point of terminal velocity. Terminal velocity is the point at which the discharged air decreases to a given velocity (generally 50 feet per minute [fpm]).

In a VAV system, a standard diffuser may develop distribution problems at reduced airflow rates. Cool air from a standard ceiling diffuser can be dumped downwards when the volume and velocity are simultaneously reduced, which causes poor air distribution within the space and often creates uncomfortable drafts for the occupants. A properly designed outlet, however, will help maintain an adequate room air movement even at reduced flow conditions.

A linear slot diffuser can maintain a fairly constant throw in a wide range of airflow modulation. The aspirating effect from this moving airstream maintains good room air circulation even at reduced flow rates. Slot diffusers are typically available in lengths of 2 ft, 2.5 ft, 4 ft and 5 ft. The capacity of a slot diffuser is about 50 cfm per foot per slot.

The "Coanda effect" (8), which helps achieve proper room air distribution, maintains the moving airstream close to adjacent parallel surfaces, such as the ceiling or a wall. The discharge air velocity pressure is increased, which creates lower static pressure at the ceiling than in the room. This causes the supply air to attach to the ceiling because of the negative pressure generated between the surface and the airstream. When the airstream leaves the outlet and is directed outward, it will follow along the flat surface over a wide range of airflow rates.

The Coanda effect can be maintained by using a flat surface 12 inches to 18 inches from the outlet of an overhead diffuser. A linear slot diffuser may be placed in perimeter zones in such a way that the discharged air flows perpendicularly to the perimeter wall. Figure 8-2 shows a comparison between a standard diffuser and linear slot diffuser. The velocity of the air striking the wall should be about 150 fpm at full volume. In a heating-cycle, any velocity significantly less than 150 fpm will cause a loss of the Coanda effect along the outside wall. Care should be exercised in placing diffusers. Because of the aspirating effect, a linear slot diffuser may be set at lower minimum air quantities than other types of diffusers. The use of no minimum stops results in energy efficiency. During the year when "no-load" conditions exist, the VAV boxes may close completely, which could cause stagnation problems. The minimum setting of stops should ensure that each zone has some air movement even during "no-load" periods. However, the need to reheat this minimum air quantity may result in higher energy consumption. If care is not taken when selecting a diffuser, the use of the minimum stop setting can also cause dumping of cold conditioned air into an already ventilated space.

In a single-path VAV system, airflow must be generally throttled back to perimeter zones during a heating cycle to eliminate excess reheat losses. The resulting reduced velocity of the higher than ambient temperature air from a ceiling-type distribution has little effect in offsetting down drafts generated by low temperature on the inner surface of outside walls during periods of design outdoor temperatures. This situation is usually handled by an entirely separate system such as a baseboard convection or constant-volume so-called "skin" air-handling system. It is recommended that the supply air temperature from the overhead diffuser be no more than 20°F to 30°F above the ambient room temperature during the heat-

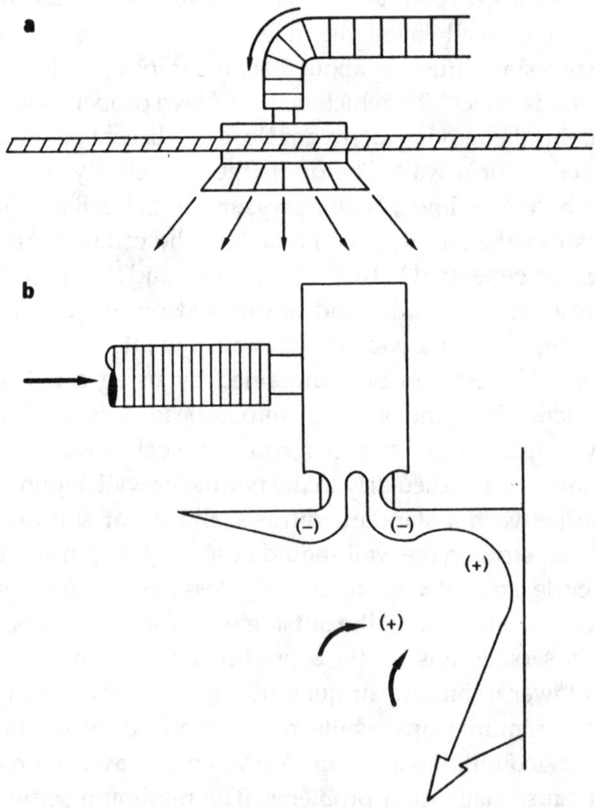

Figure 8-2. Comparison of Standard Diffuser (a) and Linear Slot Diffuser (b).

ing-cycle to avoid buoyancy. In addition, the discharge velocity from the diffusers should be such that the velocity at the intersection of the air-stream and the outside wall is approximately 150 fpm.

Where an under-window type of distribution is used, vertical throw outlets with a nonspreading pattern should be used. To prevent cool air from dropping back into the occupied space at minimum flow conditions, the outlet discharge velocity should be a minimum of 500 fpm. Modular ceiling slot-type diffusers are most desirable when large air volume reduction in a space can occur.

VAV systems utilizing blower-powered mixing boxes can employ standard diffusers without reducing airflows. While the main system may vary the volume, the zone terminals act as constant-volume units. As the supply air damper modulates towards the closed position, the return air damper opens, drawing air from the conditioned space or plenum above the ceiling and maintaining constant air volume to the zone outlets.

Proper sizing and positioning of return air openings are important in establishing room airflow patterns that minimize stratification and stagnant air zones. In general, the return air opening should equal or exceed the total area of the supply air opening. To help promote adequate air distribution in a conditioned space, the return air openings should be installed in offset positions or perpendicularly to the slot diffusers. Maintaining air quantities lower than those at design conditions during most of the operating time tends to reduce draft problems. Problems are more likely to occur at high airflow rates than at low airflow rates. A high entrainment type of outlet is recommended to achieve higher air movement at minimum flow rates.

V. VENTILATION EFFECTIVENESS

Traditionally designed ventilation systems to control odor and provide thermal comfort have assumed that the air in a building is perfectly mixed. IAQ of a given space is determined by the ability of the ventilation system to directly supply ventilation air and remove contaminants before they mix with the room air. The performance of a ventilation system depends on the building geometry, contaminant sources, thermal stratification, types of diffusers and location of a duct. Effectiveness of the ventilation depends very much on local mixing of indoor air contaminants. Flow nonuniformities can produce localized areas with unusually high concentrations of indoor air contaminants, even if an acceptable average concentration for a building is achieved at a given ventilation rate

(9). Supply air entering a space should be at such velocity and angle that sufficient mixing will take place, which is not easily accomplished at all times, when the air is supplied by a VAV system. Higher ventilation efficiency can be achieved by using a fan-powered mixing box locally. A well-designed ventilation system must provide an adequate balance between the ventilation rate and contaminant source at all occupied locations for a wide range of operating conditions, and must account for the worst case rather than average conditions.

The ratio of the transit time for the ventilation airflow to the average transit time for the contaminant flow through the room is called the average ventilation effectiveness. This ratio, by definition, is also equal to the ratio of the contaminant concentrations in the exhaust air to the average contaminant concentrations in the room at steady state. Table 8-4 shows average ventilation effectiveness for several flow types. A design goal should be to attain an average ventilation effectiveness ratio of greater than unity. Thermal and contaminant stratification that can occur in an occupied space due to poor mixing may cause occupants to be exposed to much higher than anticipated contaminant levels at times. Two characteristics of the ventilation system crucial for IAQ control are the room air exchange rate and airflow patterns in a space.

Table 8-4. Average Ventilation Effectiveness.

Flow	Average ventilation effectiveness
Stagnant flow (short circuiting)	> 0
Complete mixing	1
Contaminant source near exhaust	∞

Inadequate air distribution in a space will impair the effectiveness of the system and require excessive air exchange rates to compensate, resulting in increased energy consumption and nonuniform mixing. Air patterns in a room may be as important as the room air exchange rate (10). Generally, they are very complicated because of warm, and cold surfaces, unevenly distributed heat sources, supply diffuser design, air supply and exhaust locations, infiltration and exfiltration, and unevenly distributed indoor air contaminants with regard to thermal stratification.

In Figure 8-3, ventilation efficiency is a function of the stratification factor (S) and recirculation factor (R). The R can be thought of as the

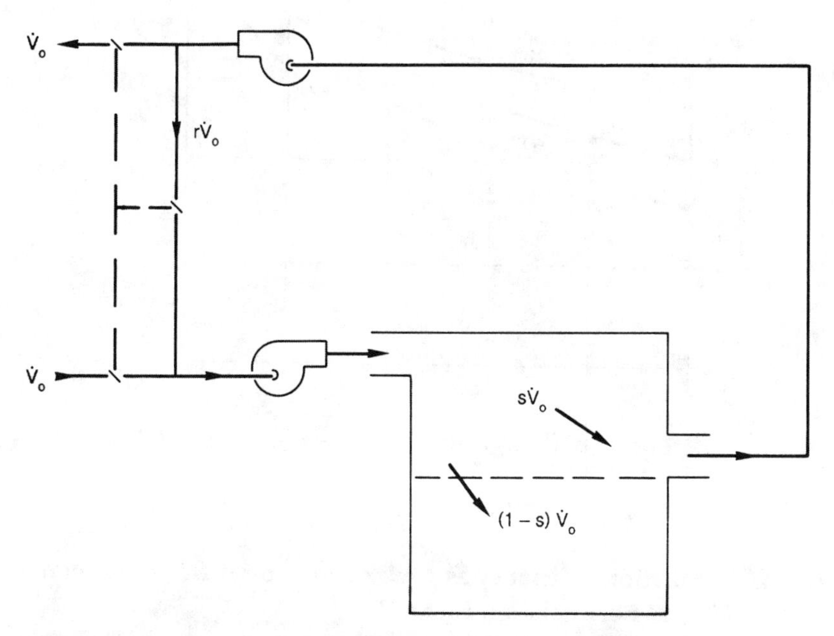

Figure 8-3. Two-Compartment Model of Recirculation System with Stratification.

percent of outdoor air entering the room that ventilates the occupied zone and is defined as (10):

$$E = (1 - S)/(1 - SR)-$$

Figure 8-4 shows typical ventilation efficiencies as a function of R and S.

VI. AIR CLEANERS

To increase the fraction of return air used, an air cleaner can be placed either in the recirculated airstream (or mixed airstream, refer to Figure 8-1). If the contaminants are all coming from the conditioned space, filtration of the recirculated air is the recommended procedure. However,

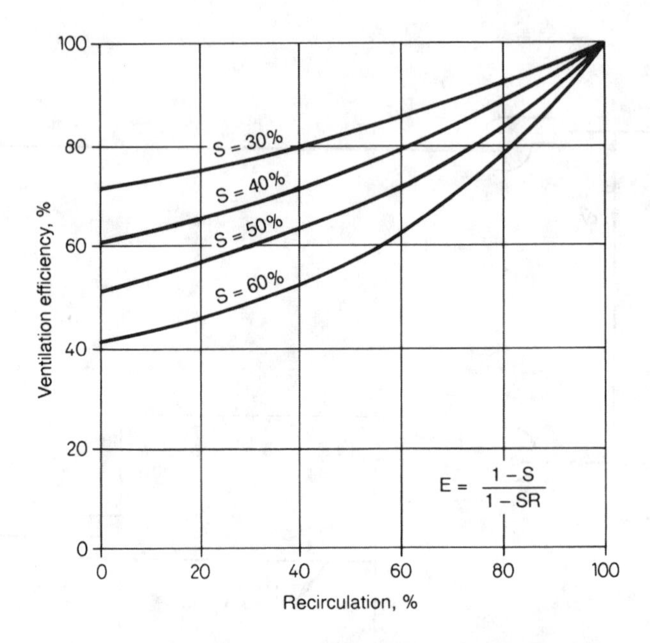

Figure 8-4. Ventilation Efficiency as a Function of Stratification Factor (S) and Recirculation Factor (R).

if the outdoor air is also a source of filtrable contaminants, this should be filtered too.

The volumetric flow through an air cleaner and its effectiveness determine the amount of filtrable material that can be removed. Consequently, when the flow is reduced in a VAV system, the contaminant removal capacity is also reduced. This is recognized by the flow reduction factor (F_r). The filter effectiveness (E_f) is best obtained from the manufacturer, but will be affected by the flow through the filter. Thus, the increased effectiveness of the filters at reduced flow rates may compensate, to some extent, for the reduced airflow. In general, effectiveness increases as velocity through the air cleaner is reduced. The R is the fraction of the return air that is recirculated. As the ventilation load increases, the amount of outdoor air needed to dilute occupant-generated carbon dioxide increases. The recirculation factor will then decrease.

Among the most commonly used air cleaners are media filters, adsorbers and absorbers, centrifugal separators, electrostatic air cleaners and air washers. New and remodeled buildings require HVAC systems that will ventilate, filter and dilute indoor air contaminants when an unsafe level of contamination is reached. It is also necessary to eliminate areas that breed bacteria and mold such as standing water in ducts. Particulates can be removed effectively by available filtration and removal technology such as electrostatic filters, ionization devices and high-efficiency particulate air (HEPA) filters effective in the micrometer range, that are commercially available. In addition, some lithium chloride desiccant cooling (air washer) systems are known to have bacteriostatic effects. Thus, the use of air recirculation in combination with adequate filtration now provides the HVAC designer with a viable cost-effective alternative to increase outdoor air rated in accordance with IAQ Procedure in ASHRAE Standard 62-1989 (1).

VII. REFERENCES

(1) ASHRAE Standard 62-1989: Ventilation for Acceptable Indoor Air Quality. Atlanta, ASHRAE, 1989.

(2) Meckler, M. Ventilation/air distribution cure sick buildings. Consulting Specifying Engineer, 1985.

(3) Fisk, W.J.; Binenboym, J.; Kaboli, H.; Grimsrud, D.T.; Robb A.W.; and Weber, P.J. Multi-tracer system for measuring ventilation rates and ventilation efficiencies in large mechanically ventilated buildings. Berkeley, University of California, LBL-20209. Lawrence Berkeley Laboratory, 1985.

(4) Janssen, J.E. Ventilation for acceptable indoor air quality: Operational implications. Proceedings of International Facilities Management Association Conference Houston, 1987.

(5) Offerman, F.J. Ventilation effectiveness and ADPI measurements of forced air heating system. ASHRAE Transactions, 1988, 94.

(6) Meckler, M. Computer design, monitoring, control and modeling of HVAC systems. Sol *AIR, 1986, 30.

(7) Meckler, M. Design for integrated IAQ thermal comfort zoning. Proceedings of ASHRAE/SOEH Conference, IAQ '89: The Human Equation: Health and Comfort, San Diego, 1989.

(8) Meckler, M. and Janssen, J.E. Use of air cleaners to reduce outdoor

air requirements. Proceedings of the ASHRAE Conference, IAQ 88: Engineering Solutions to Indoor Air Problems, Atlanta, 1988.

(9) Meckler, M. (ed): Indoor Air Quality Design Guidebook. Fairmont Press, Lilburn, GA, 1991, Part 2, Chapter 9 (by R. Anderson).

(10) Meckler, M. (ed): Indoor Air Quality Design Guidebook. Fairmont Press, Lilburn, GA, 1991, Part 2, Chapter 10 (by H. Goodfellow).

Chapter 9

Estimating Demand-Control Savings on Retrofit Projects

Milton Meckler, P.E.

I. INTRODUCTION

Since ventilation with outdoor air is essential to replace the oxygen consumed and dilute the indoor air contaminants to acceptable levels, indoor air quality (IAQ) is directly related to outdoor air quality. In conventional heating, ventilating and air-conditioning (HVAC) designs, the arbitrary increase of outdoor air has been a standard practice for many years. However, total reliance on dilution air is neither cost-effective nor energy efficient and, indiscriminate use of outdoor air may result in undesired contaminants in higher concentrations brought into an occupied space.

Table 2 of ASHRAE Standard 62-1989 recommends outdoor air requirements for ventilation for a variety of occupancies. It is necessary that at least these amounts of outdoor air be delivered to a conditioned space whenever the building is occupied except as modified in paragraph 6.1.3.4 of the standard: Intermittent and Variable Occupancy. In most cases, indoor air contamination is assumed to be directly proportional to the number of people in the space; in other cases, it is assumed to be mainly due to other factors, and the ventilation rates are based on other appropriate parameters. These ventilation rates are listed for well-mixed airflow conditions (or a ventilation effectiveness approaching unity).

The outdoor air requirements have been chosen to control carbon dioxide (CO_2) and other contaminants with an adequate margin of safety and to account for health variations among people, varied activity levels, and moderate smoking (1). For example, for an office occupancy, the minimum outside air flow rate and maximum occupancy are 20 cubic feet per minute (cfm)/person and 7 people/1000 ft^2, respectively. When occupant density differs from what is specified in Table 2 of the standard, the "per-occupant ventilation rate" for the anticipated occupancy load may be used.

The rate at which oxygen consumed and CO_2 generated depends on the physical activity of the occupants. As the ventilation load increases, the amount of outdoor air needed to dilute occupant-generated CO_2 increases. Therefore, CO_2 is now widely recognized as both a convenient surrogate for odor and an indirect measure of the adequacy of supply (mixing outdoor and recirculated) air to a conditioned space (2). ASHRAE Standard 62-1989 recommends a minimum of 15 cfm per person to control occupant odors and ensure that the concentration of CO_2 will not exceed 1000 parts per million (ppm). If CO_2 is controlled by any method other than dilution, the effects of a possible rise in other indoor air contaminants must be considered.

To establish the amount of outdoor air needed, we have constructed an interactive algorithm capable of estimating time-varying CO_2 concentrations to ensure that ventilation rates selected are capable of satisfying minimum ventilation requirements (a "worst case scenario") by investigating each zone on a real-time basis (2). In this chapter, we will present the results of the application of this dynamic modeling methodology to a 10-story office building assumed to be located in five representative U.S. cities. Calculated hourly outdoor air flow rates at preset CO_2 concentrations of 800 ppm or 920 ppm will be compared to a conventional approach in which a constant outdoor air flow rate of 20 cfm/per person during all occupied hours is supplied. The outdoor air flow control strategies utilize CO_2 sensors commercially available today and take advantage of actual "variable occupancy."

II. A BUILDING STUDY

The study building used to demonstrate the use of the methodology mentioned above for demand-controlled systems, is a 10-story high rise office building with reinforced concrete and steel frame. The walls are mainly glass with spandrel glass covering the inter-floor slabs. Two-foot-wide concrete columns are located around the perimeter and throughout the building in a grid pattern of approximately 20 ft by 20 ft. The floor is arranged with perimeter offices of an average 150-ft^2 floor area with the interior areas used mainly for open office spaces with five-foot-high movable partitions, forming a part of the furniture layout. Several conference and meeting rooms are provided for each floor as well as a lounge area. The toilets are in a central service core and are not a part of the rental space.

The entire building is served by a central cooling and heating plant having two centrifugal water-cooled chillers with interconnected cooling towers, and a gas-fired hot water boiler for heating. The hydronic system is a four-pipe arrangement with both cooling and heating allowed year-around. Each floor is served by two separate variable-air-volume (VAV) air-handling units (AHUs), one unit serving the perimeter areas and the other serving the interior spaces as shown in Figure 9-1. A common outside air (OSA) AHU serves all floors by means of a shaft located near the central core as shown in Figure 9-1. OSA is introduced to each equipment room containing both AHUs serving each floor by means of a separate OSA AHU set to deliver either a constant (Alternative 1) or variable-air-rate (Alternatives 2 and 3). All alternatives contain a 2-position damper interconnected with their corresponding ventilation air opening (also located in each equipment room) from a combination shaft interconnected through a common wall opening with relief air exiting through a separate horizontal opening located above the ceiling line into the shaft void to relieve excess air to atmosphere as needed. In this way, slightly positive floor and building pressurizations are maintained at all times (refer to Figure 9-1).

CO_2 sensors located in the return-air ductwork to each AHU in Alternatives 2 and 3 monitor only the CO_2 concentrations to corresponding levels indicated above and are directly interconnected with the building direct digital control (DDC) system. Floor-to-floor AHUs are served by a roof-mounted cooling tower capable of providing a wet economizer free cooling cycle. In this way, valuable space for delivering 100 percent outdoor air is avoided, maximizing tenanted space and minimizing comfort and IAQ problems due to uncontrolled space humidity and high seasonal pollen loading often accompanying "all air" economizer cycles.

The OSA AHUs are equipped with integral heating and cooling coils to deliver pre-conditioned outdoor air supplied to each floor AHU at respective season return-air temperatures to each equipment room so that their respective floor VAV units only provide for space heating and cool-

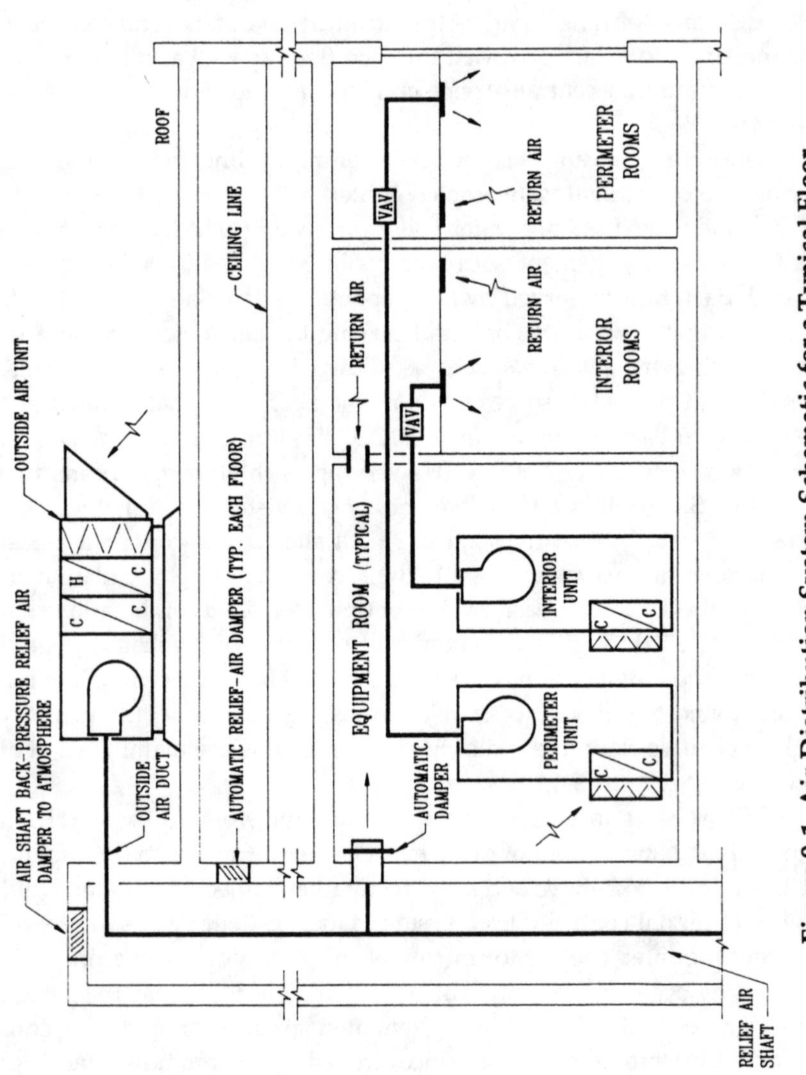

Figure 9-1. Air-Distribution System Schematic for a Typical Floor.

ing. Supply air is delivered to their respective perimeter or interior conditioned spaces through two medium-pressure duct loops, as shown in Figure 9-1. The cooling-only VAV units serve each zone equipped with preset minimum air stops for an ensured ventilation air supply to each space. The heating system consists of separate baseboard convection units that are supplied with hot water from central gas-fired, hot water boilers. The baseboard units for each zone, however, are controlled by the same thermostat which controls the VAV units of the perimeter zone thus eliminating the possibility of simultaneous heating and cooling in any given temperature control zone.

A. Building and System Modeling

The study office building was modeled and its energy use was analyzed by means of a proprietary (TRACE 600) computer energy analysis program. This program allows simulation of the building envelope behavior as well as the HVAC system throughout a typical year. The weather data for the simulations is derived from the NOAA weather tapes for a typical day of each month. Actual design schedules were employed for establishing lighting and variable occupancy profiles. The building has 180 people on each floor and a total zone volume of 231,354 ft^2 (2). Note that the outdoor ventilation air flows are slightly higher than the values listed in Table 2 of ASHRAE Standard 62-1989.

Three ventilation demand-control strategies were considered:

1. **Alternative 1: Constant Outdoor Air Requirements for Ventilation at 20 cfm/person.** This alternative is based on the recommended outdoor air requirement of 20 cfm/person for an office occupancy in accordance with Ventilation Rate Procedure (VRP) in ASHRAE Standard 621989. The calculated maximum CO_2 concentration is 920 ppm provided a two-hour daily purge cycle is incorporated. Accordingly, the OSA ventilation unit was operated during all occupied hours with the two-hour purge cycle split between 6:00 pm to 7:00 pm and 6:00 am to 7:00 am.

2. **Alternative 2: Demand-Controller Set at 800 ppm.** In this alternative, the outdoor air required for ventilation was controlled by individual CO_2 sensors located in the return-air ductwork and connected to a DCC set to maintain a present maximum CO_2 concentration of 800 ppm in each conditioned space. The outdoor air required for ventilation is thus directly proportional to actual occupancy needs for each representative AHU and the OSA ventilation unit operates only when

the office building is occupied (6:00 am to 7:00 pm).

3. **Alternative 3: Demand Controller Set at 920 ppm.** This alternative is similar to the one described in B. above, except that the maximum CO_2 concentration to be maintained is 920 ppm. This alternative closely approximates the conditions attainable with the constant outdoor air requirement of 20 cfm/person in Alternative 1.

Figure 9-2 shows the calculated hourly CO_2 concentrations at a constant outdoor airflow rate of 20 cfm/person. This is in accordance with Table 2 of ASHRAE Standard 62-1989 which recommends a minimum outdoor air flow rate of 20 cfm/person for an office occupancy provided well-mixed airflow conditions (ventilation effectiveness approaching unity) exist. Referring to Figure 9-2, note that the CO_2 concentration increases rather sharply between 7:00 am and 11:00 am, reaching a peak of 904 ppm by 11:00 am, and a maximum of 920 ppm at 5:00 pm. It does not exceed 1000 ppm for variable occupancy conditions recommended by ASHRAE. As the space use decreases due primarily to occupants leaving their offices between 5:00 pm and 6:00 pm, the CO_2 concentration experiences a rather sharp decline and then levels off at around 400 ppm until 8:00 am the next morning, when the office building is again occupied. Alternative 1 will be used as a base case to which Alternatives 2 and 3 will be compared.

Figure 9-3 shows the calculated outdoor ventilation air rates in Alternative 2 at a preset CO_2 concentration of 800 ppm in a typical floor. Note that the outdoor airflow rate increases sharply between 7:00 am and 11:00 am, and 1:00 pm and 5:00 pm. In both time intervals, the maximum outdoor air flow rate corresponds to 4770 cfm peak outdoor air rate (or approximately 27 cfm/person).

Finally, Figure 9-4 shows the calculated outdoor ventilation air rates in Alternative 3 at a preset CO_2 concentration of 920 ppm in a typical floor. Note that the outdoor airflow rate again increases sharply between 7:00 am and 11:00 am, and 1:00 pm and 5:00 pm. The maximum outdoor airflow rate at this time is 3672 cfm (or approximately 20 cfm/person).

An economic analysis was performed for Alternatives 1 through 3 using the TRACE 600 computer energy analysis program. Table 9-1 shows the estimated annual energy consumption (electricity and natural gas) for Alternative 1 for five U.S. cities (representative of different regions) as a "base case." These cities are Miami; Atlanta; Washington, D.C.; New York and Chicago.

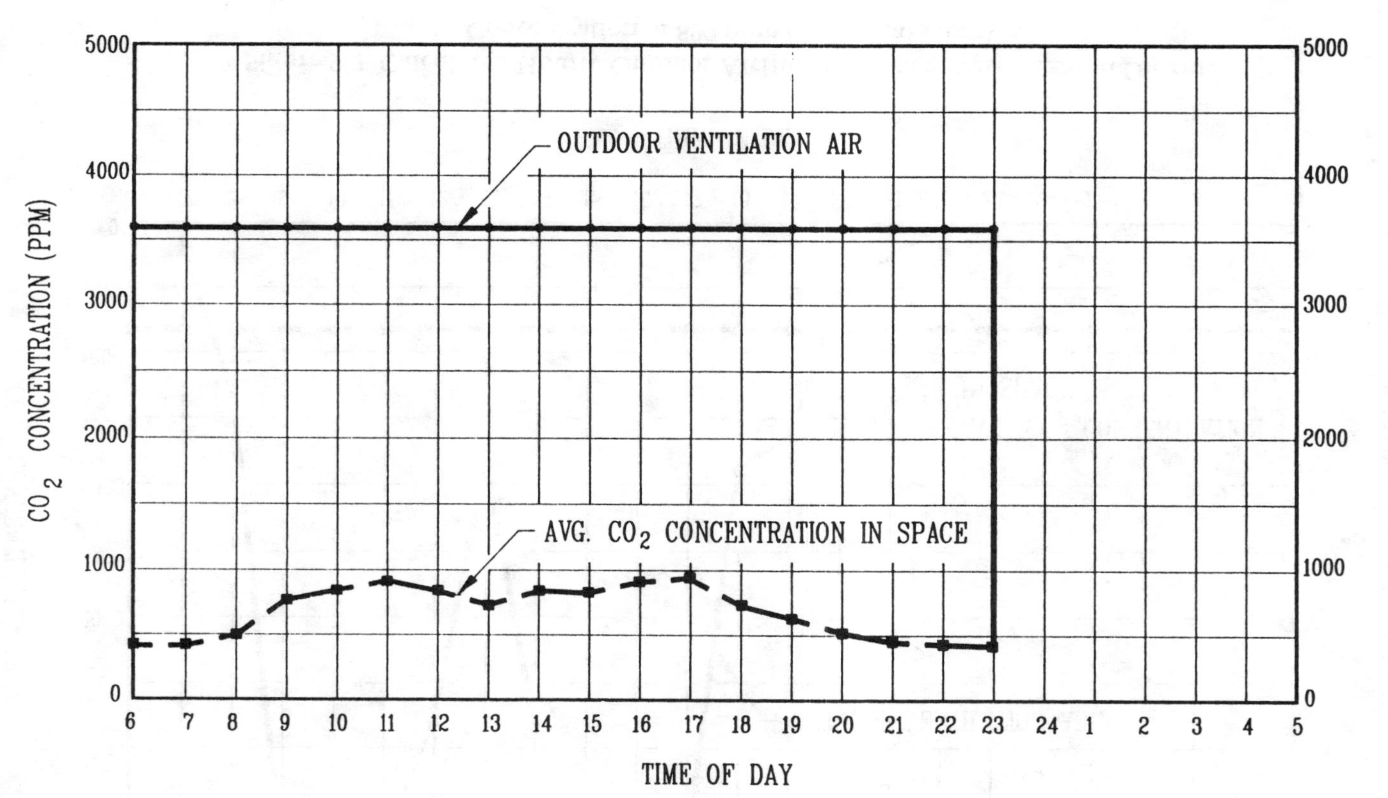

Figure 9-2. Calculated Hourly Carbon Dioxide Concentrations at a Constant Outdoor Airflow Rate of 20 cfm/person (A Typical Floor).

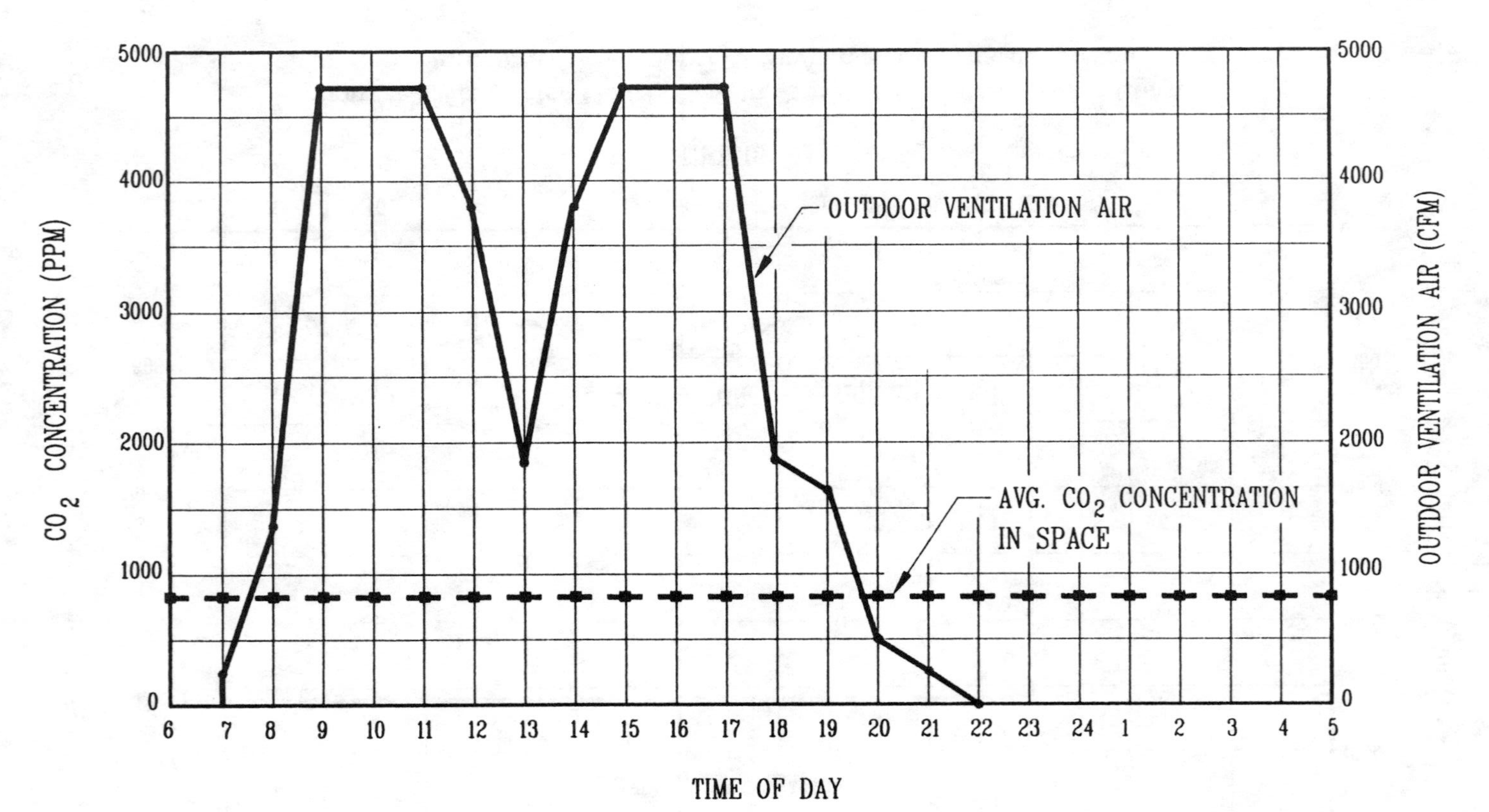

Figure 9-3. Calculated Hourly Outdoor Airflow Rates at a Preset Carbon Dioxide Concentration of 800 ppm (A Typical Floor).

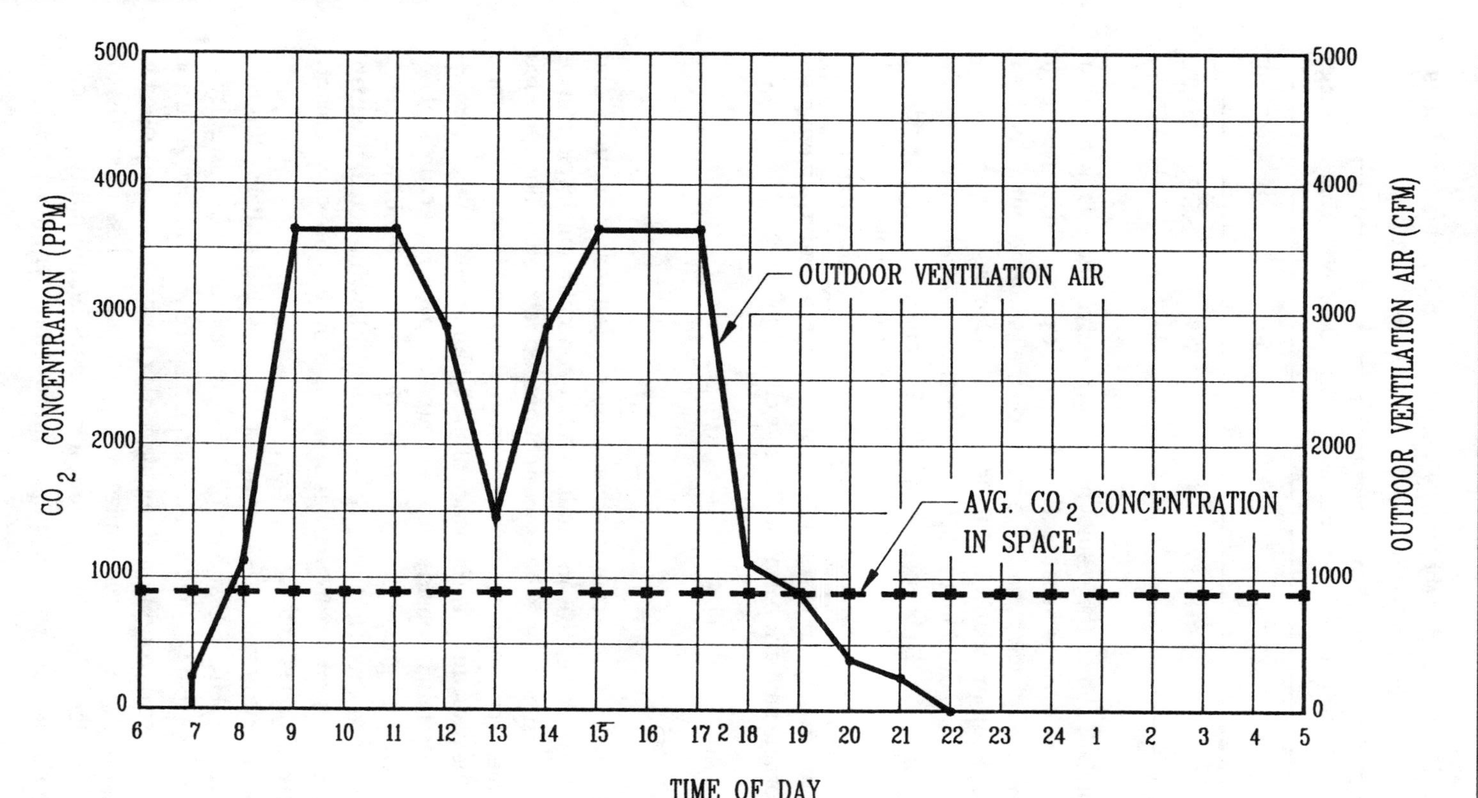

Figure 9-4. Calculated Hourly Outdoor Airflow Rates at a Preset Carbon Dioxide Concentration of 920 ppm (A Typical Floor).

Table 9-1. Estimated Hourly Energy Consumption for Five U.S. Cities (Alternative 1).

	Miami	Atlanta	Wash., DC	New York	Chicago
Electricity (kWh)	815,725	814,808	717,380	634,991	656,707
Gas (therms)	22,600	22,600	31,362	35,672	51,315

Table 9-2 shows the results of a comparative analysis in terms of annual energy savings and associated payback periods with respect to Alternative 1. Payback periods are based on the estimated initial differential cost of equipment and installation of the CO_2 sensors that was estimated to cost an additional $9,000 in comparison to Alternative 1, for the entire building. As can be seen from Table 9-2, energy savings on natural gas for both Alternatives 2 and 3 increase as the city climate gets colder. It can also be seen that the payback periods for Alternative 2 and 3 range from approximately 1.5 years to 2.0 years, with Alternative 3 having the most favorable payback period for all five cities evaluated.

III. CONCLUSIONS

The results of this comparative study suggest that continuous and accurate measurement of maintained CO_2 concentration levels can be cost-effective in matching the amount of outdoor air ventilation required for variable occupancy needs rather than fixed outdoor airflow rates based on the peak design occupancy of the building. Such rates are excessive since they fail to consider actual occupancy patterns. Accordingly, it can be shown that greater levels of comfort as well as significant energy savings can be realized as a result of accurately following actual occupant ventilation needs on a real-time basis, which is the essence of a demand-control ventilation strategy.

Properly designed CO_2 demand-control systems utilizing the IAQ Procedure of ASHRAE Standard 62-1989 can provide an opportunity to maintain acceptable IAQ within an occupied space without excessive and unnecessary use of conditioned ventilation air beyond that prescribed in VRP. Also use of a wet economizer cycle (3) permits less wasted shaft space and better control of uncontrolled factors i.e., humidity, pollens, VOCs and other outdoor air contaminants often present in higher levels

Table 9-2. Comparison of Alternatives 2 and 3 to Alternative 1, Estimated Annual Energy Savings and Payback Periods for Five U.S. Cities.

Energy Use	Miami		Atlanta		Wash., DC		New York		Chicago	
Parameters	Alt. 2	Alt. 3	Alt. 2	Alt. 3	Alt. 2	Alt. 3	Alt. 2	Alt. 3	Alt. 2	Alt. 3
Electricity (kWh)	15,907	24,004	15,862	24,012	9,260	13,597	2,277	558	8,765	11,369
Gas (therms)	4,372	4,903	4,370	4,900	6,048	6,719	6,769	7,468	8,965	10,285
Total ($)	4,964	5,934	4,033	5,042	4,256	4,913	5,865	6,337	4,543	5,291
Payback Period (yrs)	1.81	1.52	2.23	1.79	2.11	1.83	1.53	1.42	1.98	1.70

than permitted by Environmental Protection Agency (EPA) in urban environments. In addition, maintaining controlled ventilation at all times may help reduce the spread of common viruses and bacteria that lead to temporary sickness, downtime, loss of productivity, and exposure of HVAC designers to IAQ liability (4).

IV. REFERENCES

(1) ASHRAE Standard 62-1989, Ventilation for Acceptable Indoor Air Quality, American Society of Heating, Refrigerating and Air-Conditioning Engineers, Inc., Atlanta, Ga, 1989.

(2) Meckler, M, "Carbon dioxide prediction model for VAV system part-load evaluation," Heating/Piping/Air Conditioning, January, 1993.

(3) Meckler, M, "Integrated fire sprinkler and thermal storage systems," Heating/Piping/Air Conditioning, May, 1 9 9 2.

(4) Kirsch, L.S, "Legal implications of indoor air quality," Proceedings of IAQ '92: Environments for People. American Society of Heating, Refrigerating and Air-Conditioning Engineers, Inc., Atlanta, Ga, October, 1992.

Chapter 10

Commissioning for Retrofit Projects

Milton Meckler, P.E.

I. INTRODUCTION

Building commissioning and operations are both important factors in maintaining acceptable indoor air quality (IAQ) in new or existing buildings. Acceptable IAQ (as defined in ASHRAE Standard 62-1989) is indoor air that contains no known contaminants at harmful concentrations as determined by cognizant authorities and with a substantial majority (80% or more) of the people exposed do not express dissatisfaction (1). The commissioning process includes procedures and methods for verifying and documenting the performance of heating, ventilating, and air-conditioning (HVAC) systems to ensure proper operation in accordance with the original or subsequent reconfigured design intent. The commissioning process includes documentation and verification that owners/tenants design criteria have been adhered to, a description of installed HVAC systems and their intended operational modes, and performance goals (refer to Figure 10-1).

Although most buildings are generally considered to be safe and healthy working environments, appearances can be deceiving. The manner in which buildings are delivered to their owners/operators often

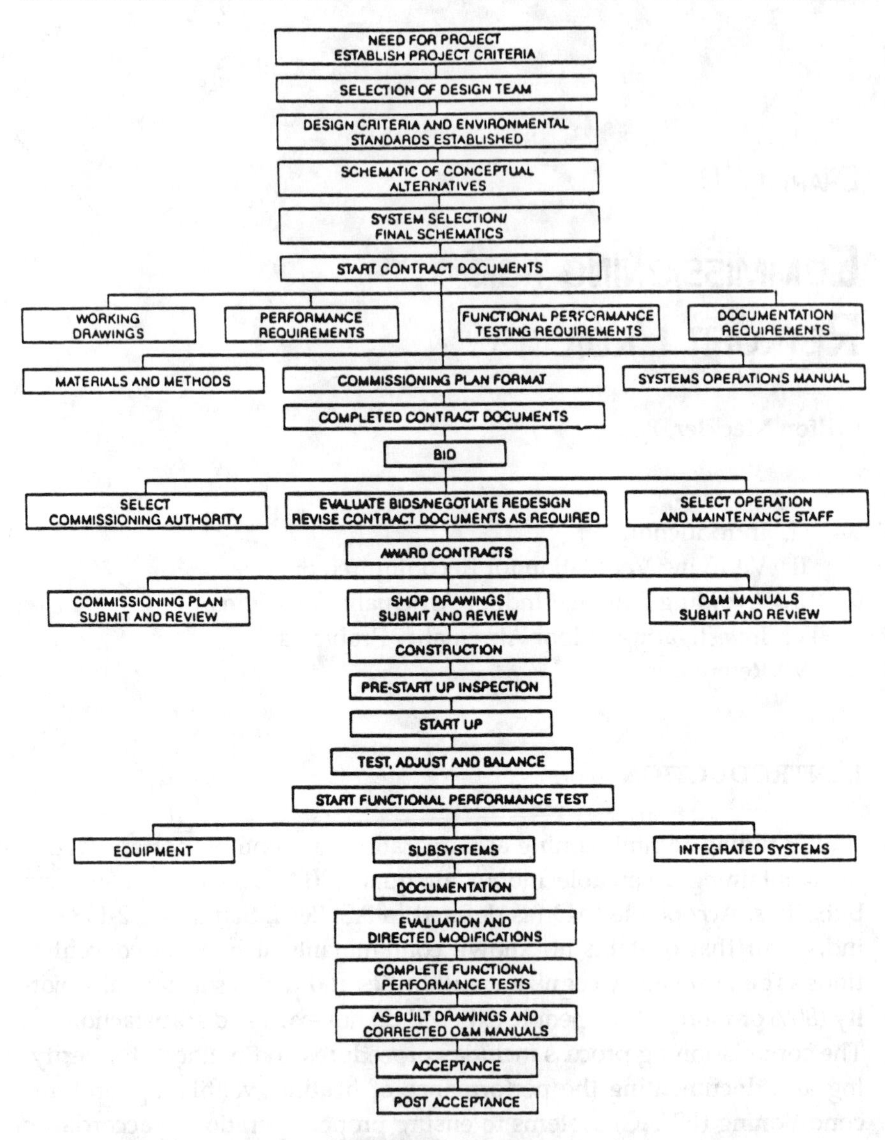

REFERENCE : ASHRAE STANDARD, ASHRAE GUIDELINE 1 - 1989 : GUIDELINE FOR COMMISSIONING AND HVAC SYSTEMS.

Figure 10-1. A Sample Commissioning Process.

makes them prone to potential IAQ problems. Important factors that contribute to the buildup of indoor air contaminants involve the placement of interior (synthetic) materials, the outgassing of volatile pollutants (e.g., formaldehyde, volatile organic compounds [VOCs], and semi-volatile organic compounds [SVOCs], energy conservation measures that minimize the infiltration and introduction of outside air, tightly sealed building envelopes (now demanded by the building owners, tenants, and energy use/construction codes, etc.), inadequate design, unsafe operations, and poor maintenance practices. To help avoid subsequent (and costly) IAQ problems, designers must now look to the manufacturers of interior building finishes and treatments to furnish product data concerning chemical composition, possible emissions, and manufacturing test methods prior to their selection, specification, and approval. The American Society of Testing and Materials (ASTM) recently adopted a guide, Standard Guide for Determination of VOC Emissions in Environmental Chambers from Materials and Products, which may serve as the basis for more specific standards for the testing of emissions from various interior building materials.

Proper building commissioning (2) can play a major role in the actual performance of otherwise properly designed and installed HVAC systems. It is difficult to determine the full design intent. To avoid performance defects, the designer should document the design intent clearly in a suitably written form (3); make it available throughout the design/construction process to architects, owners, contractors, and commissioning authorities; and ultimately employ it as a vehicle to assist in the training of building operation & maintenance (O&M) personnel.

Early delegation of a "commissioning authority" to oversee the commissioning process is often the key to achieving expected performance goals. Early involvement of the various material manufacturers, suppliers, and key subcontractors in the commissioning process is essential for identifying potential construction or installation problems. The commissioning authority must also have direct responsibility for training building operators (prior to turning over the building to its owners) by employing the above-referenced documentation as training tools and helping to familiarize building operators with overall HVAC system features and controls.

A detailed commissioning plan (2)(3) should contain a complete listing of the responsibilities each team member is expected to assume in such things as the work schedule, documentation for construction, operations and training verification procedures for installed operating systems, staffing requirements for ensuring adequate commissioning, and O&M,

etc. When commissioning becomes an integral part of the design/construction/move-in process, most potential IAQ problems are more easily identified and, therefore, more easily prevented.

II. VERIFYING VENTILATION FOR RECONFIGURED SPACES

Ventilation requirements originally intended for a given space use may not be adequate if, subsequently, the same building area becomes a reconfigured space. Reconfiguration of a space is defined as "the change in purpose of an originally intended use." When older office space is reconfigured, the net effect of different interior finishes, carpets, housekeeping procedures, and increased thermal load (due to computers, higher population densities, after-hours use, etc.) can result in a build-up of harmful indoor air contaminant levels not experienced earlier that may call for rebalancing or adjustment of the ventilation air rates.

The following general steps are suggested when undertaking reconfiguration of any building space:

1. Develop independent design criteria addressing potential IAQ concerns in accordance with the latest applicable ASHRAE filtration, IAQ, and comfort standards rather than the minimum code requirements necessary for a plan check of the building permit approval. Such standards currently include ASHRAE Standard 62-1989, ASHRAE Standard 55-1981, and ASHRAE Standard 52-76.

2. Determine and verify the capacity of existing HVAC systems based on the original design intent (refer to Table 10-1 for a sample HVAC subsystem). This can be accomplished through design documentation, a clear description of O&M tasks that must be performed on a regular basis, and HVAC test and balance reports filed prior to the building owner's final acceptance of the construction. Where such documentation does not exist or where HVAC systems are determined not to have been tested and/or balanced recently, request an updated test and air balance report. Also determine if there have been any HVAC system modifications in the interim and, if so, the effect such modifications may have on current HVAC system operational capabilities.

3. Reconfigure the HVAC air-distribution system as needed to accommodate the occupancy of new space layouts by providing air distribu-

Table 10-1. Input/Output Summary.

BUILDING SAMPLE
SYSTEM CHILLER

DESCRIPTION	INPUTS: ANALOG: Room Temperature	Duct Temperature	Water Temperature	O.A. Temperature	Static Pressure	K.W.	Flow (Totalization)	BTU	INPUTS: BINARY: Flow Switch	Switch Closure (Status)	Static Pressure
Chiller											
Chiller Output							X	X			
Chiller Input						X					
CW Supply			X								
CW Return			X								
Condenser Supply			X								
Condenser Supply			X								
Tower Fan										X	
Chiller Pump									X		
Cond. Water Pump									X		
Make-Up Water							X				

DESCRIPTION	OUTPUTS: ANALOG: Modulating Actuator	E/P Transducer	OUTPUTS: BINARY: Control Relay	Solenoid	Start/Stop	2-Position Damper Mod.	OUTPUTS: 3-POINT FLOATING: Actuator	Mercury Relay	Feed/Bleed Sav
Chiller					X				
Chiller Output									
Chiller Input									
CW Supply									
CW Return									
Condenser Supply									
Condenser Supply									
Tower Fan					X				
Chiller Pump					X				
Cond. Water Pump					X				
Make-Up Water				X					

DESCRIPTION	ALARMS: High Limit	Low Limit	Run Time	Maintenance Message
Chiller			X	X
Chiller Output				
Chiller Input	X			
CW Supply	X	X		
CW Return	X	X		
Condenser Supply	X	X		
Condenser Supply	X	X		
Tower Fan			X	X
Chiller Pump			X	X
Cond. Water Pump			X	X
Make-Up Water				X

DESCRIPTION	FEATURES: Scheduled Start/Stop	Temperature Setback/Setup	Optimized Start (Adapt.)	Demand Limiting	Duty Cycle	Ventilation Delay	Economizer Cycle (D.B.)	H.W. Reset w/O.A. Temperature	VAV Control	Remote Read/Reset	Tenant Override	Color Graphics	Point Lockout	Trend Log
Chiller														
Chiller Output														X
Chiller Input														X
CW Supply														X
CW Return														X
Condenser Supply														X
Condenser Supply														X
Tower Fan														
Chiller Pump														
Cond. Water Pump														
Make-Up Water														

REFERENCE : ASHRAE STANDARD, ASHRAE GUIDELINE 1 - 1989 : GUIDELINE FOR COMMISSIONING AND HVAC SYSTEMS.

tion to ensure maximum ventilation efficiency. Analyze the overall new system capacity to determine its adequacy for new thermal loads associated with the new reconfiguration.

4. Thoroughly test balance the reconfigured system and document the results in a suitable Associated Air Balance Council (AABC) or equivalent report format.

5. Verify the performance of the HVAC system (anticipated full- and part-load conditions) to comply with the new tenant plans and specifications developed by the architects and/or space planners and check that all applicable codes have been followed in accordance with commissioning guidelines (ASHRAE Guideline 1-1989).

6. Where the performance of the reconfigured HVAC system is determined to be unfavorable, repeat the above procedure.

III. AVOIDING POTENTIAL INDOOR AIR QUALITY PROBLEMS

Unfortunately, many IAQ problems are not easily foreseen and cannot be corrected without substantial rework. Therefore, early prevention involves sensitivity to occupant complaints, avoidance of build-out construction odors on partially tenanted floors, and application of common sense in selecting the most appropriate initial corrective measures. Fortunately, the major causes of sick-building syndrome (SBS) or building-related illness (BRI) are known and, therefore, prolonged unacceptable IAQ can be avoided. Although preventive measures may, in some cases, increase construction and occupancy costs, failure to maintain adequate IAQ can ultimately be more costly as a result of loss of productive time and subsequent modification and/or replacement of troublesome HVAC components and/or interior finishes and materials (i.e., carpet, paneling,) that may be needed to correct the problems. The following procedures are recommended during applicable phases:

A. Design

1. Choose building materials, finishes, and equipment carefully and review those selections made by an employed outside interior decorator.

2. Design an energy-efficient HVAC system that provides a reasonably contaminant-free interior environment. HVAC systems (otherwise adequately designed for peak thermal loads) may not be suitable for meeting ventilation needs at all part-load conditions. Additionally, questionable energy conservation measures affecting air supply rates should not be allowed to jeopardize IAQ or occupant health concerns without a thorough review of all foreseeable operating modes and related consequences.

3. Allow proper access to all HVAC systems for ease of maintenance or replacement of potential microbiological contaminant sources.

B. Construction

1. Do not substitute materials or modify the design where such changes are known to result in increased indoor air contamination.

2. Check carefully to ensure that the entire HVAC system is installed in accordance with the original design. It is not uncommon for entire ductwork runs to be omitted or dead-ended; for all moving parts not to be in operation; for fans and motors to be omitted or in reverse relation; or for supply and return ducts to be omitted, interchanged, or interconnected. Attendance of the owner's O&M staff during the construction phase is recommended.

C. Close-Out

1. Thoroughly clean the HVAC system and especially remove all dust and debris from all interior equipment surfaces prior to start-up.

2. Allow emissions from new interior construction materials to vent to the atmosphere before gases permeate other building materials, thus lengthening the outgassing period. This can be done by ordering the materials and receiving them at the construction site early enough to allow unpacking and venting potential odors outdoors long before actual installation is planned.

3. Thoroughly test balance the HVAC system in accordance with the applicable testing and air balance industry practices. For example, the ratio of outdoor make-up air to clean recirculated air, particularly in variable-air-volume (VAV) systems, cannot be allowed to change during part-load (i.e., variable supply air rate) conditions. One way of accomplishing this is to supply outdoor air by a separate supply fan

capable of maintaining a predetermined flow rate. Another way is to utilize an automatic return-air damper that is controlled by return airflow and is therefore capable of adjusting to a lower system pressure loss at part-loads. Be sure to take appropriate airflow measurements as part of the HVAC system balancing process. Consideration should also be given to increasing make-up air considerably for the first few weeks or even months of building operations. Occupants should be advised as to potential causes of stress related to moving, odors due to new construction, and individuals to contact in case of complaints.

4. The following test procedure is recommended as a minimum and should be modified, where required, based on the complexity of each building to obtain the best possible test results:
   - Inspect the HVAC and electrical systems to verify proper installation/interconnection and to ensure that all corrective measures have been finalized in accordance with construction contract requirements.

   - Clean the interior of the building to ensure that it is free of standing water prior to testing.

   - Install moveable screens, furniture, and other fixtures prior to testing.

   - Set the HVAC system to operate at a maximum heating mode for at least 24 hours with full illumination (i.e., lighting) on. The "bake-out" time required to heat some heavy structural masses may require more than 24 hours, so consider the effect of construction mass in setting this appropriate time frame.

   - Set the HVAC system to operate at a normal mode for 12 to 24 hours after a predetermined temperature has been reached.

   - Set the HVAC system to operate at a maximum heating mode for 12 to 24 hours with all space lighting on after a predetermined temperature has been reached.

   - Set the HVAC system to operate at a normal mode for a minimum of 24 hours after a predetermined temperature has been reached.

- In buildings with repetitive (i.e., typical) floors and separate HVAC systems, it may be economically advisable to measure the concentrations of indoor air contaminants on the first floor for a week or two to determine if it is effective or if additional bake-out time is necessary. This will be of invaluable use in case of a subsequent litigation.

D. Occupancy

1. Provide heating or cooling capability early in the morning (i.e., before normal office hours usually begin) on a daily basis with a 100% outdoor air only for the first few weeks (or longer if necessary) followed by normal heating or cooling operations with maximum ventilation provided (for purging). During occupancy hours of the first few months, operate the HVAC system below normal thermostat settings to reduce adverse outgassing effects.

2. For the purpose of evaluating IAQ conditions (temperature, humidity, stuffiness, etc.), designate a regular employee on each floor to record these conditions a few times each day.

3. While monitoring employee comments very closely, decrease the cooling/heating/ventilation requirements prescribed above gradually for three to six months until normal design cooling/heating/ventilation levels are established.

E. Operations and Maintenance

1. Avoid outside sources that will contaminate air intakes. These sources may be such items as bird nests or feathers (located near outside air intakes, etc.), standing water, and exhaust air from adjacent buildings, parking garages, or streets. Replace outside air filters regularly.

2. Use adequate ventilation air to avoid the build-up of indoor air contaminants. A decrease in lighting load to conserve energy may result in a decreased heating load and air-change requirement. Use of inadequate outside make-up air can cause excessive recirculation of indoor air contaminants and should be avoided at all costs. Also, inadequate recirculation can create stagnant air.

3. Maintain all parts of the HVAC system to the state of being operational at all times.

4. Train O&M personnel thoroughly on complete system operation and acceptable tolerances for system adjustments, emergency procedures, use of O&M manuals, etc. Maintain a standard method of recording HVAC operation complaints.

IV. INVESTIGATING INDOOR AIR QUALITY PROBLEMS

Investigation of IAQ problems poses a great challenge even to the most experienced health professionals, since the complaints in buildings can be rather diverse and non-specific. For example, the SBS has been known to cause a host of nonspecific symptoms such as headache, dizziness, nausea, eye irritation, etc. Of the 346 indoor air investigations conducted by National Institute of Occupational Safety and Health (NIOSH), 179 (or 51.7%) were attributed to inadequate ventilation.

Should building owners, managers, operators, or others with to investigate IAQ problems, they may gather the necessary information by using questionnaires that also promote the efficient use of investigative time during subsequent site inspections. These questionnaires should include inquiries about occupants, building and building environments, and building HVAC systems.

VOC levels in new and existing buildings may be measured by using screening devices, such as a survey meter, and levels in excess of 1 parts per million (ppm) may call for an in-depth investigation. Increased ventilation rates may not always reduce the concentrations of VOCs. Other control techniques, such as a "bake-out," in which the air temperature in a building is increased for a certain period of time while providing adequate ventilation, may be used. High temperatures, maintained for a sufficiently long time, increase the VOC emissions and eventually drive them off from furnishings and building materials. Necessary precautions must be taken to avoid material damages due to low relative humidity and high temperature. The high costs of conducting a bake-out and delayed occupancy may limit the use of this process in all buildings.

Failure to identify the cause of an IAQ problem following the above-referenced inspections may warrant a more extensive investigation. Should that occur, the actual airflow rates of outdoor and recirculated air should be measured and compared to the design and recommended ASHRAE Standard 62-89 airflow rates. Other measurements also recommended to be taken are the indoor air temperature, relative humidity, and outdoor wind velocity and direction. These measurements will help to ascertain whether symptoms coincide with specific outdoor weather conditions consistent with transmission of seasonal pollens, dust, etc. Low relative humidity (less than 30%) can also cause dryness of the eyes, nose, and throat, while high humidity (greater than 70%) can cause proliferation of microorganisms.

The use of air-sampling to confirm prevailing IAQ levels can be expensive, and the results may not always be reliable indicators of off-normal operating conditions. Permissible exposure limits (PEL) and threshold value limits (TVL), developed by Occupational Safety and Health Administration (OSHA) and American Conference of Governmental Industrial Hygienists (ACGIH), respectively, are based primarily on industrial environments. The procedures are used for sampling an 8-hour time-weighted average (TWA) and a 15-minute short-term exposure limit (STEL). The sampling procedures may not be capable of measuring concentrations of contaminants much lower than those normally found in industrial environments and, therefore, may not be applicable to all indoor environments. Some of the most commonly collected samples and instrumentation (detector tubes or direct-reading devices) are carbon monoxide (2 ppm to 200 ppm), carbon dioxide (0 to 2000 ppm), nitrogen oxides, ozone, radon, and particulates (as low as 2000 particles per cubic centimeter of air (4).

In determining the nature of a specific IAQ problem, additional information, not available through questionnaires, may be necessary. This information should be collected by a "walkthrough" process (i.e., visual inspection of a building) to look for possible indoor air contaminants in building materials, finishes, furnishings, equipment, and supplies. In some cases, the walk-through process may require several site inspections and a more intensive IAQ investigation, including environmental monitoring. Prior to a walk-through, it is recommended that the blueprints of the building and HVAC system, the HVAC modification and O&M records, and the current employee list be thoroughly reviewed.

V. REFERENCES

(1) ASHRAE Standard 62-1989, Ventilation for acceptable indoor air quality. American Society of Heating, Refrigerating, and Air-Conditioning Engineers, Inc., Atlanta, GA, 1989.

(2) "Ventilation design for reconfigured office space." Air conditioning, Heating, & Refrigeration News, 1990.

(3) Meckler, M. "Role of commissioning and building operations in maintaining acceptable indoor air quality." Proceedings of the 5th International Conference on Indoor Air Quality and Climate, Indoor Air '90, Toronto Canada 1990.

(4) Indoor Air Quality Design Guidebook. Chapter 15: Investigation of Indoor Air Quality in Buildings, by Meckler, M. The Fairmont Press, Lilburn, GA, 1991.

SECTION 3
ENERGY MANAGEMENT AND LIGHTING STRATEGIES

CHAPTER II

ENERGY AUDITS AND ENERGY MANAGEMENT

Larry W. Bickle, Ph.D., P.E.

I. INTRODUCTION

Energy conservation is a popular subject—so popular, in fact, that the federal government appropriates millions of dollars every year to conduct energy conservation programs in public schools and hospitals. Energy audits play a major role in these and other federal programs; but so much attention has been given to energy audits that they have become almost synonymous with energy management. This is a dangerous trend. Energy audits do not in themselves produce energy savings. They are simply one of several essential steps in an energy management program.

Energy audits do have a useful role and should be conducted. This chapter, however, takes a critical look at the subject. The goal is to raise important issues and limitations for both owners and auditors to consider. It is hoped that this will allow more effective use of energy audits within the framework of a meaningful energy management program.

II. TYPES OF ENERGY AUDITS

A wide range of activities are loosely described as "energy audits." At one end of the scale, simply collecting utility bills, calculating energy consumption per square foot, and comparing this consumption to "normal" or "average" values might be termed an audit. At the other end of the scale, the audit of a professional engineer and/or architect might include spending several man-days inspecting a building, testing HVAC equipment, measuring lighting levels, computing theoretical performance, determining life-cycle costs, and preparing retrofit construction documents.

The exact level or type of energy audit is not pertinent to the issues raised in this chapter. In general, however, most of the points relate to audits in which a professional makes site visits and performs technical calculations. In the terms used in various federal programs, this would be a Class A type audit and would include some aspects of the Technical Assistance Program (TAP) type of technical and economic calculations.

III. LIMITATIONS OF ENERGY AUDITS

With this general background, consider the following:

A. Energy audits do not save money by themselves. It costs money to conduct any kind of energy audit. Unless the findings of the audit are used, the audit itself will produce no savings. Conducting an audit before there is top-level management commitment to implement the results can be a serious waste of money.

B. It is possible to do an energy audit too early in the program. If a detailed energy audit is conducted too early in the overall energy management process, there is possibility for misdirection. Many of the most important first steps in the energy management program, such as changes in administrative policies, cannot be easily quantified. Unfortunately, neither the true cost nor the mathematical relationship can be easily determined. However, there is evidence suggesting that these actions may have benefit/cost ratios 50 to 100 times greater than capital improvements.
While most energy audits do identify low- and no-cost actions, the energy audit with its calculations and "precise" numbers for capital improvements can divert attention from these more important early action areas. The net result may save energy but not be the most cost-effective program.

C. There are significant limitations in the engineering techniques used to compute energy consumption in buildings:

1. Engineering analysis techniques to analyze the long-term average impact of small actions are not readily available. For example, it is not possible accurately to estimate the yearly savings that would result from replacing a specific piece of weatherstripping or from installation of edge seals on outside air dampers.

   But many of these so-called minor capital improvements can, when taken collectively, produce substantial energy savings. These "minor" actions also involve considerable cost, and thus it is important not to apply them indiscriminately in every case. In technical terms, the result of each action is smaller than the uncertainties in the calculation techniques themselves.

2. Even if there were not uncertainties in the engineering techniques, there would be uncertainties in the input data. The input data that are difficult to obtain precisely include a description of the building, environmental variables (such as temperature, wind, and solar radiation), and internal loads (such as occupants and lights).

   In an older building there are usually uncertainties about wall insulation, control system set points, *in situ* efficiencies, and other "details." Even when this is not the case, reducing a real physical building to a set of idealized nodes, conductances, and terms in a calculational model introduces simplifications and loss of precision.

   Shade trees, obstructions at ground level, local ground reflectants, small lakes, etc., change the specific microclimate for an individual building. In most cases, engineering calculations will use macroclimate information from nearby weather stations, with resultant uncertainty and imprecision regarding the weather data specifically applicable to the building being studied.

   How many people are in the building at the same time? During what hours is it operating? Are exhaust fans switched off at certain times? Do occupants turn off lights when they leave the room? How are drapes used? These and other important questions about the interaction between occupants and the building affect the precision of the energy consumption calculations. These occupant use patterns are extremely difficult to determine with any precision because they change from hour to hour, day to day, month to month, and year to year.

Assumptions are added to assumptions, and so on and on. Even the most complex, comprehensive computerized methodologies such as DOE-2, BLAST, TRACE, and AXCESS can rarely predict actual energy consumption in a building within 10% to 15%. Discrepancies of 25% to 50% between theoretical calculations and actual consumption are not uncommon. An occasional difference of 100% or greater is not unknown. It is unclear how much these differences are due to calculational techniques, and how much they are due to errors and lack of precision in input data. What is clear is that the overall precision of the process is not much better than the total combined savings of possible modifications.

3. Another engineering problem is that few simplified methods exist for predicting the interaction between energy conservation methodologies, and generally modifications are not cumulative; two modifications that each would save 10% probably would not save a total of 20% if both were implemented. In the simplest case, the first modification saves 10%, and the second saves 10% of the remaining 90% or 9%.

   But the situation can be worse: both modifications can compete to save the same energy. Consider, for example, the combination of double-glazing and a night setback thermostat. Both modifications reduce heat conduction through the windows, but the interaction is highly complex. If the night setback temperature is 55°F (13°C) and the nighttime outside air temperature is 55°F, there is no temperature difference to cause a heat flow. Thus, the double-glazing has zero added benefits at that particular instant. Clearly the savings obtained by making both of these modifications is not the sum of the savings that would be calculated for each modification individually.

   These complex interactions can be modeled to some extent using sophisticated computer simulation programs such as DOE-2. However, there need to be simpler ways to evaluate the interactions between potential modifications.

D. Another major limitation of energy audits is cost estimating. Most of the standard cost estimating methods and data files are designed for use with either new construction or major remodeling. Much of the energy conservation retrofit work is really "odd jobs" that are handled by small independent contractors. These costs tend to be highly localized and difficult to predict. While no single retrofit project is large, there can be a large percentage error in cost estimating. These cost estimating errors can accumulate to produce large errors in the total project cost.

IV. ENERGY MANAGEMENT PROGRAM

In spite of the limitations raised in this chapter, energy audits do have an important role to play in an overall energy management program. However, an energy management program needs to focus on more basic issues.

Exactly what is basic varies from one client to another. Based on our own past experience, and a review of available literature, we would propose the following as building blocks for a successful energy management program. Whether you agree or not, an internal discussion of these fundamental issues will help focus energy management efforts for maximum results.

A. Energy conservation should be viewed as an upper-level management responsibility. Energy conservation involves improved operations and maintenance and investments in hardware. These are but pieces of the broader management problem of controlling energy costs (refer to Figure 11-1). A well-balanced program will cut across internal divisions and require policy changes, integrated administrative practices, improved operations and maintenance practices, public relations programs, and, finally, capital investments.

B. An energy conservation program should be financially sound. Energy conservation actions cost money. Weigh these costs carefully against potential savings so only cost-effective actions are taken. The definitions of cost-effective must be formulated at the policy level using opinions from as many interested parties as possible.

Administrative and policy changes tend to be the least costly changes to implement, are the most cost-effective, and should be undertaken first. Improvements in operations should be undertaken next. And capital improvements should be deferred until last (refer to Figure 11-2).

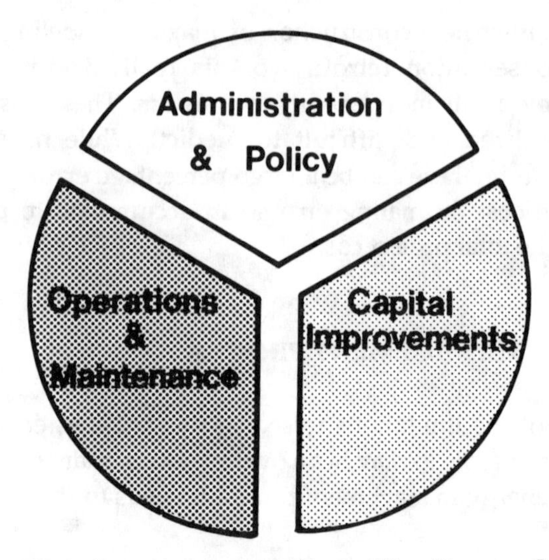

Figure 11-1. Components of Controlling Energy Costs.

By progressing from less expensive changes to funding the more expensive improvements out of proven future savings, a client could have a self-financing program.

C. Establish goals and priorities. State precisely the goals of the energy conservation program. Is the object to reduce costs? Improve the public image? What are the relative priorities of the various goals? Insofar as possible, these goals should be quantitative.

D. Energy conservation is a team activity. Whatever program evolves must be a cooperative program among all participants. Recognize that energy management is a sensitive activity. The program defined for a particular client must be unique and responsive to local needs, conditions, personalities, requirements, and restraints and the broader goals of the client's organization.

E. Focused accountability and responsibility are critically important. While energy conservation is a team activity, one person must be in charge. This person must provide clear, strong leadership. The energy management program must be well defined, and specific responsibilities and authority agreed upon by all participants.

F. Motivation and evaluation of progress are vitally important. Energy conservation is really a collection of many small actions, like turning

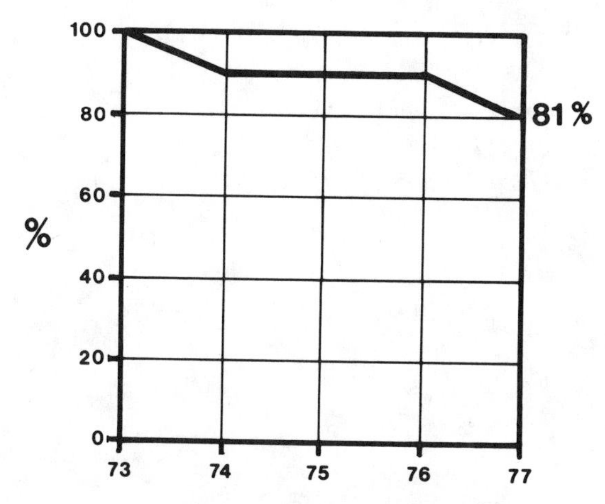

Figure 11-2. Savings Achieved through Administrative, Policy, and Operations Changes.

out the light when leaving an unoccupied room. Motivation and constant feedback are essential to a cost-effective energy conservation program. One of the best ways both to motivate and to provide feedback is to implement an effective, easily understood, and highly visible "scoring system" for measuring the success of the energy management program. An added benefit of this scoring system is that energy conservation actions—policy changes or capital investments—can be evaluated, and quantitative estimates of actual savings can be made.

G. Energy conservation programs are site-specific. They must be tailored to specific climatological conditions and building types.

H. A Master Plan for Energy Management (MPEM) should be developed. This plan should identify specific responsibilities, establish evaluation criteria, and systematically rank actions into logical priorities for implementation.

Clearly these basics are not absolute, but they do illustrate how the common-sense management techniques of commitment, motivation, education, and evaluation can be integrated into a meaningful program to reduce energy costs. Perhaps more important, implementation of an effective energy management program provides the proper perspective for interpretation and use of energy audit results.

CHAPTER 12

DISTRIBUTED DIRECT DIGITAL CONTROL ENERGY MANAGEMENT FOR THE 1990S

John J. McGowan, CEM

I. Introduction
II. Direct Digital Control Architecture
III. Distributed DDC for Equipment and General Purpose Applications
IV. Distributed Zone Level Control
V. Future Systems
VI. References

I. INTRODUCTION

The building automation business, with its latest distributed direct digital control (DDC) technology, is expected to exceed $725 million by 1995. What began as one of the many technologies evolved to respond to the oil shocks in the 1970s is fast becoming an essential component of any modern building. In other words, the energy management system (EMS) has grown up.

Conventional control systems over the years have evolved from pneumatic and simplistic electro-mechanical systems. Lower cost microprocessors and electronic technology has allowed for microprocessor-based control products, such as the EMS, to be widely applied. The results of distributing control are new benefits for the customers through precise control, technical capabilities and remote communication. These benefits have allowed DDC selling points to extend beyond just energy and cost savings to occupant comfort and less building and equipment downtime due to remote monitoring. Other benefits come from intelligent building type features such as after hours tenant billing for space conditioning. Keeping pace with technology developments and new trends can be a significant challenge, because changes are rapid.

Energy management has always required knowledge of heating, ventilating and air-conditioning (HVAC) and basic controls, and now one must also understand computer and networking technology. This knowledge is critical because of the growing number of systems installed each year. Networking and sophisticated controls have led some to call DDC a "local area network (LAN) building automation system (BAS)." Some of the drivers for market growth may be summarized as:

- energy costs and security,
- lower cost systems and replacing conventional controls,
- quality building operation including maintenance, and
- natural evolution to DDC technology.

The distributed DDC market has evolved extensively in recent years. Significant growth has occurred through the late 1970s as a result of the energy shortages of 1973 and 1977. Control systems had been continually evolving yet the increased demand for technology that would aide building owners to save energy led to new features. A new name was given to these devices: energy management control (EMC) systems.

American Society of Heating, Refrigerating and Air-Conditioning Engineers, Inc. (ASHRAE) defines DDC as closed-loop control process implemented by a digital computer. Closed-loop control simply means that a condition is controlled by sensing the status of that condition, taking control action to ensure that the condition remains in the desired range, and then monitoring that condition to evaluate whether the control action was successful. A simple closed-loop example is a proportional zone temperature control. The zone temperature is sensed and compared to a setpoint. If the temperature is not at a setpoint a control action is taken to heat or cool the zone, and then the temperature is sensed again.

Combining highly accurate controls with networking is an essential feature of digital distribution for building control. Building owners realize that real energy cost reductions which continue to be attractive as well as the benefits of better control. These benefits include longer equipment life, more satisfied tenants, and a host of new features available due to expanded building information. The communication features available through these systems has allowed for managers to track their buildings more effectively and respond quickly and intelligently to problems. Inteligent response involves remotely diagnosing a problem via system telecommunication and sending a service person knowledgeable of the task at hand to the site. These features are extremely desirable particularly in such applications as office buildings. In summary, the DDC market continues to grow based on the on-going importance of energy and cost savings, and the expanded benefits available to the building owner through automation technology.

Architecture defines the distributed system structure. Architectures may be viewed as a group of control related devices with common communication networks. This chapter will provide the reader with a basic understanding of the control architecture necessary to implement a distributed system. DDC "nodes" are intelligent devices capable of equipment control, networking and remote communication. Unlike the control systems of the past which employed one central processing unit (CPU) to monitor, communicate and control, distributed systems rely on the networking of smart independent controllers. These are typically on a LAN. Quite simply, a network covers a limited geographic area and every node on the network can communicate with every other node. Computer LANs usually do not need a CPU; control LANs, on the other hand, use a single network interface or management unit (island host).

There are four architecture levels in a DDC system: sensor/actuator, distributed controller, island host, and master system operator interface (SOI). Of these levels, the most important ones are the distributed controllers and the island host. The sensor and actuator technology at this point is hardwired, and uses electrical signals rather than networking technology. Industry standard signals like 4 to 20 milliamps that are wired to a controller are most common. Controllers must then use analog to digital, or A to D, conversion technology to convert the signal to an engineering unit for use with the control sequence. At the upper level of architecture the SOI is a critical system component, yet it is ancillary to control. The most common implementation of distributed systems is to have both a zone level controller, tailored to simple applications like the VAV terminal unit, and an equipment level, or general purpose controller. These two controllers will have different requirements and designs to accommodate the given application, yet they both fit within the level of a distributed controller. Figure 12-1 indicates a more typical distributed DDC system architecture with the island host and two levels of distributed controllers.

II. DIRECT DIGITAL CONTROL ARCHITECTURE

A. Island Host

Island hosts are custom panels that integrate building control, manage distributed networks and handle remote communication interface via

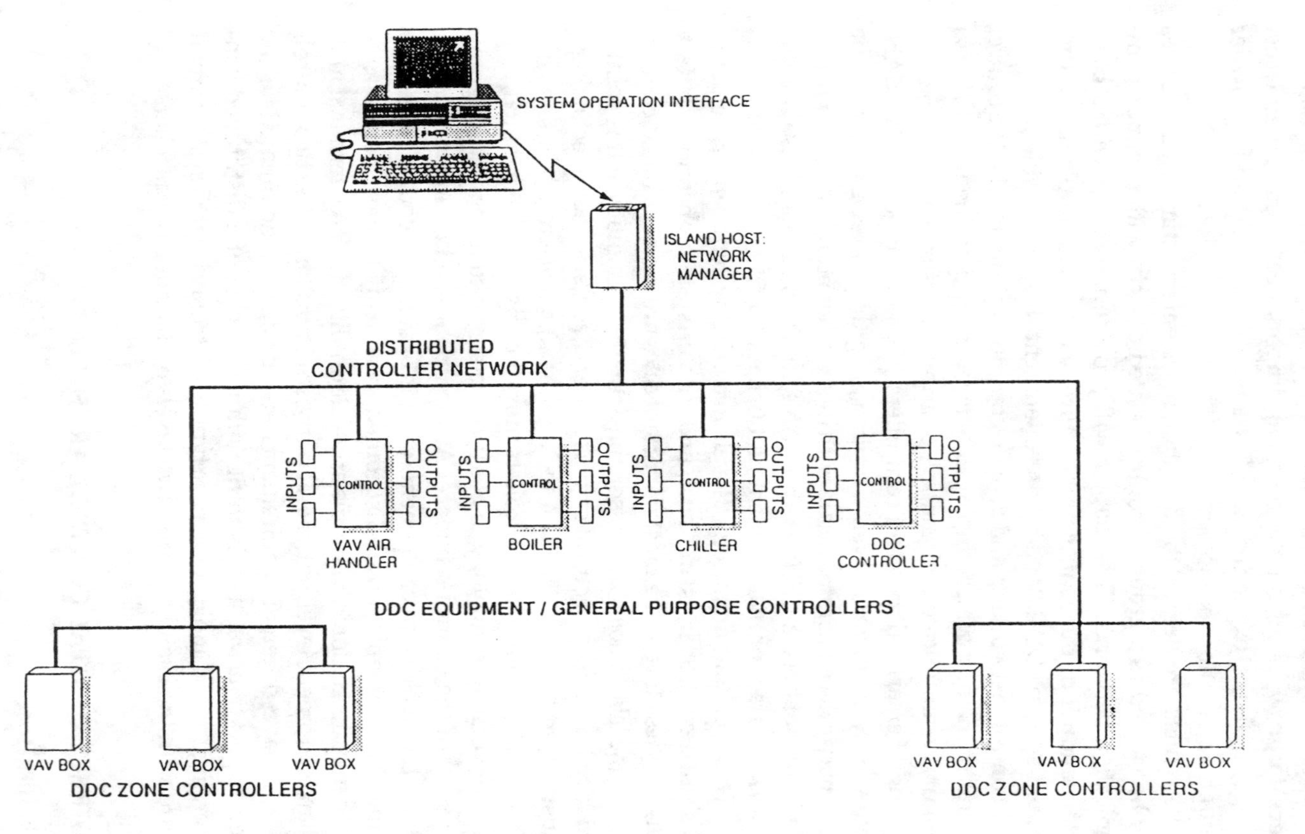

Figure 12-1. A Typical Distributed DDC System Architecture.

modem. Key topics to this discussion will be hardware requirements, building coordination which provides a sequence of operations for the entire system, and trends at this level of control.

Hardware in the case of the island host varies greatly. Approaches have been taken which involve personal computers (PCs) as well as custom microprocessor panels. The custom-microprocessor-based island host must be viewed primarily as a vehicle for managing the network, integrating building-wide control among all the distributed controllers and serving as a central focus for outside communication with the network. These concepts will be expanded significantly under sequence of operations, yet they are mentioned here to indicate the hardware complexity required at the island host level.

The island host is typically focused on providing communication and networking functions, and therefore often does not have on board input and output (I/O). In some cases these devices will have a limited number of I/O, but the trend is certainly toward use of system-wide data in the form of pseudo points. A pseudo point is a piece of information that physically resides elsewhere in the system but is held in a "data field" at the host or other distributed controllers.

From a hardware perspective the island host is the focal point for communication. It provides outside access through a modem and telephone lines, allowing access to all system data. The island host must also provide some overall network management functions, though in many cases these systems are self-policing. Self-policing means that each network member, or controller, has responsibility for ensuring communication and sequences that are automatically enabled when a failure occurs.

The island host hardware as a powerful microprocessor-based system which is focused primarily on two levels of system communication: networking and remote interaction. The host must also coordinate building-wide control functions and availability of system-wide data sharing. These functions are typically implemented in one piece of hardware, however there may be other functions required such as life-safety, fire and security systems.

Building wide coordination with a distributed system will be viewed in three key areas: (a) control integration, (b) building wide monitoring and history, and (c) remote communication with an SOI. The term "integration" is used because of the need to accomplish control via coordinating, or more specifically integrating the functions of multiple microprocessors. In other words integrating control means the accomplishment of complex strategies that are implemented in part by more than one device. There are many functions that fit into this category and are easy to accomplish with a CPU-based system. The examples include demand limit control or optimal start/stop for the building.

CPU-type systems initiate all control from one central "all knowing" device. Distributed systems, on the other hand, must monitor building-wide functions at one device, and then integrate this information to carry out control sequences with other devices. There are many design approaches with distributed systems. For simplicity, this discussion will cover one of the more common approaches, island host integration. In this approach the island host serves as a CPU-type device for building-wide functions. This device monitors the building status, and based on that data, initiates control sequences. For example, the island host monitors the building power consumption for the network, and based on demand consumption and a setpoint, initiates demand shed.

The DDS control approach will not allow loads to be shed without integrated control sequences implemented via distributed controllers throughout the building. These controllers must physically disable loads to avoid exceeding the setpoint because the island host may not have access to actual control devices that can be shed. The key to control integration is the ability for the island host to either "request" the implementation of local control sequences, or to override local control and force a specific sequence. In any case, an island host must track the actions of distributed controllers and be capable of taking appropriate steps if sequences are not implemented.

The final topic in control integration is the system data. In essence the island host serves as the vehicle for ensuring that key information is available throughout the network. There are two types of system data: global and shared. Global data is the information needed by many or all of the controllers in the same format. Typical examples of global data are outside air temperature, time and enthalpy. These data are provided by one controller which has a physical point for monitoring, though for system integrity purposes back-up locations may be provided. The data from that hardware point are then "broadcast" to the network and available to all controllers.

The second type of system data is shared data. This is a much more custom approach to using system data for control. A simple example is a temperature reset. Let us assume that Controller A is responsible for maintaining a discharge air setpoint. A reset temperature control provides for the setpoint to be adjusted based on an indication of the load. Controller B monitors the space temperature for local control, but this is also a good indication of the load. The shared data function provides a vehicle for controller A to get access to the space temperature with the same frequency as if it were physically connected to that device. Data sharing requires that a specific point be accessed by the island host at intervals of 5 to 30 seconds, which accommodates the need for control, and that the data be transmitted to Controller A. Functions such as these are critical to building-wide control.

Another building-wide function of the island host is to serve as a vehicle for monitoring and operating historical data. Monitoring functions will be discussed in the remote communication portion of this section, yet the most important note is that a "window" to the network must exist. This window must allow for communication with all network devices via a single modem and telephone line in the building. Through that window a variety of monitoring and alarming features must be provided.

Historical system operating information is also of critical importance. Typically the island host serves as a central repository for data files containing trend data on various system points. This is generally the most efficient approach to ensure that the overall cost of the system is not detrimentally impacted by the need for data from all network devices. A key topic that will only be explored here is the cost per controller, or "node," to have network capability. To ensure that this cost does not impose an undue cost burden on the overall system, the general approach is to charge one network device, the island host, with responsibility for history. This requires that the host access data from network devices and provides data storage, or memory, for later retrieval.

The final island host function is to allow for remote communication with an SOI. There are three key functions which must be provided: programming, monitoring and alarming. Programming requires that each network member including the island host be commissioned via a local or remote SOI. Commissioning includes programming of control sequences, network communication parameters and alarm limits. Monitoring includes current status on all controller points as well as access to system historical data.

Island host alarming is perhaps one of the most critical features of any building control system. This feature requires that critical conditions be defined in addition to default alarms, such as sensor or controller failure. The system designer may then note whether such alarms should be logged for availability when the system is polled or should initiate a dial-out. The dial-out function will use the modem and communication capability to initiate communication with one or more predefined SOIs, and send the alarm data.

B. Island Host Trends

Among significant island host trends is standard communication. Driving forces for this trend are (a) end-user demanded remote communication with multiple manufacturer systems via an SOI, and (b) freedom to bid system expansion without replacing major system components.

There are many other trends with island hosts that will simply be mentioned in passing, but represent significant technological and market strides. These trends include:

- movement from minicomputers to microprocessor based systems,
- expanded control integration within the building system between the island host and controllers,
- easy-to-use (program and monitor) systems,
- non-volatile memory storage to avoid program loss,
- cost-effective and higher speed remote communication, and
- expanded PC based SOIs with simple user interface (1).

III. DISTRIBUTED DDC FOR EQUIPMENT AND GENERAL PURPOSE APPLICATIONS

There are many features and capabilities available with distributed controllers for more complex applications. Microprocessor-based control has become common quickly with large HVAC equipment and general purpose control. The term equipment level is derived from the use of these controllers with such equipment as centrifugal chillers and air-handling units. General Purpose controls use similar panels with different control sequences such as start/stop or on/off control of HVAC, non-HVAC loads, lighting and other electrical equipment. Equipment level controllers (ELCs) and general purpose controllers are cost-effective. ELCs provide application-specific control sequences, and integrate these with island host and zone level devices for building-wide functions.

The topics which will be covered in this heading include:

- equipment level hardware,
- sequence of operation,
- control integration,
- networking, and
- equipment level control trends.

A. Equipment Level Hardware

Perhaps the most distinguishing feature of equipment level hardware is the on board input and output. Devices applied at this level typically provide a mix of point types: analog and digital input and output. The I/O requirements tend to be more demanding at this level due to the expanded applications. There is generally a demand for at least eight of each point type. Since these controllers may be applied as general purpose devices or application-specific controllers, it is not uncommon for devices to be expanded up to as many as 64 total points. This may be done through add-on modules, cards, etc.

The equipment applications generally require control through digital output and analog or pulsed modulating output, as well as monitoring of both analog and digital values. Analog values are generally used to monitor temperature or pressure, while digital points provide contact closures typically indicating an alarm. Device output then respond to the commands of the controller based on the appropriate input value and the sequence of operations. These output will either be enabled or disabled, or may be modulated to a commanded position.

A common trend with this equipment/general purpose hardware is for manufacturers to develop one hardware device or "platform," and use this with all applications. The platform will provide a base I/O capacity, and in many cases, allow expansion through any of several techniques. Application of the controller is accomplished by changing software in the device. Software issues related to this approach will be discussed further in sequence of operations. Hardware design issues to be resolved also include the requirement for a full-line of ancillary component products. These products include controller, sensors, transducers, electric relays, motors, actuators transducers, and output devices.

A key related issue with these products is whether the manufacturer will "fix" the ancillary components that may be used with a device or allow flexible application of a wide variety of industry standard devices. The benefit of flexibility is reduced cost of system retrofit because existing field devices can be reused.

There are two additional points that should also be mentioned in the category of hardware: controller size and ambient operating requirements. Controller size is typically not an issue with general purpose devices because they are often mounted in a mechanical room. The assumption is that these devices can be acquired with enclosures that meet standards, agency approvals and local codes. With equipment controllers however, there is an on-going trend to preintegrate DDC devices within the control panel of HVAC equipment. Preintegration imposes a size constraint in smaller units, and may be a consideration if the building owner is buying equipment without controls. Operation under outdoor temperature conditions is an absolute requirement for equipment level control devices.

The need for extended temperature operation is evident with a variable-air-volume (VAV) rooftop air handling units. It is typical for this equipment to be provided in a fully self-contained manner. A single enclosure containing the HVAC equipment and controls is provided in one package mounted on the roof. The package will be exposed to the full extreme of outdoor temperature and humidity, and may be shipped to any location in the world, further adding to the potential extremes.

B. Sequence of Operations

The basic hardware platform outlined above makes it possible for that controller to be applied in any number of applications. The sequence of operations is the specific program that tailors this controller to that application. In this section, we will briefly explore the techniques used to accomplish that programming. Basically all control products on the market today use one of two programming approaches: equation-driven or library-loop control parameter approach.

ELC control sequences will vary dramatically by type of equipment. These devices require complex and sophisticated control sequences tailored to the specific applications. One approach to solving this problem is to have the controls designer, contractor or integrator write the sequence with an equation-driven-language. This is feasible, though time consuming, and it allows for custom application. A common variation on this theme is to use a controller that allows flexible programming, and maintain a custom library. This allows the designer to start with a reasonably complete control sequence and edit that sequence to meet the new requirement.

The equation-library approach works well for equipment-level control because sequences may be maintained for chillers, boilers, etc. The designer is able to quickly modify the library, yet a key consideration is control reliability. In such cases each new control sequence must be tested and verified prior to field application. Testing can be extremely time consuming and does not always verify every aspect of control operation. Library-loop-control parameter type programming basically involves controllers that employ sequences that have been programmed, tested, verified and hard-coded by the manufacturer. Hard-coded means that these sequences are held in a portion of memory that may not be accessed to review or change the sequence. Obviously, the control user will sacrifice a portion of the control flexibility available with equations. Yet the engineering time and job support required to deal with testing is virtually eliminated. This is a strong selling point for such devices and has led to many products on the market that offer preprogrammed control sequences of various types.

Another example of the library-loop approach is with general purpose control. In this case general purpose is defined as simple control loops that provide either start/stop or modulating control. Some examples include time-of-day (TOD) scheduled sequencing of loads such as lighting and/or other electrical loads such as start/stop programs.

To be more specific about the technology is difficult in this section, however there is another important concept. That is the complex nature of equipment level control. To illustrate this point, consider a VAV air-handling unit. The control for this piece of equipment requires several major control loops including:

- mode of operation, heat or cool,
- discharge air temperature control (heat/cool)by sensing temperature,
- outdoor air damper control (free cooling and ventilation) sequences,
- discharge temperature reset,
- duct static pressure control, and
- space/building pressure control.

One of the benefits to equipment level intelligent control is the ability to monitor a variety of alarm and safety conditions. For the most part these become simple control loops that monitor the condition of a contact closure or an analog value. Based on a change in state of the digital point, or an analog point exceeding a set of range limits, an alarm is initiated.

C. Control Integration

The concept of control integration was briefly discussed in the island host. One reset approach involves installing a representative sensor somewhere in the space, and making control decisions based on that single point of information. A more integrated approach is the monitoring of the space temperature at any number of distributed zone controllers. The equipment level control then averages those values or resets to the highest or lowest value. This function is too equipment-specific to be handled by an island host, yet it is ideal for the ELC. Control integration at the equipment level is not likely to be limited to interaction with lower level devices. As noted in the island host, these ELCs will be integral to carrying building-wide functions such as demand limiting and optimal start/stop.

D. Networking

The final topic in equipment level controls is networking control. The functions discussed in the island host for programming, monitoring and remote communication necessitate that full access is available with the ELC. This device must have complete communication capability. Given that statement however, it bears noting that the devices at this level may not require the same level of communication available at the island host.

E. Equipment Level Control Trends

Short term standard ELC communication will occur through the island host. Systems with multiple manufacturer controllers will need island hosts from each manufacturer. A major reason for this is control integration complexity, especially with multiple manufacturer devices. Integration becomes difficult with function like demand limiting, because everyone takes a different approach. So architectures will be left in tact from the island host down, but future controller communication standards are likely.

There are a number of other trends which include:

- application specific ELCs with expanded I/O capacity,
- implementing multiple control sequences with a single ELC,
- control integration with island host level,
- easy-to-program, object-oriented and parameter driver rather than IF, THEN, ELSE,
- non-volatile memory storage to avoid program loss, and
- lower cost controllers.

IV. DISTRIBUTED ZONE LEVEL CONTROL

Although zone level control (ZLC) devices are relatively new, they have rapidly become common with applications like VAV terminal units, water source heat pumps and small packaged single-zone units (refer to Figure 12-2). ZLCs introduce new features and new complexity to the systems. ZLCs will be viewed in the following categories: (a) hardware, (b) sequence of operations, (c) networking and control integration, and (d) zone level control trends.

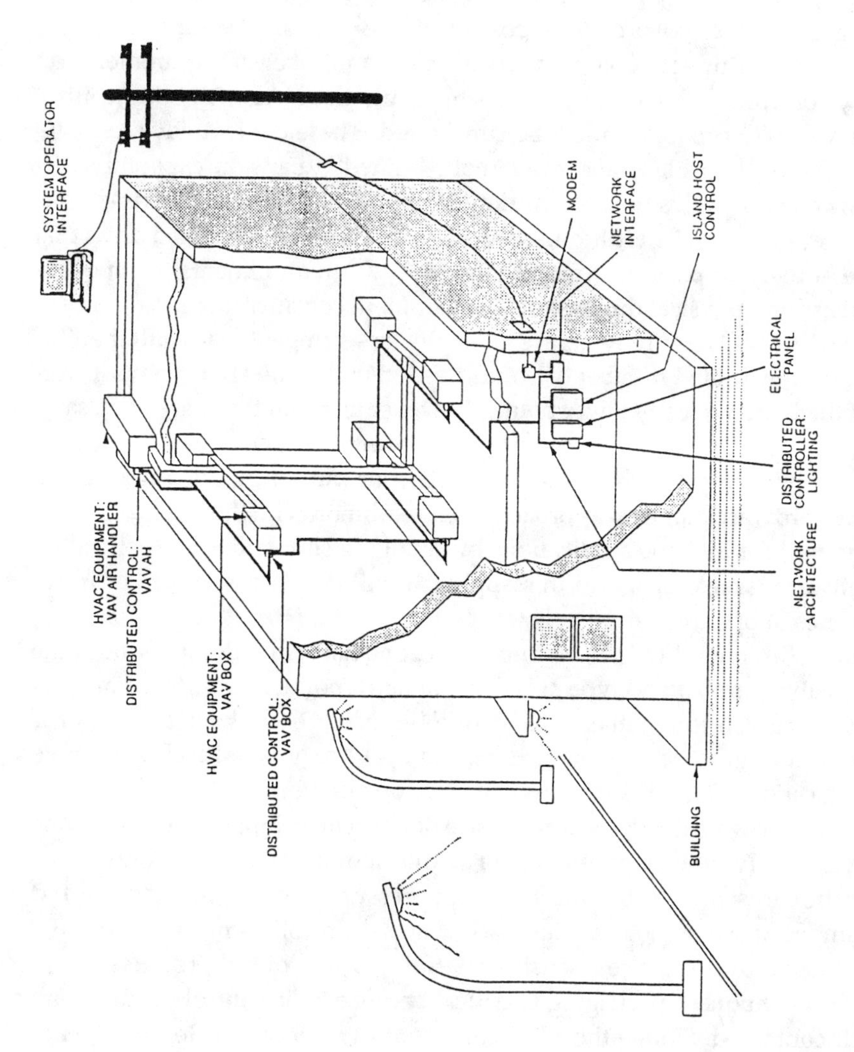

Figure 12-2. Zone Level Controllers.

A. Hardware

Lower cost microprocessors and electronic technology have made distributed zone control possible. ZLCs are critical to the evolution of DDC, because they provide control that is tailored to a specific application, and is located as close as possible to the controlled equipment.

The nature of zone products imposes a number of requirements on any controller which is applied to this equipment. As with other control devices I/O capability must be considered. These devices have less I/O capability than ELCs and offer more limited expansion capability. Yet they must be well suited to a variety of applications, and at the same time, be cost-effective. In addition, there are several hardware related areas that are equally important. Among these are I/O requirements, component packaging, physical dimensions, and ambient considerations.

The I/O requirements are generally less stringent than with the ZLC due to the limited number of points needed with zone applications. Overall there are typically fewer points, however all point types are necessary. Zone applications usually require both monitoring and control which are accomplished through digital and analog or pulsed modulating output. The hardware platform approach is also common with the ZLC, and these are used for all zone applications by changing software in the controller. Software issues related to this approach will be discussed further in sequence of operations.

With regard to I/O, an important zone hardware design issue is the available number and type of I/O channels or points. Consideration must also be given to whether a fixed number of I/O will be available, or if expansion is to be provided. Expansion I/O may be critical if multiple applications are to be accommodated by one platform.

Hardware design issues that are of particular importance at the zone level involve the requirement for a full-line of products including the controller, sensor or thermostat, static pressure measurement device and damper or valve motors. The costs of these controllers must be carefully evaluated due to the cost sensitive nature of zone control products.

Component packaging is critical because zone control products are self-contained. Unlike the ELC, zone control applications tend to be applied by the original equipment manufacturers (OEMs), each offering a variety of control products. There are actually two distinct issues: (a) packaging as it relates to easy application of zone products, and (b) packaging as it relates to physical design. Easy application is an important facet of packaging because zone HVAC equipment is also self-contained. In addition, zone OEMs generally install or mount a large percentage of the controls for this equipment at the factory. Using VAV as a component packaging example, there is a growing demand to provide one easily installed package containing the controller, sensor or thermostat, duct static-pressure measurement device and output devices for dampers and valves.

Unlike the ELCs applied with larger equipment, OEM controls may be highly discretionary and require that the OEM be able to install and verify operation on any device. The concept of component packaging is targeted at satisfying customer needs by offering a fully-packaged control product incorporating all the components in a tested hardware product. This is an important option to reduce the installation time at the factory.

Physical design and packaging address whether the control meets the space constraints of the application. This is actually a hardware design and selection issue. VAV boxes, for example, will vary greatly in size based on the zone air flow rate requirement and the type of box. Further VAV boxes are not the only application that may be controlled by a ZLC. As a result, use of any ZLC will be dictated to some degree by its ability to meet space requirements for any intended application.

Ambient considerations is another important hardware consideration. Key to the discussion of ZLC products is that the variety of applications for these devices require that they meet a range of operating conditions. In nearly all cases, ZLCs which are mounted with VAV boxes need only to operate within the standard interior ambient temperature conditions while single-zone-packaged units, etc. are likely to be exposed to exterior ambient temperature and humidity conditions. This dramatically increases the requirements, and perhaps the cost, for these devices because of potential applications in Alaska or Florida. Since the controls business continues to be more global, conditions must be considered for the entire world, therefore a ZLC could conceivably be applied at the arctic or the equator. These are valid issues for consideration by any end-user selecting a DDC system. Because of the above-mentioned hardware considerations, it is highly unlikely that the larger and more complex ELC panels can be applied effectively in this market.

B. Sequence of operations

Control sequences vary dramatically by type of equipment, but there are added distinctions between the ELC and ZLC. At first impression, a reasonable case could be made that ELCs must maintain more complex and sophisticated control sequences than would be encountered in zone applications. In many cases this may be true, yet the zone device must be able to satisfy the requirements of all applicable equipment. If we were to consider the VAV box as an example again, there are several control implementations ranging from pressure-dependent to more sophisticated approaches such as pressure-independent dual duct applications. For example, a pressure-independent single duct system requires five control loops including fan, space temperature, reheat, duct-pressure and space-pressure.

Considering that zone applications typically involve smaller and less sophisticated equipment, and less I/O, it is easy to assume that sequences are also simplistic. Sequences tend to be complex and there are also interactive sequences to be considered. The same VAV box control requires networking communication to allow integration with the air-handling unit for simple functions like unoccupied override to start the air-handling unit. The full power of a microprocessor is more than warranted in the ZLC to provide control as well as communication requirements.

C. Networking And Control Integration

The need for networking was mentioned earlier in connection with control sequences that are carried out between several modules. Also ZLC are distributed processors, and networking is essential to distribution. In island host and ELC, these topics were viewed separately, but with the zone discussion they must be considered together. This is because ZLCs are highly reliant on network interaction (more than any other network member). There are two reasons for networking with these systems: (a) remote communication, and (b) control integration. Remote communication is essential for all those functions discussed in the island host.

Control integration is key to the ZLC because it is not possible to terminate every necessary point to this panel. Also it may not be feasible to build every control function into these devices such as time-of-day. In some cases a clock does not reside in the controller. Time (or time and a ZLC schedule) is maintained elsewhere and the network broadcasts this information to the controller. These and other data points fall into the system data category. However, it may also be necessary to implement building-wide control such as optimum start or demand-limiting by issuing commands or requesting sequences through the ZLC. All of these functions are completely reliant on a distributed-network.

D. Zone Level Control Trends

It is very difficult to discuss trends in this equipment because the existence of ZLCs is in fact a trend. Perhaps the most important trend that may be noted is the growing implementation of this technology. In fact the same trend noted in the island host and ELC has relevance here as well, standard communication. Though to an even greater degree than the ELC, ZLC level devices are not likely to be integrated as standard network nodes for some time. The evolution of standard communication is expected to be the island system first, then at the ELC and finally at the ZLC. Following the ZLC, it is possible that intelligent sensors and actuators will evolve.

There are also some specific trends within this technology that are of interest. It is important to note that the growing trend of HVAC OEMs shipping product from their factory with preintegrated ZLCs. Further, this technology also makes use, by necessity, of several trends previously mentioned. These trends include:

- application-specific ZLCs,
- more control integration with other architecture levels,
- nonvolatile memory storage to avoid program loss, and
- lower cost controllers (1).

V. FUTURE SYSTEMS

It is critical to focus on the significant industry issues, protocol and control networking. One of the undisputed characteristics of future systems will be distributed networking controllers with substantial control flexibility. The communication capabilities of these devices will be equally important to the control sequences.

The basic architecture discussed with regard to current DDC technology is setting the stage for the future. Without question, networking controllers will continue to expand their penetration into all types of equipment, and at the same time, network communication capability will continue to expand. In fact these future systems may be viewed in two categories: (a) control, and (b) communication.

Among the future control capabilities of distributed systems modular expandable systems made up of smaller, faster and more distributed controllers that are applied at each control process or local unit (2). These controllers will require adaptive PID, or expert system types of, control capabilities, and networked communication is an absolute essential.

Communication characteristics of future systems will be also explored briefly in this chapter. Perhaps one of the most exciting, and undoubtedly one of the most challenging, future trends is standard communication. This industry has only passing familiarity with data communication through experience with proprietary control system development. Also because it is ancillary to our control products, there are some real questions as to the appropriate level of internal research and development that should be funded. At the same time, investment is hampered because users have stated that they are willing to pay only a modest premium for standard communication (2).

Without a doubt, networking is essential to support distributed control, and it has been verified that the industry wants this standard for networking. Prior to availability of that standard though, there will be transitional products. Transition refers to the attempt by users and others in the industry to find a short-term solution while the standard is being developed. This demand has led to a number of related trends and products in the industry that will not be discussed here. These transitional efforts are likely to change the concept and implementation of future systems. Standard communication protocols for systems networking will not evolve until the mid to late 1990s. In the interim these transitional approaches will be implemented to provide short-term solutions to non-standard communication issues.

In addition to communication standards, there are some other specific communication trends that are likely to evolve with future systems. These include:

- faster data transmission speeds over the network and via modem;
- voice communication for monitoring and alarming;
- expanded network data sharing capabilities; and
- multi-functional integrated networks for building control, security and other systems.

There are also a number of ancillary products that will likely be integrated with control systems such as PC-based LANs for data access throughout a building and an entire organization. These future systems must also be "easy-to-use." Ease of use is extremely important point because the expanding power of microprocessors and software has allowed for "user-friendly" control products. Control industry users are being exposed to a wide variety of PC and computer-related products that shape their expectations of the technology available. This industry must satisfy those expectations with control products that allow an average user to achieve sophisticated building operations without complex programming.

A PC-based SOI is likely to be an integral part of any future system. The SOI will be the window through which the user is able to view the system. It will also serve as the focal point for ease of use through a high technology simplistic user interface. The SOI will probably be available as two distinct packages. In the past SOI packages were provided that allowed programming, monitoring and a host of utility functions for interface with control systems. The on-going trend toward easy-to-use systems will result in specialized packages designed for the intended user. One of these SOI tools will be focused to the contractor or system integrator that must design, install and start-up the control system. A second tool will be specialized with features for the building owner who must maintain the system over time.

VI. REFERENCES

(1) "Survey on Energy Management Systems #99," Future Technology Surveys, Inc., Lilburn, Ga.

(2) Abramson, A., P.E., Honeywell, Inc. Report on intelligent buildings institute focus group survey of control industry users. Proceedings of World Energy Engineering Congress, Atlanta, Ga., October, 1990.

Chapter 13

Demand-Side Management and Energy Services Industry

James P. Waltz, P.E., C.E.M.

I. INTRODUCTION

The purpose of this chapter is to share valuable experience on the subject of demand side management (DSM) with the reader. While the following material is not a comprehensive dissertation of the subject of DSM, it does provide significant insight into some of the very important issues and pitfalls to be dealt with and/or avoided in implementing energy efficiency programs, whether they are under the banner of DSM or not. While DSM professionals may well be interested in this chapter, the material contained herein should be most informative to building owners.

While it might seem like old history, the Arab Oil Embargo of 1973 was a very practical demonstration of the law of supply and demand. Faced with a manyfold increase in the cost of energy commodities, the U.S. responded by reducing its energy consumption and, ultimately, increasing its energy efficiency. Real improvements in the energy efficiency in our existing inventory of buildings took quite a while to work themselves into place—due to some very real and perhaps unfortunate characteristics of our building management and operating infrastructure.

The plant services or facilities engineering department in many organizations face a similar problem. The owning and operating of facilities are frequently viewed as necessary in order to conduct the organization's main business. It could even be said that paying the cost of maintaining facilities is somewhat akin to paying taxes in the eyes of management. Furthermore, when ready to spend money on facilities, management can generally be expected to be more excited about wall and floor coverings than mechanical or electrical equipment, for example.

Interestingly enough, there is even a "pecking order" for the funding of facilities projects. For example, if the air-conditioning system has broken down, the funds will be provided to repair it. By contrast, improvements and upgrades to systems that are apparently in satisfactory working order are viewed much differently. These projects are much less likely to be perceived as real needs. Compounding this situation is the fact that the profession of plant/facilities engineering has a relatively bad image. The situation in the electronics industry illustrates this point well. Many of the high technology firms in the Silicon Valley were started by an individual or a small group of individuals with very high-level technical skills, perhaps unique and unparalleled in the industry. Starting small, these firms hardly had any need for facilities engineering staffs and, as they grew, seem to have hired a handyman who was put in charge of facilities.

As the organizations grew, it became abundantly clear to management that the handyman could not perform the job. Frequently, the owners added a layer of management to the facilities staff, generally someone from their own high technology field of endeavor who, often times, would have otherwise had been laid off. The end result (common in many industries) is that facilities engineering is perceived as very low technology and possibly even trivial in nature. The net effect of this image problem is that the facilities manager who brings a high-cost and somewhat high technology facilities upgrade project (for example, a building automation system) to management for funding is frequently rebuffed—even though the project will provide a return on investment equal to other capital investment projects available to management. Indeed, the energy services industry came into being in the early 1980s as a way to solve two facilities problems:

- scarcity of capital, and
- low credibility/competence image of facilities engineering.

The business proposition at the core of the energy services industry is a simple and frequently compelling one. On a turnkey basis, the energy services company will identify, finance, install, and maintain energy efficiency improvements to a facility and guarantee to the owner that the energy costs thereby saved will exceed the cost of the improvements. This business proposition allowed many building owners in the 1980s (and 1990s) the confidence to proceed with expensive, complicated, energy efficiency improvements to their facilities. Many companies were founded to serve this marketplace and others in allied fields diversified into the energy services business. Unfortunately the collapse of Organization of Petroleum Exporting Countries (OPEC) in the late 1980s resulted in a rebound of energy commodity prices and a stagnating of the industry. However, the well-founded firms survived, and today represent the bulk of the DSM implementation resource as viewed by the utilities (indeed 9 of the 13 finalists in Pacific Gas & Electric (PG&E) 20-Megawatt pilot bid were energy services companies, and many of the other building owner finalists will likely be utilizing ESCo's to implement their work). As a result, throughout this chapter, the terms "energy services" and DSM will be used as though they are synonymous.

Just as "energy conservation" was a "catch" phrase of the energy crisis of the late 1970s and early 1980s, it seems that "DSM" has become the catch phrase these days. Our lawmakers and public utilities commissions have come to realize in recent years that a kilowatt-hour of energy can be created through energy efficiency at a much lower cost than by building a power plant to produce it. As a result, utility companies around the nation have been mandated to create and implement DSM programs. As practicing energy engineers well realize, there is very little new in terms of energy efficiency technology but the enthusiastic support of energy efficiency by the utilities and their willingness to implement DSM programs.

While the utilities and utilities commissions tend to take rather global points of view relative to the implementation of such programs, the core of any building energy efficiency improvement program has always been, and will always remain, a matter of product technology and application engineering. Unfortunately, the rush to implement global programs frequently had a tendency on the part of the bureaucrats and program managers towards oversimplification of the problems at hand as well as a tendency to treat the solutions as commodities. For example, the Emergency Building Temperature Restrictions of 1978 attempted to implement a short-term solution by instructing building owners to raise the summer thermostat settings on terminal reheat HVAC systems. This action, rather than saving energy, actually increased the energy consumed.

Unfortunately wholesale order-of-magnitude increases in the level of energy conservation investment can only be accomplished, and is apparently being accomplished, by a wholesale creation, virtually from scratch, of a whole new cadre of energy auditors and project implementors as well as new auditing and project implementation techniques and procedures. Unfortunately this forcing the rate of energy conservation market penetration has largely not made use of the value of the experience and expertise of the energy engineering cadre that "paid their dues" during the energy crisis of the 1970s and early 1980s. As we enter the 1990s, we find that those who are probably the least qualified are in possession of the "keys" to the energy conservation. The result has been the creation of a new highly theoretical and academic energy conservation industry which may produce more demand and less energy savings.

As a result of this, DSM programs will yield early success, but will rapidly become more and more difficult to further. The true test of the credibility and sincerity of the present day rush to embrace energy efficiency will be whether the managers of these programs will "stay the course" as the going gets tougher and recognize that the character of their programs will have to evolve to achieve substantial efficiency improvements.

While the essence of the energy services business proposition is simple, its practice is not. Unfortunately, in order to acquire this package that includes funding and apparent significant protection against both technical and financial risks, much control over the project implementation process must be given up and this opens the door to abuse. The main problem with single-source turnkey arrangements is the fact that there is hardly ever a building or facility where the results of an energy services project can be simply and accurately determined. Changes in the occupancy, use of facility, weather, and several other factors tend to complicate the matters. This is particularly true when unscrupulous business practices are employed in energy services. Ironically, it is exactly those building owners who have the greatest need for a "low risk" method of conducting energy retrofits are the most likely to be deceived.

Not only is this a threat to a building owner who is depending on an energy efficiency project to reduce a facility's energy consumption and cost, but it is also a threat to the credibility of the utility company which has sponsored such a project and program. Indeed, the credibility of an entire industry is at stake when turnkey energy retrofit projects are implemented in a sweeping program.

The following portions of this chapter are focused on the implementation of comprehensive energy retrofit projects, exploring:

- case studies of successful and failed projects,
- the use of computerized building simulation as a critically important energy analysis tool,
- post-retrofit monitoring and measurement methodologies,
- project funding alternatives, and
- project implementation resources.

II. CASE STUDIES

The following discussion presents the results of evaluations of two similar energy retrofit projects, both of which were done on a financed and guaranteed-savings basis. In one case the project was an unqualified success. In the other case it was a complete failure. The following describes the projects and their implementation methodologies, documents the project performance, and makes recommendations regarding the selection of DSM and energy services vendors.

A. Project No. 1

This project was the retrofit of a county administration building and courthouse complex. The administration building was built in the late 1960s and is relatively modern in terms of its building construction, HVAC, and lighting systems. The courthouse building, by contrast, was built in the 1920s and shows its age in terms of its construction, the wide variety and age of its HVAC and lighting systems. Each building was separately supplied with electricity and both buildings shared a common central cooling and heating plant.

Through a competitive proposal process, the county chose a team to implement the project. This team consisted of a prime contractor (who was actually a local mechanical service contractor), a consulting engineering and a financier. The steps to project implementation included the following:

- completion of a very rigorous energy retrofit feasibility study including computer simulation of the entire facility to within 5% of its actual utility company invoices, performed collaboratively by the consulting engineer, the prime contractor and the owner's staff.

- detailed review of the feasibility study by the owner, along with owner participation in selecting the final package of retrofit work to be performed—to suit their financial criteria, building maintenance and repair concerns, and other needs.

- negotiation of a final turnkey retrofit contract.

- completion of detailed final design, including such detail as point-to-point control wiring diagrams.

- installation.

- start-up and debugging of the project and training of owner's staff.

The ultimate project implemented include the following features:

- installation of an energy management computer for time-scheduling of virtually all HVAC equipment.

- major modifications to the majority of the air-handling systems in both buildings including direct digital controls (DDCs) conversion to variable-air-volume (VAV), and the addition of outside air economizers on systems not so equipped.

- extensive lighting fixture retrofit.

1. Results of Project No. 1

The mechanism chosen by the parties to this particular contract for measuring the avoided cost produced by the project is a methodology generally referred as "stipulated calculations." In this methodology, a series of formulas are developed which utilize energy factors which are agreed to by stipulation by both parties and which formulas also contain variables, such as equipment runtime and fluid temperatures, which are monitored and recorded by the building automation system. These formulas are embodied in an automated spreadsheet which is used on a periodic basis by the energy services contractor and the owner (both parties possess the spreadsheet) to account for the avoided cost produced by the project.

In addition, as a control mechanism, our firm was asked to prepare monthly cost avoidance reports utilizing the comparison of monthly utility bills. In this mechanism, the units of energy used during the pre-retrofit base period is compared to the units of energy consumed after the retrofit is complete, and the difference in energy units is multiplied by the average unit cost of each type of energy consumed. These reports were generated on Energy Resource Associates, Inc. (ERA) proprietary Energy Accounting Report System, which is described in "Energy Engineering" annual: Directory of Software for Energy Managers and Engineers.

As can be seen in Figure 13-1, the electrical and natural gas consumptions for the building have been significantly reduced following the retrofit, when compared to the prior "base" period. As can be seen in Figure 13-2, which shows 12-month-long moving window totals, the long-term trends in energy consumption are clearly down. The 12-month-long totals are frequently used to neutralize seasonal effects in the data so that the trends can be more easily observed.

B. Project No. 2

This project was a retrofit of a full service community hospital, also located in California. Similar to the administration building and courthouse complex, this facility was constructed in a multitude of projects spanning a number of decades. While fairly modern, this hospital pos-

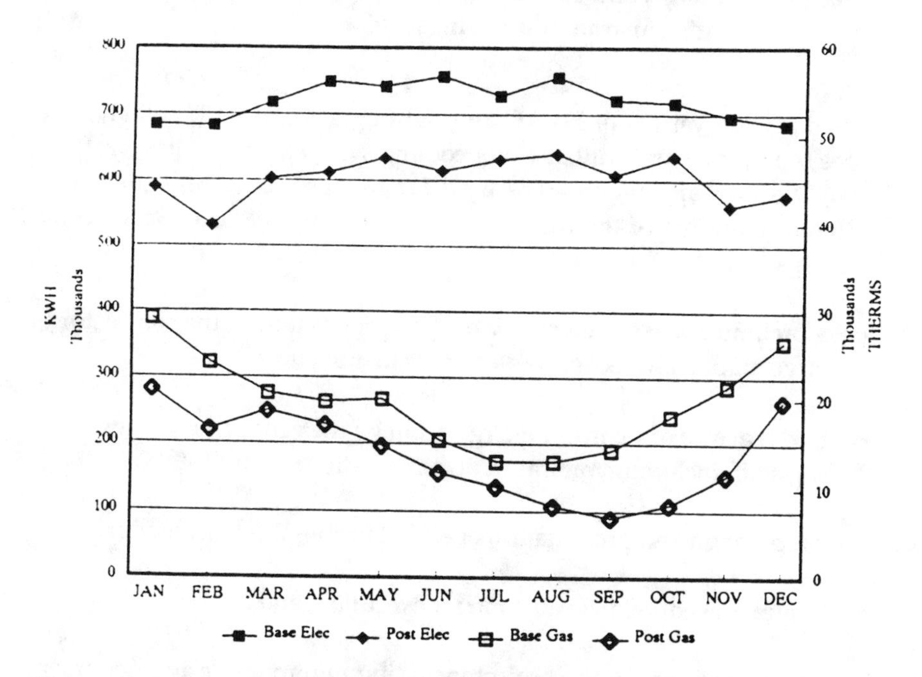

Figure 13-1. Energy Savings for County Courthouse and Administration Buildings (Base-Year vs. Post Retrofit).

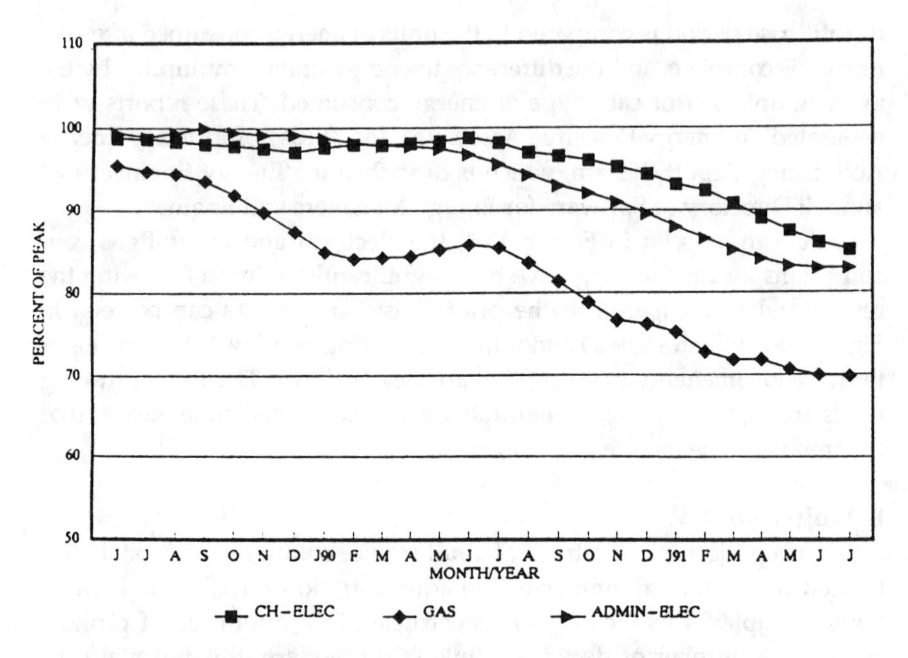

Figure 13-2. Long-Term Trend Analysis for County Courthouse and Administration Buildings.

sessed a wide variety of HVAC and lighting systems and, notably, had three separate central chilled-water cooling systems prior to its retrofit.

The implementation steps used for this project were significantly different than that of the first project described previously. Specifically, the steps included the following:

- a preliminary assessment and outline proposal from the contractor (a large, nationally-recognized controls manufacturer);
- performance of a cursory energy audit by the vendor's sales engineering staff, including minimum documentation;
- preparation and presentation of a final proposal by the vendor;
- contract negotiations and contract execution; and
- implementation of the project including minimum design documentation.

The scope of construction for this particular project consisted of two basic components. The main component was the installation of a new head-end computer on the existing building automation system. This computer was to provide automated scheduling of all equipment along with optimized control of chillers and their auxiliaries and optimized reset of the supply air temperatures of the majority of the air-handling units (AHUs). The project also incorporated field hardware necessary to provide the automated supply air temperature reset. The second component was the integration of the existing stand-alone chilled water systems. This work included interconnecting piping and the provision of automated shut-off valving for various chillers.

1. Analysis of Project No. 2

As a result of its non-performance, a detailed audit of this project was conducted and produced a wide variety of observations, as described below:

(a) The original energy audit identified savings of approximately $150,000 per year. Unfortunately, this audit, when examined in detail, revealed that little, if any, factual information formed the basis for the savings calculations, and:

- Calculations were incorrectly performed in that incorrect units were used in formulas (e.g., degrees Fahrenheit vs. enthalpy in Btu per pound of dry air).
- Savings were double-counted (e.g., cooling energy savings were counted at the AHUs and again at the chillers).
- Values in the formulas such AHU airflows in cubic feet per minute (cfm) appeared to have been guesses, as they did not correlate in any fashion at all to the airflow rates listed in the as-built drawings. (e.g., 30,000 cfm vs. 10,000 cfm).

Investigation indicated that no actual field measurements or detailed survey work had been performed. In addition, even the simple formulas that were used for estimating savings were misused in that formulas intended for calculating the savings from supply air reset on double duct HVAC systems were used for single zone, terminal reheat, and high-pressure induction systems.

In short, the engineering feasibility study, or energy audit, which was performed by the vendor was little more than a marketing ploy used to make the customer feel comfortable and thereby close the sale.

(b) As the retrofit contract provided for a guarantee of savings, the contractor agreed to provide monthly audit reports which would document the avoided cost produced by the project. These reports provided simple tabulations of numbers which purported to show that the energy savings guarantee was being met. However, none of the formulas used in these audit reports were ever documented to the customer, but apparently embody the same erroneous formulas used in the original estimates of savings. Furthermore, the formulas were to employ variables such as the actual supply air temperatures at AHUs as reset by the automatic controls, and monitored by the building automation system. However, the audit reports simply show the same constant supply air temperature which was assumed for the original savings calculations. As a result, the monthly audit reports were little more than the original savings calculations presented in a new format.

(c) Not only were the original estimates of savings erroneously optimistic and the audit reports bogus, but further investigation of the modified building automating system ultimately revealed that, in fact, it was not working. While a new head-end computer was indeed installed, and the optimization software installed in this computer, the system had sufficient difficulties in its operation that the installing vendor ultimately removed the software and left the computer in place. This left the system in virtually the same condition it was prior to the retrofit. That is, reset of supply air temperatures, starting and stopping of chillers and pumps, etc. all had to be initiated by the system operators, as opposed to being automatic.

(d) Because a detailed engineering feasibility study was not conducted and therefore a detailed technical plan for the project implementation was also not created, the project evidently went straight from sales to installation without really passing through a design stage. The results of this omission, for example, were such things as two air handling systems (one serving an interior zone while the other serves an exterior zone) having their supply air temperatures reset by means of a single control point from the building automation system. Clearly, the needs of these two spaces vary greatly and, therefore it would be impossible to optimize the supply air temperatures simultaneously for systems having dramatically different cooling and heating needs. During the project audit, no documentation whatsoever was found for the modifications to the building automation system. Construction documentation was discovered for the modifications to the chilled-water piping, valves, etc. However, this documentation was found to be significantly inaccurate when compared to the actual installation.

2. Results of Project No. 2

As can be seen in Figure 13-3, there was not any noticeable change in electrical energy consumption of the facility following the project implementation. In addition, Figure 13-4 (which again shows 12-month-long moving window totals) clearly demonstrates that the long-term trends in energy consumption are also unchanged since the project was implemented. No natural gas savings was estimated for this project and, therefore no natural gas data are presented here.

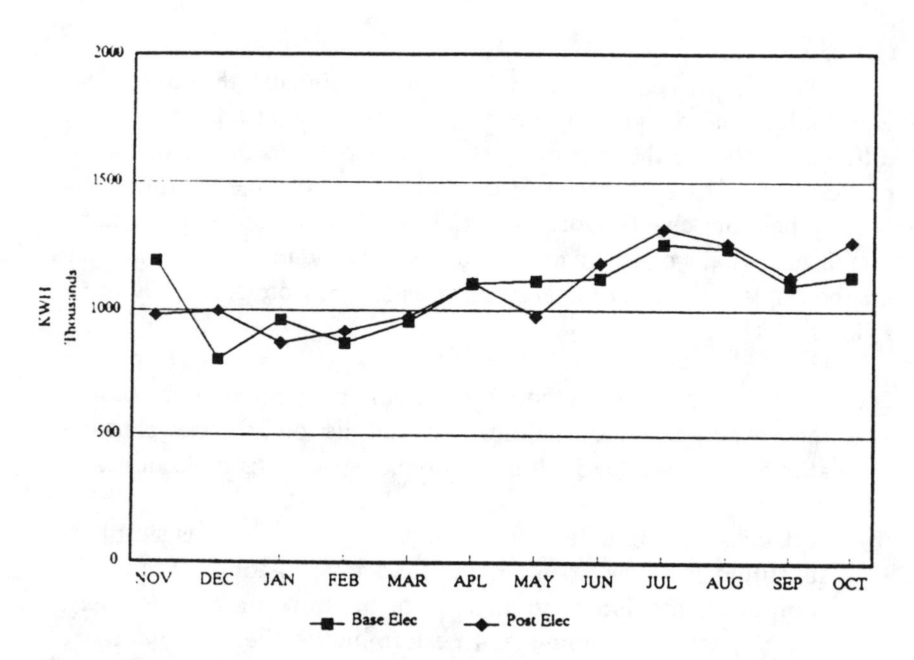

Figure 13-3. Energy Savings for a Community Hospital (Base-Year vs. Post-Retrofit).

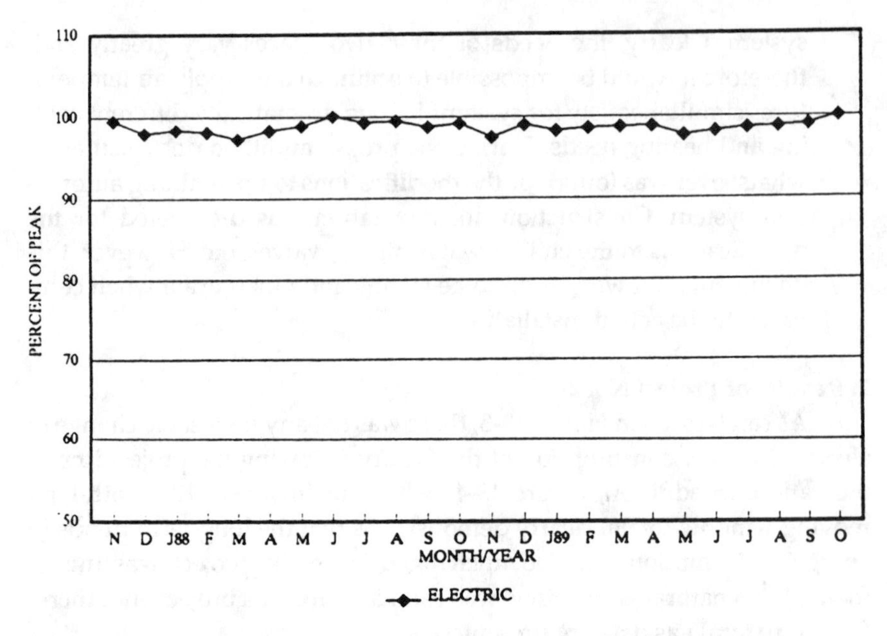

Figure 13-4. Long-Term Trend Analysis for a Community Hospital.

C. Conclusions

The two projects described here were nominally the same. They were both financed and guaranteed turn-key energy retrofit projects. The difference between the two projects lie principally in the nature of the participants and the character of their approach to project implementation. To help our clients avoid the pitfalls chronicled above, our firm has established what we believe to be a good set of fundamental ground rules for the implementation of energy services or DSM projects. They are as follows:

(a) The firm or team to do the project should be selected on the basis of their experience and qualifications, not just on financial considerations, optimistic (and admittedly enticing) savings projections.

(b) A detailed feasibility study is always required. There is simply no substitute for time spent in the field investigating systems and equipment, time spent in analysis of the building and its energy using systems and time spent performing detailed calculations of the potential energy savings that might be achieved by implementing certain energy conservation measures. Generally speaking, computer simulation or other means to achieve an energy balance, (i.e., a total of all the sources and uses of energy in the facility) is essential.

In addition, estimates of savings should be obtained from the comprehensive energy use model so as to prevent double-counting or optimistic estimates of savings. As a specific example, if a building only spent $60,000 per year operating its cooling equipment, an estimated annual savings of $50,000 for this function is probably not reasonable, even if the total annual energy bill is $500,000 or more.

(c) The engineering feasibility study, its source data and the bulk of the assumptions and calculations should be documented for review by all parties.

(d) The intended energy conservation work should be identified by means of detailed scopes of construction work so that the installing company as well as the buyer can have a reasonable means by which to measure whether or not the project has actually been implemented.

(e) Extensive construction documentation should be developed, both to guide the installing contractor's craftsman, but also for the owner to see and concur with the detailed installation work planned, and to use as a troubleshooting tool once the work is complete and/or the contract term has run out.

(f) Whatever means is agreed to by the parties for accounting for the avoided costs produced by the project, these means should be clearly defined and well documented and implemented in a way that both parties can track avoided cost when starting with the same periodically measured source data (unit costs of energy, system operating parameters, equipment run times, etc.).

II. COMPUTERIZED BUILDING SIMULATION

As we have seen above, state legislatures, Public Utilities Commissions (PUCs), and utility companies are pushing DSM as a less costly and environmentally superior alternative to building power plants. In addition, the current economic climate has brought renewed emphasis on controlling operating costs. Along with this renewed emphasis on energy conservation by our public agencies comes a greatly increased scrutiny of energy savings projections and post-retrofit verification of savings. Added to this is the fact that as more and more of the more straightforward energy retrofit work is performed (such as lighting, high-efficiency motors, etc.) the market for energy conservation must shift to the more difficult energy retrofits, such as HVAC system and control system modifications.

The result of this evolution of the demands of the energy retrofit marketplace, energy engineers, contractors, building owners, government, and utility company employees must find more accurate ways to predict the future energy savings performance of energy retrofit work. The technology rapidly being employed to meet this challenge is that of computerized building simulation. This technology works for a number of reasons:

(a) An accurate energy balance of all the sources and uses of energy in a building is an integral part of building simulation. By properly assigning energy consumption to end uses, estimates of end use savings is dramatically improved.

(b) When considering multiple energy retrofit measures, building simulation is the only technique which can accurately and completely account for the interaction between measures to avoid double-counting of savings.

(c) The effects of changes in a building's size, occupancy, or hours of use following retrofit can only be properly accounted for by remodeling the pre-retrofit building to create a new pre-retrofit baseline for comparison with post-retrofit energy use of the changed building.

While clearly the "technology of choice" for energy retrofit engineers, computerized building simulation suffers from a number of problems at the present time. These include:

(a) Most of the building simulation programs that exist were created to address the question of new construction, not retrofit. As a result of these programs there are many gaps and weaknesses which must be overcome by their users when employed in the evaluation of energy retrofit.

(b) The only instruction generally available on computer simulation of buildings is provided by the vendors of individual computer programs. Unfortunately, their instructions generally are limited to the use of their program, not the overall process of building simulation and its quality control. As a result, this instruction tends to be very academic in nature, and therefore incompletely effective.

(c) Existing common practice and tradition tends to treat building simulation programs as "black boxes" with an attendant unwarranted "blind faith" in the validity of the computer generated program output.

As observed by the author in the process of auditing and reviewing energy savings calculations and building simulation work performed by reputable and experienced professionals in the field (both as an expert witness for building owners and as a quality control consultant for PG&E all of the above has resulted in the frequent inaccurate estimation of savings for energy retrofit projects, to the embarrassment and consternation (and occasional financial downfall) of the parties involved in major retrofit projects.

A building simulation training seminar has been developed by us to improve the skills of engineers and energy analysts in performing computerized building simulation, regardless of the specific computer program in use. We have developed a concise set of procedures and practices which, when followed, regularly produce very accurate building simulations as well as very accurate estimates of potential energy savings—as demonstrated by numerous projects producing savings as projected. The following discussion presents some of our observations and experience in developing computerized building simulation as an accurate and essential tool for use in energy retrofit projects.

A. Historical Perspective

Starting in mid-1970s, computer simulation of buildings was developed as a practical tool for the engineering of buildings. The tool has found significant use during the design phase of a building to develop load estimates and optimize combinations of building features. The tool has perhaps found even better use in the analysis of existing buildings for energy conservation retrofits. By simulating retrofit options on the computer, reliable estimates of potential energy savings may be achieved, assuming that the initial modeling and subsequent modeling of retrofit measures have been well implemented. In fact, we have used a wide range of computer simulation tools over the past 10 years to prepare savings estimates for a large number of comprehensive energy retrofits in both large (1.8 million ft^2) and small (25,000 ft^2) buildings with great success.

In at least one case, a major energy services company declined to accept the conservative savings generated by computer modeling of the building, implemented a project based on their own optimistic estimates of savings, and is now paying for more than $100,000 annually for the project's "shortfall" in savings from their optimistic estimates. Their optimistic estimates increased total estimated savings by more than $200,000 per year—fortunately for them, the energy services company did not guarantee 100% of their estimated savings.

B. Accuracy Defined

Accurate modeling of existing buildings is clearly critical to the business of energy services, DSM, or any form of energy retrofit. For a computer simulation of a building to be of value in evaluating energy retrofit opportunities, it must be accurate. To be accurate, the model should account for essentially all of the sources and uses of energy in a building. Such a model should include the calculations of total energy consumption that is close to the building's actual annual energy consumption (e.g., within 5%). Such a model would also closely mimic the actual seasonal variations in energy used by the building. Finally, such a model should allocate energy consumption by function in a faithful fashion. This last virtue is particularly important if the model is to be used to evaluate the effectiveness of specific energy retrofit measures, such as lighting controls, outside air economizers, etc.

Accuracy in computer simulation of buildings, in our experience, is founded in three basic areas:

(a) a thorough understanding of the simulation tool being used;

(b) a thorough understanding of the building being simulated, through its physical and operational characteristics — in essence, in existing buildings, the quality of the survey or audit determines the quality of the simulation;

(c) careful analysis and critique of output data—the comments made elsewhere in this chapter generally apply to mainframe programs, though they also apply to other simulation tools.

By utilizing the above techniques, we have found it possible to regularly model buildings within 5% of their actual annual energy consumption with a high-degree of confidence in the simulation of each energy-using system and functional use of energy in the building. It should be noted that, in buildings where weather is a strong energy-use factor, modeling to less than 10% variance from the actual energy use may be of limited value—as our ability to predict weather for a given future year may not even be that accurate.

C. Importance of Knowledge of the Tool

The first foundation of accurate building simulation is knowledge of the tool to be employed. While the above statement may seem obvious, the computer simulation tools available to consulting engineers are very complex and have a "reality" of their own that cannot be ignored or violated if accurate models of buildings and energy retrofit measures are to be accomplished.

It is fairly common to utilize return air troffer lighting fixtures to reduce the in-space load on supply air, thus allowing a lower supply air quantity and a raising of return air temperature that allows selection of a smaller cooling coil for the same cooling capacity. In one project during the design development stage, the engineer was modeling an office building that had a large amount of core space that served mostly as traffic ways, secretarial space, and file/storage space. As a result, the main cooling load was created by the lighting systems. Unfortunately, the design engineer assigned virtually all of the lighting load to the return air (which is not physically possible) and specified a minimum outside air quantity in cfm/ft^2. The result was that supply air was calculated by the program at 0.2 cfm/ft^2, the outside air was 0.1 cfm/ft^2 and the computer calculated a return air temperature of around 500°F When half of this return air was discarded, roughly half the cooling load went with it, for substantial savings in energy consumption. Upon detailed examination of the computer output, we were able to point out the fallacy of the simulation and got the project back on track. The experience did make the point, however, that a lack of detailed understanding and familiarity with the calculation methods of the simulation program can easily lead the modeler to failure.

Another example is the capability or lack of capability to handle desired simulations by the program. Before variable-flow chilled-water pumping was commonly employed, few programs had the ability to simulate such a system. In order to do so, a series of "dummy" chillers were described to the program in such a manner that the program selected each in turn as loads increased. Associated with each of these dummy chillers was a constant-speed power pump. The effect of each chiller/pump combination was to sequentially simulate the overall pump power curve that would be produced by a single variable-speed pump. For accurate estimation of savings, only the energy consumed by the pumping systems was compared from one run to the next, thereby eliminating unwanted secondary impacts such as changes in chiller efficiency.

To be knowledgeable about a simulation program, the user must understand how the input data are utilized by the program, the calculations/algorithms employed by the program, the flow of input and calculated values through the program, and the precise effect various program controls exert on the calculations performed by the program. The bottom line here is that an inferior simulation tool in the hands of an engineer well versed in its features and capabilities is superior to the best simulation tool in the hands of an engineer unfamiliar with it.

D. Importance of Knowledge of the Building

Perhaps the single most important factor in developing accurate computer models of existing buildings is developing an intimate knowledge of the physical and operational characteristics of the building to be modeled.

1. Envelope and Weather vs. Operators and Controls

While many practitioners of computer simulation of buildings work toward more detailed time-related simulation of weather and its effect on building structures, those who are well acquainted with the practical aspects of building operation know that the effect of operating engineers and temperature control systems are manyfold more dominant in affecting a building's energy use. Perhaps one or two anecdotes would be illustrative of this point.

In one study of a major high-rise office building in San Francisco, it was learned that the watch engineer was putting the outside air economizer control for the entire building on "automatic." Further investigation revealed that it was this engineer's nightly practice to override these controls to place all operating HVAC systems (a few terminal reheat systems serving the entire core of the building) on 100 percent outside air. In this way, the supply air for the engineer's office in the basement was return air from the core of the building and, by overwhelming the reheat coils with 100 percent outside air, the building core temperature dropped a few degrees and, in turn, cooled the engineer's office a few degrees. Modeling the building with automatic control of outside air would not have produced an accurate simulation. in fact the building was modeled using an average outside air percentage of 70%. The very first output for the mainframe program simulation of this building showed a calculated energy use that was within 5% of the building's actual energy consumption.

In another downtown San Francisco high-rise, the chief engineer utilized a variety of electro-mechanical timeclocks to start and stop the building's various HVAC systems. As he explained, he was then using the 7 am to 6 pm timeclock. Late night observation, however, backed up by review of building electrical demand recordings, revealed that he had inadvertently "patched" himself into the timeclock set for 6 am to 7 pm, resulting in a 10%- to 12%-increase in the building's HVAC energy consumption.

Not only are operational practices manyfold more dramatic in their effect than the effects of changes to the building envelope (which influence weather-related loads), but the whole issue of weather data are greatly misunderstood. Some building simulation programs have been criticized in the past for not providing 8,760 hours of actual weather data for simulations. The well-known mainframe program developed by Department of Energy (DOE) and its various affiliations provide 8,760 hours of simulation by means of a weather data source known as the test-reference-year (TRY). Other programs, such as one developed by a well-known air-conditioning equipment manufacturer, provide a weather "trace" consisting of an average 24-hour profile for each month of the year, for a total of 288 hours of simulation. There is not much difference between these methods for two reasons. The first reason is that the TRY is not an actual year's weather data. It is, in fact, an amalgam of 12 actual month-long data. These months of real data are selected for incorporation into the reference year by a process that effectively chooses the mean month out of the months of data available. Given the continuous nature of our solar system and the statistical difference between an "average" and the "mean," the true difference between 8,760 hours of simulation and 288 is difficult to discern, except in the run times of various programs (which vary according to the number of hourly calculations that must be made).

The second reason, which applies to new or existing buildings, relates to the purpose of performing a building simulation in the first place. The general thrust of any simulation is to project the future so as to make technical and economic decisions regarding building design features. All of this presupposes that the weather that will actually occur in the future period under consideration (generally 3 to 10 years) will be essentially equal to the weather data being used for the simulation. Since this cannot be known for a certainty, any decision that would turn on the small effects in the calculations caused by the difference between 8,760 and 288 hours of simulation would be a decision of dubious wisdom at best.

2. Observational Surveys

As a result of experiences similar to the above, it has become our practice to perform two specific types of surveys in the buildings we study. The first of these surveys is observational in nature and includes careful observation of the functioning of the building's temperature control systems—as opposed to simply reviewing the temperature control as-built drawings. We have found that controls frequently were not installed as designed, have been overridden (known as "auto-manual" control), or have simply failed in one fashion or another. This observational survey generally includes a sample measurement of the system operating parameters (supply air temperature, mixed-air temperature, space discharge air temperature, etc.) as a means of observing the actual performance of the control system. The results of the inspection are frequently invaluable.

The observational survey also regularly includes a "late night" tour of the facility and its HVAC systems to identify actual operating schedules (frequently at odds with what is reported by the operating engineers) and control system performance during this period. In one building surveyed, the control air compressor was off at night but the fans and pumps were still running, resulting in extreme overheating of the facility at night, which also made the chillers work hard in the morning to bring the building temperature down when the controls came back on. This late night survey is also invaluable in confirming the operating schedules for lighting systems which are frequently under the control of the custodial crew.

3. Electrical Load Surveys

The second type of survey we find essential is an electrical load survey. Where great accuracy is desired, such as in the modeling of large high-rise office buildings, every electrical panel and piece of equipment should have its instantaneous power draw measured. This can be done with a hand-held power factor meter, with the data recorded and entered into a spreadsheet developed just for this purpose (refer to Table 13-1). It should be noted that simply reading voltage and current is insufficiently accurate, as induction motors especially have very wide ranges of power factors (depending upon their loading) that can cause volt/amp readings to be in error by 50% or more when compared to true power draw.

In addition to the instantaneous measurement of electrical loads, it is also important to look at specific large loads (chillers, elevators, computer rooms, etc.) overtime, using a power-recording instrument. This instrument can also be used to observe the total power demand profile for the entire building if the building is small and time-of-day metering is not employed by the utility company. Frequently, particularly for large buildings, the utility company records the building's power demand overtime (utilizing magnetic tape or bubble memory meters) and the information from these meters is almost always available from the utility company (refer to Figure 13-5, which shows a 24-hour plot of utility company demand interval records).

As large buildings, even in cold climates, spend most of their time in a cooling mode of HVAC system operation, electrical energy use makes up the vast majority of the building's energy use. This being the case, it is important to compare the sum of the various instantaneous load measurements with the recorded peak demand for the building, as shown in Table 13-2. If the individual measurements do not equal the total demand, then any attempt at modeling will fail. Furthermore, a building's energy use is the product of connected loads and hours of use. By utilizing the data from the operational survey and checking it against the record of electrical demand over time, a high level of confidence can be achieved as to the actual operating schedules of the various energy-using systems in the building.

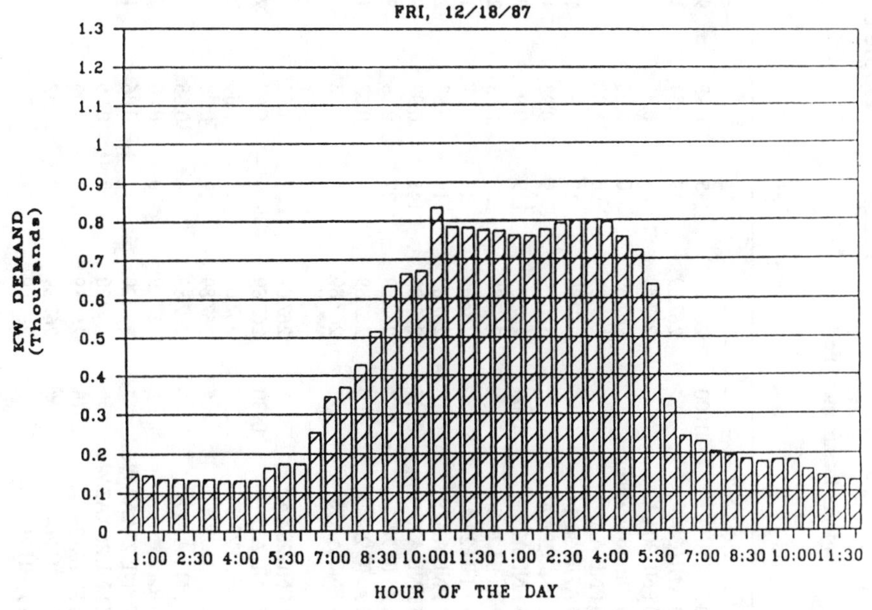

Figure 13-5. Sample of Utility Company Demand Interval Records.

Table 13-1. Sample of Electrical Load Survey Data.

Power Measurement Form | Date: 12-14-83 to 1-12-84 | Recorded by: JPW, JLH, MFS, RW

Load	Time	Ave Volts	L-1	P.F.	L-2	P.F.	L-3	P.F.	kW	Remarks
-----North Tower--------							----------------			Note: Floor 31 is at bottom
30 PNL 30A	1000	267.89	115	0.99	91	0.98	88	0.96	77.0	
30 AHU 30-3	"	267.89	19	0.84	19	0.91	17	0.91	13.1	
30 PNL 30AA2	"	267.89	3	0.82	5	0.63	9.9	0.88	3.8	
30 Window washer	"	267.89	0	0.00	0	0.00	0	0.00	0.0	
30 PNL AA30	"	267.89	8	1.00	9	0.81	3	0.81	4.7	
30 AHU 30-4	"	267.89	15	0.81	17	0.82	17	0.73	10.3	
30 PNL 29AA2	"	120.09	9	0.93	11	1.00	23	1.00	5.1	WMR @ bottom of 30AA2
29 PNL A29	1000	267.89	129	0.99	95	0.96	85	0.98	81.0	
29 AHU 29-3	"	267.89	23	0.83	23	0.75	21	0.83	14.4	
29 AHU 29-4	"	267.89	19	0.77	17	0.83	19	0.85	12.0	
29 PNL AA29	"	267.89	19	0.88	23	0.52	11	0.98	10.6	
28 PNL A28	0945	267.89	141	0.97	93	0.94	99	0.94	85.0	
28 AHU 28-3	"	267.89	27	0.88	29	0.81	25	0.86	18.4	
28 AHU 28-4	"	267.89	19	0.62	15	0.65	17	0.76	9.2	
28 PNL AA28	"	267.89	3	0.72	3	0.99	7	0.82	2.9	
						Date:		Recorded by:		
28 PNL 28AA2	"	267.89	3	1.00	11	0.83	10	0.64	5.0	
27 PNL A27	0930	267.89	103	0.97	71	0.96	81	0.95	65.6	
27 AHU 27-3	"	267.89	21	0.84	23	0.76	19	0.82	13.6	
27 AHU 27-4	"	267.89	19	0.78	16	0.83	14.9	0.89	11.1	
27 PNL AA27	"	267.89	3	0.85	5	0.87	7	0.93	3.6	
27 Radio	"	267.89	0	0.00	0	0.00	0	0.00	0.0	
27 PNL 27AA2 West	"	120	17	0.96	17	1.00	10	1.00	5.2	West Mech Room
27 PNL 26AA2 West	"	120	9	0.96	15	0.71	9	0.97	3.4	Read @ bottom of 27AA2
26 PNL A26	0915	267.89	133	0.99	90	0.97	65	0.97	75.6	
26 AHU 26-3	"	267.89	19	0.87	19	0.78	17	0.82	12.1	
26 AHU 26-4	"	267.89	19	0.76	15	0.81	18	0.83	11.1	

Table 13-2. Sample Comparison of Field-Measured Electrical Loads to Peak-Demand Recorded by Utility Company.

County Courthouse Measured Demand

Electrical Load	kW
Misc Power & Lighting Combined	330.9
Mechanical Equip.	
Air Handlers	82.0
Return & Exh fans	32.9
Pri, Hot Water Pump	2.2
Control Trans	0.6
Comp. Rm.	2.0
Oil Burn PNL #4	6.1
Comp. Rm. PNL #4	4.4
MCC SF #1 MEZZ	10.3
Air Control Compressor	4.2
Boiler Rm. Exh fan #3	0.4
Pump Panel #2	4.5
Pent PNL #15 & 16	2.7
Amp PNL (Chill Rm)	0.2
Rm. PNL 17	3.3
New PNL Rm. 8	0.7
3rd Flr Fans	23.7
N. Courtroom Fans	14.5
S. Courtroom Fans	15.7
Panel #1S1S2	39.4
9-11th Fan	22.4
Sub Total	603.1
Chry Chiller Load from Dranetz	175.0
Chw Pump	6.0
Cw Pump	6.0
Tower Fan	9.0
Sub Total	196.0
Total Measured kW Demand	799.1
Utility Co. Demand (Winter)	800.0
Percent Variance	0.1

E. Output Critique

One of the hardest things to do in performing a building simulation is to honestly critique the computer output. After spending hours or even days preparing the input data, it is easy to fall into the trap of believing that the output must be correct. However, as our mistakes prove to us, it is critically important to critique the computer output with a skeptical attitude. Three specific techniques are valuable in regard to critiquing simulation program output.

1. Annual Energy-Use Profile Comparison

The first technique is a gross, year-long evaluation of the modeled energy use in comparison to actual energy use. While the totals may agree, seasonal variations may not agree well with each other, indicating that weather influenced systems are not modeled well. Graphic comparison of modeled and actual energy use is most valuable in this evaluation, as can be seen in Figures 13-6 and 13-7.

In addition, since computer simulations generally utilize weather data that are a composite of multiple years' data (including National

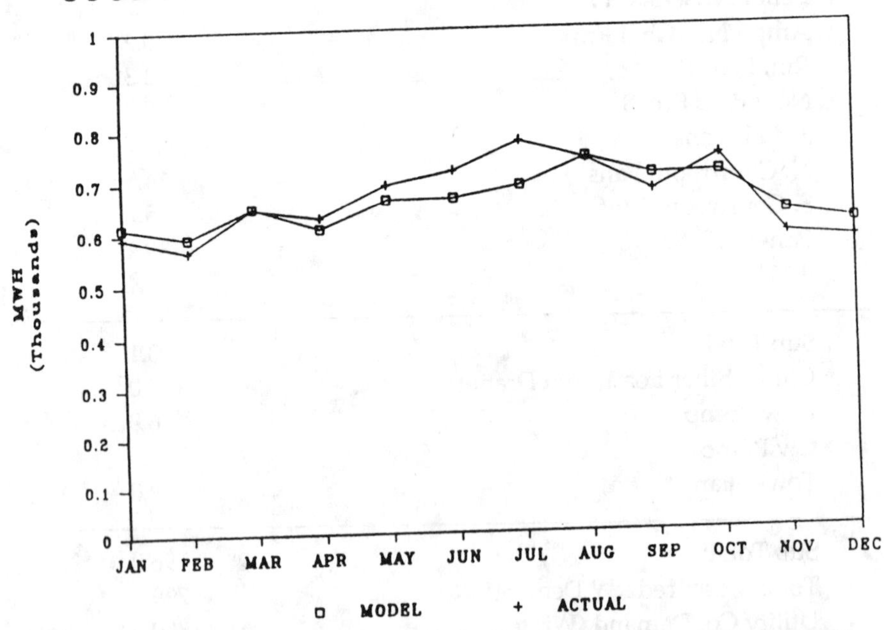

Figure 13-6. Comparison of Modeled and Actual Annual Electrical Consumption Profile.

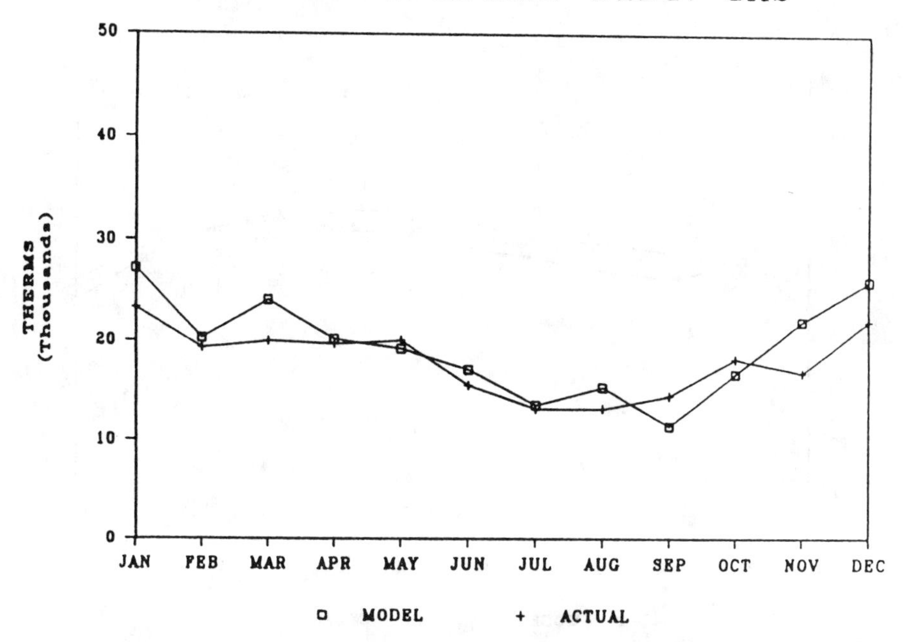

Figure 13-7. Comparison of Modeled and Actual Annual Natural Gas Consumption Profile.

Oceanographic Atmospheric Administration (NOAA) TRY tapes as previously discussed), it is valuable to contrast the actual weather data for the year being modeled to the weather data employed in the simulation, as shown in Figure 13-8. When modeling a building using a year's worth of actual energy use for validation, it is more important that the modeled energy use vary according to the changes in the model weather for the same period rather than absolutely agree with the actual utility data being used for comparison. For example, if the model shows higher than actual electrical use for cooling in a given month and both the actual electrical use and actual temperatures are lower than the model, then this lends credence to the model and means the model is meaningful for evaluation of multiple future years' potential for energy savings.

2. Peak Load Comparison

The second of these techniques is to evaluate peak modeled loads against known values. From the utility company's data, the building's peak electrical demand is known for all seasons of the year. Generally, computer models will provide a monthly peak electrical demand for the

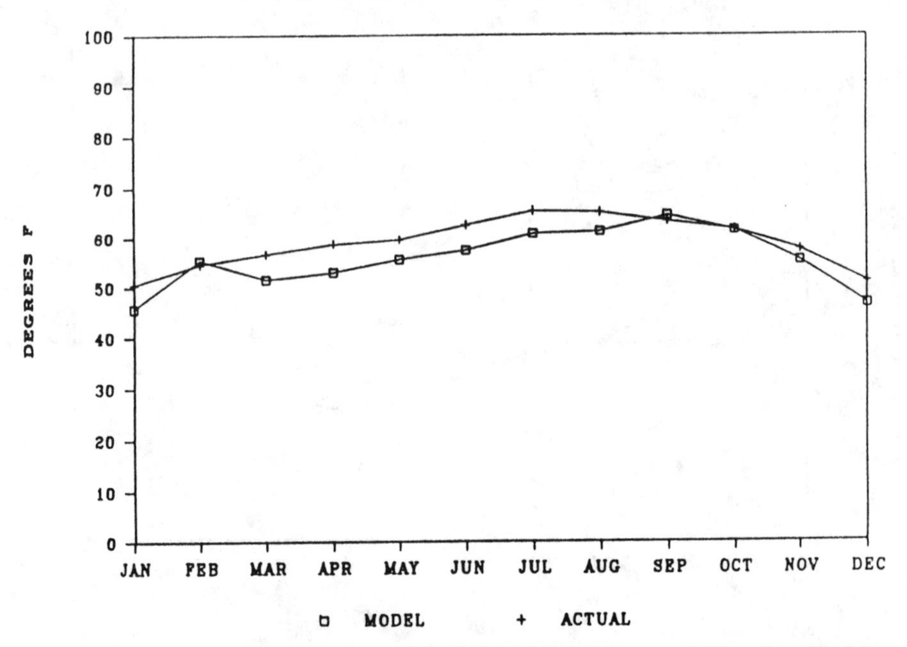

Figure 13-8. Comparison of Modeled Actual Annual Weather Data.

various components of the model. By comparing the main seasons (summer, fall, spring and winter), it can be observed whether all of the loads measured during the survey found their way into the model and whether the seasonal modeling of cooling loads is correct. Furthermore, the building's peak cooling load is probably known from operating engineers' observations and/or operating logs, and this too, can be used as a scale of measure for evaluating the accuracy of the computer model.

Again, the issue of the weather data employed for the simulation must be taken into consideration. Generally, all weather data used for simulations are missing the hottest and coldest days of the year. Accordingly, the actual demand data used for comparison would best be selected as a day experiencing the same, or nearly the same, temperature extremes as present in the weather data used for simulation. Interestingly enough, one building we modeled had one of its chillers fail and was quite short of capacity. As a part of assessing the comfort risk caused by the failed chiller, the mainframe simulation output was reviewed and we identified the ambient temperature at which the simulation would predict "losing" the building on a hot day. In fact, within a few weeks of completing the modeling process, an unseasonably hot day was encountered with a peak temperature exceeding our predicted "lose the building" temperature by a few degrees. Indeed, the chief engineer reported that he had "lost" the building on that one day.

3. Detailed Output Analysis

The third technique is primarily oriented towards evaluation of energy retrofit models. In order to develop savings estimates for energy retrofit measures under consideration, the retrofit is modeled and then contrasted with the original model, thus showing the savings that might be achieved. Since it is very easy to make small errors in editing the input data for a computer model and cause an unintended result, a useful quality control technique has been to analyze the computer model in detail (by functional use, i.e., lighting, cooling, fans, pumps, etc.) and develop a specific figure for the savings estimated for each retrofit in each functional use area.

As can be seen in Table 13-3, a very detailed analysis of the output from a mainframe proprietary computer model is possible. The analysis allows a "plausibility" check of the savings from a particular retrofit. For example, if a variable-air-volume (VAV) retrofit is under consideration, it is possible to develop a specific estimate for the savings to be achieved by the fan alone. This savings can then be compared to the original energy used by the fan and the plausibility thereof evaluated. If a simple inlet vane conversion is anticipated and the system operates a single shift per day during weekdays, a savings in the neighborhood of 30% to 40% may be anticipated. If the detailed analysis indicates a savings of 70% or 80%, then review of the model input is warranted to determine the error in the input or determine the reason that a savings much higher than the engineer's is reasonable. For example, perhaps the system does, after all, operate on a 24-hour per day basis or was grossly oversized and will experience very low loads compared to its installed capacity for most of its operating hours. In any event, when the savings vary greatly from that which is "plausible," it indicates either an error in the modeling or error in the plausibility logic—either of which should be determined before using the savings numbers generated by the model.

It is theoretically possible to create a perfect model in which every small unique thermal zone in a building responds to weather input virtually the same as the actual building. However, the practicality of such modeling is doubtful as the engineering costs to prepare such a model may actually exceed the value to be created by the modeling process, particularly in smaller buildings. As a result, even the best modeling tools and reasonably constructed models will be limited in their ability to

Table 13-3. Sample of Detailed Analysis of Output from Computer Simulation.

Analysis of Run/Alternative to Determine Energy Savings:
ECM # 1 & 2
Title Admin Bldg Penthouse and Basement Double Duct to VAV

Equipment	Energy	Base	Comp to	ECM Use	Delta	% Reduction
Chiller 1	kWh	442052	442052	430574	11478	2.6
Chlr 1 Aux	kWh	157787	157787	155471	2316	1.5
Chiller 2	kWh	390069	390069	337348	52721	13.5
Chlr 2 Aux	kWh	70208	70208	64355	5853	8.3
Chiller 3	kWh	44762	44762	25963	18799	42.0
Chlr 3 Aux	kWh	38332	38332	24441	13891	36.2
Boiler	Therms	191586	191586	137840	53746	28.1
Boiler Aux	kWh	65831	65831	60051	5780	8.8
Sys 1 Sf	kWh	397881	397881	129122	268759	67.5
Sys 1 Rf	kWh	183582	183582	59574	124008	67.5
Sys 1 Ef	kWh	13690	13690	5397	8293	60.6
Sys 2 Sf	kWh	46228	46228	46228	0	0.0
Sys 3 Sf	kWh	38082	38082	38082	0	0.0
Sys 4 Sf	kWh	147921	147921	147921	0	0.0
Sys 5 Sf	kWh	48861	48861	48861	0	0.0
Sys 5 Rf	kWh	1972	1972	1972	0	0.0
Sys 6 Sf	kWh	16701	16701	16701	0	0.0
Sys 6 Rf	kWh	3336	3336	3336	0	0.0
Sys 7 Sf	kWh	83202	83202	83202	0	0.0
Sys 8 Sf	kWh	67196	67196	67196	0	0.0
Lights	kWh	1827783	1827783	1827783	0	0.0
Base Elec	kWh	3652192	3652192	3652192	0	0.0
Base Gas	Therms	39909	39909	39909	0	0.0

Total Electric Savings: 511898 kWh
Total Gas Savings: 53746 Therms

predict the effect of retrofit measures. Therefore, in some cases, it is an appropriate engineering step to derate or discount the savings figures for engineering conservatism (refer to Table 13-4).

Table 13-4. Sample of Conservative Derating of Energy Savings from a Spreadsheet Simulation Program.

Savings from best computer models:

ECM #1	Run Description	kWh	Therms
	Penthouse Unit Baseline	284146	32712
	Penthouse Unit Opt S/S	265684	28471
	Savings	18462	4241
	Plausibility Factor	1.00	1.00
	Net Savings	18462	4241

(Note: kWh includes only cooling and fans)

ECM #2	Run Description	kWh	Therms
	Penthouse Unit Opt S/S	265684	28471
	Penthouse Unit VAV	133832	4156
	Savings	131852	24315
	Plausibility Factor	0.75	0.60
	Net Savings	98889	14589

(Note: kWh includes only cooling and fans)

ECM #_	Run Description	kWh	Therms
	N/A	0	0
	N/A	0	0
	Savings	0	0
	Plausibility Factor	1.00	1.00
	Net Savings	0	0

(Note: kWh includes only cooling and fans)

A good example of this is the fact that many computer models that utilize hourly heating and cooling load calculations as part of their modeling (not all do, as we will see below) are unable, without micro-zoning of the model, to avoid the sharing of internal heat gain with external zones needing heating, thus underestimating the actual heating requirements of the building. Similarly, tall buildings in central city locations often have large vertical exterior zones, part of which need cooling and part of which need heating at any given time, primarily due to solar exposure and shading from adjacent buildings. These perimeter systems can be difficult to model and sometimes will show optimistic results from even the most conservative attempts at modeling, thus necessitating an engineering discounting of savings. The bottom line here is that even the best models still have limits to their capabilities.

4. Plausibility Check

Finally, by summing all the savings for all retrofits, a gross plausibility check can be performed, based on engineering judgment regarding whole building energy-use levels that are reasonable for the type of building being evaluated. This is a gross measure, but it is an excellent final check on the entire process, as shown in Table 13-5. Even such a simple check can be effective in catching unreasonable optimism in energy savings estimates that may have slipped through all the other quality control measures in this very complex process of building simulation. Had such a macro check been part of the project documentation associated with the project mentioned in the introduction, that energy services company would not have the problem they currently face.

F. Simulation Tools

It is likely that a wide range of opinion exists in the energy engineering field as to what constitutes building energy simulation. Our view is a rather broad one and encompasses a wide range of calculational strategies as being appropriate to specific project goals and project environments.

1. Mainframe Programs

The high end of the practice are programs that have traditionally run on mainframes (minicomputers). Both proprietary and public-domain programs are in common use. The availability of such programs to run on high-end personal computers is becoming fairly commonplace. In general these programs have similar, if not common, ancestry and are founded in hourly heating and cooling load calculations that are then applied to the HVAC systems and equipment described to the program. These programs

Table 13-5. Sample of "Gross" or "Overview" Check of Savings Calculations.

Energy Savings Summary

Energy Conservation Measure:		kWh	Therms	$
1. Admin penthouse double duct to VAV & 2. Admin basement double duct to VAV		511898	53746	$62,246
3. Admin courtroom multizones to VAV		37124	3589	$4,415
4. Convert jail multizones to VAV		53579	0	
7. Large courthouse multizones to VAV		127761	36682	
Sub-total		181340	36682	$27,696
5. Supervisor's AHU control mod		12372	332	$1,195
6. Lighting retrofit		448987	–1946	$38,888
8. Courthouse small Mz's to VAV & 9. Courthouse small Mz to VAV		148703	43771	$27,093
10. Summer steam shutdown		0	10287	$3,292
11. Energy management computer		74974 63146	6190 0	
Sub-total		138120	6190	$14,135
12. Variable flow chilled water		15647	0	$1,377
	Btu/sf/yr			
Total (Excl ECM#12)	43680	1478544	152651	$178,960
Plausibility factor		0.95	0.95	
Net savings	41496	1404617	145018	$170.012
Existing consumption	104035	7896022	214272	$763,417
Percent reductions	39.9	17.8	67.7	22.3
Retrofit Btu/sf/yr	60355			

Note:

1. Electricity average unit cost for 12 mo. ending Oct. '88 was $0.0796/kWh, plus approx 10% rate increase in Jan. '89 equals $0.088/kWh used above.
2. Natural gas unit cost used is $0.32/therm.

are very powerful simulation tools in that they allow for detailed input of both the envelope and the lighting and HVAC systems employed in the building, and produce excellent results as shown in Figures 13-6 and 13-7).

In addition, these programs also provide extensive output data for use in output critique. While very powerful, these programs require significant engineering labor to prepare the data necessary for input (often 40 to 80 engineering labor hours) and are sometimes too costly for use on smaller buildings or for use in the qualification of sales prospects in the energy retrofit business. To meet the need for less costly simulation methods, we developed some spreadsheet-based simulation tools that have proven to be very effective.

2. Complex Spreadsheet Simulation Tool

One of the spreadsheet-based tools developed is a complex spreadsheet that allows time-related loads to be scheduled by hour, by three-day types (weekday, Saturday and Sunday/Holiday), by type of energy used, or by type of functional energy use (cooling, fans, lighting, etc.). In addition, the calendar of day types for the model year can be customized to cover virtually any situation. With respect to weather-related loads, the model takes a totally different approach than mainframe programs. In this case, the program accepts peak loads as input and distributes the loading over the period of a year according to the differential between the modeled ambient temperature and user-input "no-load" temperatures for heating and cooling. Other variables include heating and cooling lockout temperatures, minimum loads, and daily and seasonal operating schedules. The model calculates hourly ambient temperatures for application of the loads by using a near-sinusoidal model and varying the temperature up or down from the average temperature by half the average daily range. The model utilizes as input degree-days and average daily range by month, or average maximum and minimum temperatures by month. The model provides hourly heating and cooling loads for typical day types each month and hourly time-related loads for typical day types each month.

As can be concluded from Figures 13-9 and 13-10, this modeling tool can produce simulations of high accuracy and requires only a few hours for input generation and model runs. In addition, because there is great control over the model, many different retrofit measures can be modeled and custom simulations can be produced by modifying the code or extracting output from the base building model and performing subsequent calculations thereon. This tool is most effective on smaller or simpler buildings, where a high level of confidence in energy-savings is desired while engineering costs are kept to a minimum.

3. Simple Spreadsheet Simulation Tool

Another tool developed is a one-page simulation spreadsheet. Its purpose was to provide an extremely quick and inexpensive simulation tool for use where limited accuracy is acceptable and simulation costs are of greater importance than accuracy. Two versions of this model exist, one for HVAC systems that mix heating and cooling (e.g., terminal reheat) and one for nonmixing systems. As shown in Table 13-6, this simulation tool has very simple input and basically views a building as having lighting, heating, cooling, HVAC accessories, domestic hot water, and two types of miscellaneous energy use (electrical and heating fuel). Input are generally in units per square feet (e.g., lighting input is in watts per square foot) and percentage of operating hours. In addition, provision is made for reduced summer operation (primarily for schools) and "off hours" loads in all functional areas. Time-related loads are calculated based on "hours on" times input loads, similar to the spreadsheet described above, without the ability to customize day types or the annual calendar. Weather-related loads assume a linear, directly proportional relationship with degree-days, which are input to the spreadsheet.

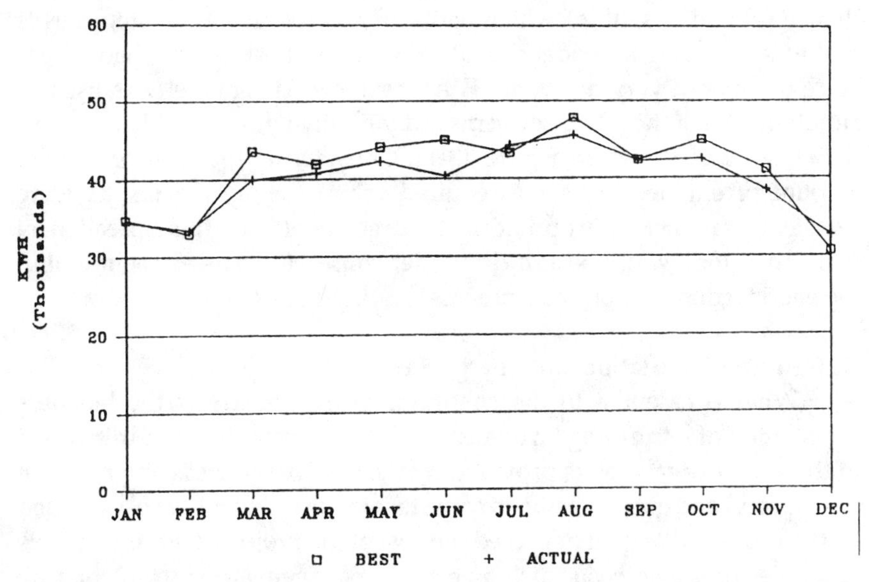

Figure 13-9. Sample of a Spreadsheet Simulation Program (Electricity).

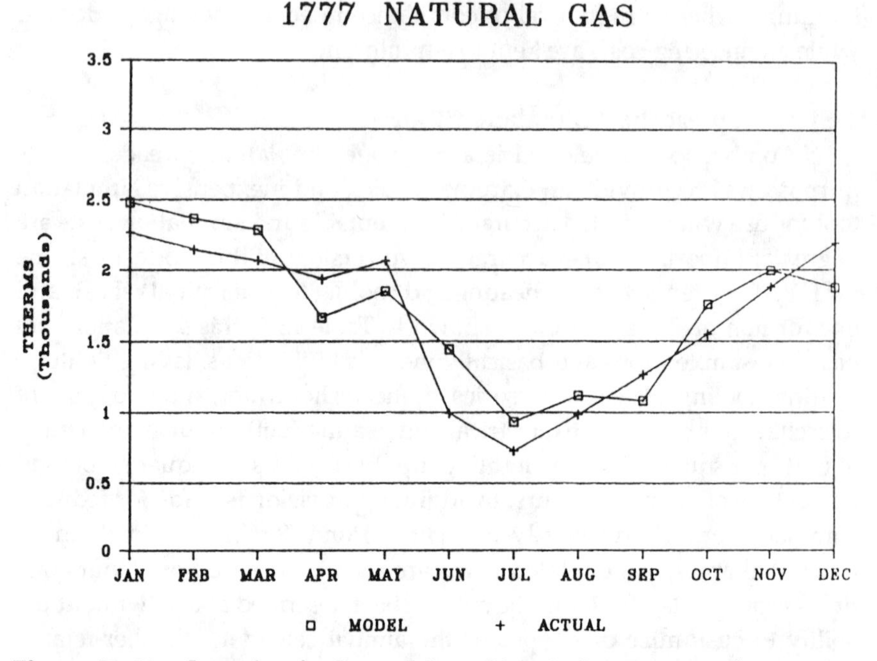

Figure 13-10. Sample of a Spreadsheet Simulation Program (Natural Gas).

This model was developed to simulate a college campus of more than 100 buildings, all of which had fairly simple HVAC systems. This tool was also used to model a small community hospital that had a very large number of very different HVAC systems. This model was used to simulate each of the HVAC systems individually with the modeling accuracy results as shown in Figure 13-11. Considering the relatively small amount of engineering effort required for modeling, the results were excellent. Another appropriate and attractive use of this spreadsheet simulation tool would be as a first-order conservation assessment tool in the energy conservation sales process.

G. Building Simulation and Energy Services

What is essential to the energy services industry is the business proposition of retrofitting an owner's building at essentially no initial cost to the owner (financing is provided by the energy services company or a third-party) and guaranteeing in some fashion that the utility cost avoided by the project will equal or exceed the cost of the project (debt service plus any other on-going costs such as project management or maintenance). Unfortunately for some projects done in this industry, the sales people

Table 13-6. Input and Output of a Simple Spreadsheet Simulation.

COMMUNITY HOSPITAL ENERGY USE TEMPLATE
FOR NON-MIXING SYSTEMS

ENERGY UNITS>>>>UNIT # TYPE

UNIT #	TYPE
1	ELEC, KWH
2	GAS, MBTU
3	STEAM,MBTU
4	CHW, MBTU

INPUT BY [M. MUNIZ] DATE [9/7/89]

AREA NAME [EMERGENCY] SYS NO. [8-2] OCCUPIED SQ. FT. [3200] YEAR BUILT [1974]

LIGHTING		COOLING		HEATING		CLG/HTG ACCESSORIES		DOM. HW OFFICE		DOM. HW RESID.		MISCELLANEOUS A		MISCELLANEOUS B	
NRG UNIT >	1	NRG UNIT >	1	NRG UNIT >	2	NRG UNIT >	1	NRG UNIT >	3	NRG UNIT >	3	NRG UNIT >	1	NRG UNIT >	2
WATTS/SF >	2.1	SF/TON >	350	BTU/SF >	20	KW/TON >	0.2	SF/PERSON >	150	SF/PERSON >	200	WATT/SF >	0.5	MBH CAPAC >	0
% HRS OCC.>	80	EQU. FCTR >	1	EQU. FCTR >	0.3	EQU. FCTR >	1	GAL/PER/DA>	0	GAL/PER/DA>	0	% OCCUP. >	60	% OCCUP. >	33
SUM LOAD %>	100	EQU. FCTR >	0.1	EQU. FCTR >	1.2	% OCCUP. >	100	TEMP FCTR >	1	TEMP FCTR >	1	UNOCC LD %>	20	UNOCC LD %>	5
UNOCC. LD%>	80	% HRS OCC.>	60	% HRS OCC.>	60	UNOCC LD %>	0	SUMMER LD%>	100	SUMMER LD%>	20	% SUMMER >	100	% SUMMER >	50
		UNOCC LD% >	60	UNOCC LD %>	100	HTG USE % >	0	assumed:		assumed:					
		% SUMMER >	100	MIN LOAD %>	0	SUMMER LD%>	100	HWS TEMP	140	HWS TEMP	140				
		MIN LOAD %>	0	% SUMMER >	100			CWS TEMP	60	CWS TEMP	60				
KW TOTAL >	7	TONS >	9	MBTU'S >	64	KW >	1.8	OCCUPANCY >	21	OCCUPANCY >	16	KW >	2		

USAGE	UNITS NRG	SEPTEMBER	OCTOBER	NOVEMBER	DECEMBER	JANUARY	FEBRUARY	MARCH	APRIL	MAY	JUNE	JULY	AUGUST	TOTALS
LIGHTING	ELEC, KWH	4,645	4,800	4,645	4,800	4,800	4,335	4,800	4,645	4,800	4,645	4,800	4,800	56513
COOLING	ELEC, KWH	4,627	2,694	1,451	94	42	672	545	2,232	3,575	5,256	6,857	5,798	33843
HEATING	GAS, MBTU	0	0	2,905	17,142	12,876	7,456	3,910	0	0	0	0	0	44289
CLG/HTG ACCESSORIES	ELEC, KWH	1,317	1,360	1,317	1,360	1,360	1,229	1,360	1,317	1,360	1,317	1,360	1,360	16018
DOM. HW OFFICE	STEAM,MBTU	0	0	0	0	0	0	0	0	0	0	0	0	0
DOM. HW RESID.	STEAM,MBTU	0	0	0	0	0	0	0	0	0	0	0	0	0
MISCELLANEOUS A	ELEC, KWH	783	809	783	809	809	731	809	783	809	783	809	809	9531
MISCELLANEOUS B	GAS, MBTU	0	0	0	0	0	0	0	0	0	0	0	0	0
EMERGENCY	ELEC, KWH	11,371	9,664	8,196	7,064	7,012	6,967	7,515	8,977	10,545	12,000	13,826	12,767	115905
SYS. NO. 8-2	GAS, MBTU	0	0	2,905	17,142	12,876	7,456	3,910	0	0	0	0	0	44289
OCC. SQ FT 3200	STM, MBTU	0	0	0	0	0	0	0	0	0	0	0	0	0
	CHW, MBTU	0	0	0	0	0	0	0	0	0	0	0	0	0

TOTAL COMBINED ETU/SF 137,424

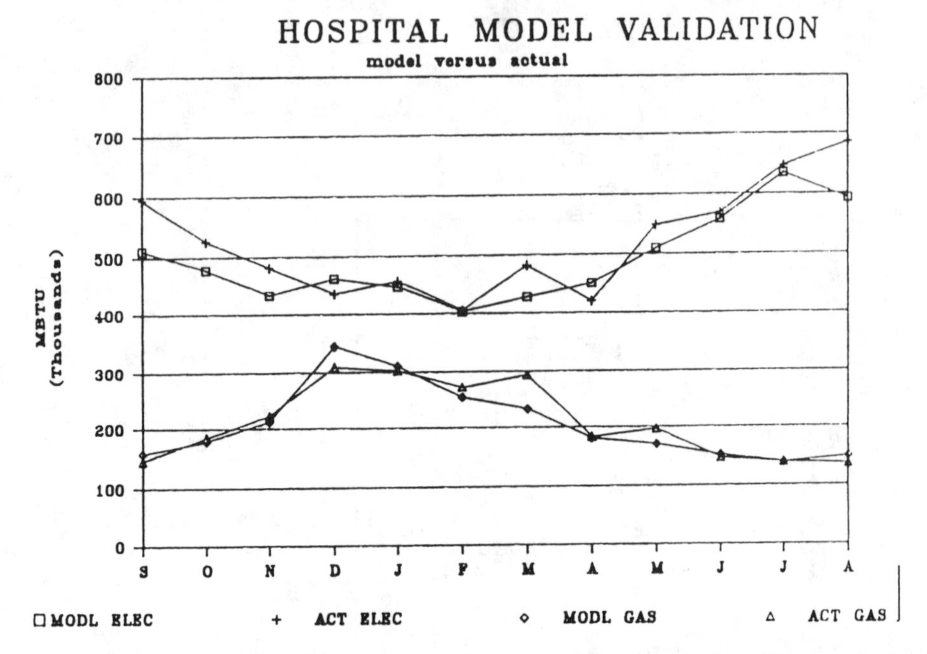

Figure 13-11. Sample of a Simple Spreadsheet Simulation (Hospital Model Validation).

involved viewed the business proposition as simply a way to make their job easier and they exhorted their technical staffs to generate savings calculations that would support a high cost value for their projects. This is unfortunate, and even frightening, because savings so generated are difficult if not impossible to achieve in reality and, if the guarantee offered is reputable, it must then come into play to cover the savings shortfall that must necessarily occur.

One energy services company with which we have worked had the unpleasant experience of having a sales engineer substitute his own savings calculations for those generated by the computer model. The result of this was a guarantee of natural gas savings on one project that actually exceeded the natural gas consumption of the building. Needless to say, his own management failed to properly consider the plausibility of such a proposition and approved the project for funding—and is presently refunding the annual savings "shortfall" ($100,000 per year). The use of cost-effective and accurate tools and methods of building simulation is an essential part of identifying and implementing successful energy services or DSM projects.

H. Conclusions

While it is a fascinating and complex engineering tool, the fundamental value of computer simulation of buildings is that it forces a quality-enhancing step in the analytical process. This step is essentially a systematic confirmation of the engineer's knowledge of where and how energy is consumed in a building. If the modeling step is done well, it is difficult to make "off target" recommendations for specific types of retrofits or "off target" estimates of savings. With such a high level of confidence established on the technical side of a project, the assessment and mitigation of project performance risk can rightly be performed on the financial side of the project evaluation, resulting in a very high probability of success for energy retrofit and DSM projects.

III. METHODS FOR PERFORMANCE MONITORING OF ENERGY SERVICES PROJECTS

The following is a discussion of alternative methods for the monitoring of cost avoidance achieved by energy services/DSM projects. This discussion is developed to provide a basis for dialogue regarding of the issues involved among the parties to future or past energy services projects/DSM projects, with the goal of creating consensus among the parties with regard to how performance/cost avoidance monitoring should be performed for these projects. The consensus thus achieved should provide a high level of accuracy and veracity while minimizing the administrative burden to both parties. The following discussion will also serve those who contemplate pursuing an energy services or DSM project and who are concerned about demonstrating the project's energy savings performance (and compensation to the contractor or activation of the contractor's savings guarantee).

A. Background

Accounting for or demonstrating the utility expense avoided as a result of the implementation of energy conservation measures in facilities is a small, but critical part of the energy services (now sometimes referred to as DSM) business. Accounting for avoided costs determines the energy services company's revenue and/or claims against cost avoidance guarantees. No method of accounting for cost avoidance is perfect or completely accurate. Because of this, purchasers of energy services have come to learn that the experience, qualifications, skill, and veracity of an energy services company is far more important in selecting a company than the apparent value of projected savings and/or savings guarantees offered during the sales process.

B. Approaches to Performance/Cost Avoidance Monitoring

Historically, a very wide range of basic methods have been employed in demonstrating the utility cost avoided through energy retrofits. In general these methods have fallen into three categories, and are sometimes a mixture of categories. These three categories are:

- utility bill comparison,
- energy-conservation-measure-specific instrumentation, and
- stipulated calculations/values.

As we go down this list of methods, values of avoided cost are less impacted by variables beyond the control of the energy services company. As we go up the list, values of avoided cost are more greatly impacted by variables which are within the control of the project-host/building-owner.

For example, should a tenant in a commercial office building decide to install a mainframe computer in their space they might well do it with the knowledge of the building owner, but without this fact being communicated to the energy services company. The result is an increase in the utility consumption in the building, and a decrease in the avoided cost determined by utility bill comparison, which has absolutely nothing to do with any energy conservation measures which may have been implemented by the energy services company. This situation is generally viewed as a risk in the eyes of the energy services company.

At the other end of this scale, for example, the installation of motion detection lighting controls could be agreed by both parties to reduce annual energy consumption by a set number of kilowatt-hours per year. This kilowatt-hour value would be the product of the stipulated wattage of fixtures to be controlled multiplied by the stipulated annual hours of reduced operation of the fixtures. Given the level of expertise possessed by both parties, the energy services company may well have superior ability to ascertain both the wattage of the fixtures and the annual reduction in operating hours. If the energy services company overstates these values or fails to install the proper equipment, actual avoided cost would be less than projected. This situation is also generally viewed as a risk in the eyes of the owner. All of the methods are workable and fair, depending on the characteristics of the facility to be retrofitted, the characteristics of the specific energy conservation measures to be implemented and the detailed mechanics of each method's employment which are agreed to by the parties.

1. Utility Bill Comparison

Utility bill comparison contrasts post retrofit utility consumption, as invoiced by the utility company, to pre-retrofit consumption (in a "base" period) and generally multiplies the difference in units of consumption by the current unit cost for each utility. Factors influencing this method include the frequency and likelihood of changes in the use, occupancy, and operation of the facility, (these can affect both the selection of a "base" period and possible adjustments to consumption figures for changes in occupancy, use or operation) and the effect of weather on the facilities' energy use. The advantages and disadvantages of this approach are shown in Table 13-7. This methodology would generally find application in simple, single-use facilities, with stable histories of facility use and energy consumption. Simple or complex retrofits can be accommodated by this method.

Table 13-7. Utility Bill Comparison.

ADVANTAGES	DISADVANTAGES
Simple, small number of data source	The effects of energy conservation measures can be "masked" by unrelated changes in use and occupancy
Since savings come "right off the bill," the figures are often considered more "real"	Adjustments for changes in facility use can be complex and an administrative burden—extensive adjustments can make the figures appear "artificial"

2. ECM-Specific Instrumentation

In this concept of demonstrating cost avoidance, temperature sensors, event sensors, flow meters, power transducers, and other instruments are applied to the equipment or loads affected by the energy conservation measures. The data gathered by these devices are then manipulated and integrated by the building automation computer system or another computer to calculate the actual energy savings produced by individual energy conservation measures. Factors that impact the effectiveness of this approach include the complexity of the energy conservation measures implemented, the accuracy and reliability of the instrumentation equipment, and the validity of the mathematical formulas used in the energy calculations. Advantages and disadvantages of this approach are shown in Table 13-8. This method would generally find application in industrial environments, large, expensive, and complex projects or project where only a very small number of simple retrofits are replicated in great quantity.

Table 13-8. ECM-Specific Instrumentation.

ADVANTAGES	DISADVANTAGES
Separates out the effect of ECMs from other changes in building use and occupancy	Prohibitively expensive except on the simplest of energy conservation measures
Data gathering and calculalations are verifiable	Requires considerable administrative manpower to maintain, calibrate, and troubleshoot the instrumentation and calculational software
Due to direct measurement, cost avoidance figures are often considered more "real"	Verification can be time consuming and costly

3. Stipulated Calculations/Values

This approach is based on both parties possessing a high degree of confidence in the effectiveness of a particular energy conservation measures in reducing load and confident knowledge of the annual hours of use of the systems/load being retrofitted. This approach consists of establishing set figures or values or precise formulas for the calculation of cost avoidance. In the case of a stipulated value, for example, a set number of watts are agreed by the parties to be saved by a particular lighting fixture retrofit, such as replacing a 200-watt incandescent lamp with a 32-watt compact fluorescent replacement unit.

In addition, the parties agree to a stipulated value for the annual operating hours for the fixture (i.e., 8760 hours for a hallway fixture in a college dorm). The calculation of avoided cost then is simply the watts saved per fixture, multiplied by the number of fixtures, multiplied by the stipulated hours per year, all of which are then multiplied by the current unit cost of electricity (which is taken from the current rate schedules or utility invoices). Similarly, stipulated calculations are formulas that have one or more variables which are measured, such as the final connected electrical load of a high efficiency motor and its hours of operation as recorded by the energy management computer. The calculation of avoided cost is the difference between the original connected load and the new connected load, multiplied by the hours of operation, all of which are again multiplied by the current unit cost of electricity. The advantages and disadvantages of this approach are listed in Table 13-9. This method would generally find use in facilities with constantly changing occupancy or facility use which would mask the use of utility bill comparison, or where a wide variety of conservation measures are being employed or where very complex retrofits are being employed (both of these make instrumentation prohibitive).

Table 13-9. Stipulated Calculations.

ADVANTAGES	DISADVANTAGES
Extremely simple—frequently the only variables are either measured once at the completion of the installation, are tracked by the energy management computer or are the current unit costs of energy—often these figures are more "real" than other methods, especially in a complex, unstable environment	Can work to reduce the motivation of the installing contractor to work carefully and accurately
Administrative costs are kept to a minimum	Savings figures are sometimes viewed as "artificial" or "unreal"

4. Conclusion

Choosing one of the above methods of monitoring actual cost avoidance achieved by an energy retrofit project varies according to the perceptive of the purchaser and, of course, the provider of services. End users such as building owners or occupants tend to view this activity as crucially important to a project. This is because building managers must budget for each year's operating costs whether they are in the form of energy purchases or energy services. A non-functioning project, as one hospital manager in Southern California and a large Northern California

County recently discovered, can create a $100,000 annual budget problem.

The method chosen needs to fit the characteristics of the building and the retrofits planned for implementation and the level of trust between the parties and the concomitant need for real verification of savings. This is why it is so important to choose an energy services company carefully, according to how they are likely to perform the needed work and follow-up after the project is completed—in contrast to selecting the company that seems to be making the most attractive financial and guarantee offers. A well-studied facility with a carefully implemented project may not need verification.

IV. PROJECT FUNDING ALTERNATIVES

Fortunately for those facility owners whose funding of a comprehensive energy retrofit program is unlikely out of internally generated funds, there are a variety of possible funding alternatives available. The most obvious, as discussed in great detail here, is that of utilizing an energy services company as a turnkey agent to fund, implement and maintain the project.

An alternative scenario is simply a lease/purchase agreement. In the process of developing an energy management program implementation plan for a large hospital, we identified a number of sources for funding including one that even offers a program of ensuring energy savings performance on their projects through a third-party underwriter. In some ways, this method of financial risk mitigation is more attractive than the turnkey energy services approach in that a third-party with a particular interest in the successful performance of the project (the underwriter) will perform a review of the project to assess its plausibility.

An additional funding source was identified in the form of tax-exempt bonds. For organizations categorized as 501-C3 by Internal Review Service, activities such as energy conservation and facility restoration qualify for funding through the sale of tax-exempt instruments. Turnkey or packaged deals tend to carry higher interest rates and numerous hidden costs and, are therefore probably not optimal for organizations with financial and project management skills. Whether turnkey funding, commercial money or the savings guarantee or tax-exempt funds are chosen is largely a matter of management's level of comfort with the facility staff's ability to administer such a program.

V. PROJECT IMPLEMENTATION ALTERNATIVES

A considerable number of alternative project implementation resources are available to facility owners to implement energy retrofit projects. These resources include:

- energy services companies,
- contractors,
- engineering consultants,
- vendors, and
- in-house staff.

Strategies for employment of these resources are chosen based on a number of considerations including:

- in-house expertise and available time required,
- mitigation of financial risk,
- technical/design control,
- construction/installation control, and
- hardware/equipment standardization.

Investment in an energy conservation project has the expectation of a financial return. Therefore, any energy conservation project that is implemented should be done in a fashion which creates either a high level of confidence that the desired end-result will be achieved or reassigning that risk elsewhere through some form of underwriting. In other words, prudence dictates either some sort of self insurance or the arrangement of insurance through another party.

Control over technical and design issues generally takes two forms. The first has to do with the owner being appropriately involved in and cognizant of design issues such as selection of system components, equipment operating philosophies, the extent of the scope of the retrofit work, the physical location of components, the arrangement of components into systems, system features of a maintenance, operations and longevity nature (which are secondary to and occasionally in conflict with energy conservation and maximizing ROI goals of typical energy retrofit projects). The second area of concern has to do with ensuring that the receiving organization receives appropriate design documentation—which is many times performed inadequately, incompletely, or not at all, particularly with respect to such energy retrofits as the installation of energy management systems and lighting retrofit projects.

Control over construction and installation matters is principally an issue of seeing that the persons installing or constructing the energy retrofit project employ good workmanship and follow appropriate practices in performing their work. Again, projects such as the installation of energy management systems and lighting retrofits are often performed inadequately and in an unacceptable fashion when not properly controlled.

Commercial buildings, multibuilding complexes, and particularly health care facilities are highly complex and becoming more so in light of increasing environmental, health, safety, and efficiency demands are placed upon them. The operations and maintenance of such facilities, especially health care facilities, is an extremely challenging proposition. Accordingly, selection of equipment, devices, components, and even entire systems which simplify, standardize or otherwise facilitate the operation and maintenance of facilities is desirable and should be achieved whenever possible. For example, when implementing a building automation system throughout a multibuilding complex, it is desirable to standardize on a single manufacturer for the essential hardware and software components so that these systems can be networked together and provide a common, universal operator interface. Doing so reduces operating training, spare parts inventory, enhances expendability, effectiveness of control, manpower utilization and still allows the majority of the field hardware to be generic or manufacturer-independent for a summary of the general character of the resources available for project implementation and the implications attendant to each (please refer to Table 13-10).

VI. CONCLUSION

There is great pressure from the various state legislatures and public utilities commissions to implement DSM programs. Because of the surge of business in the marketplace, many vendors are taking advantage of this opportunity even though they are not competent in the field. The potential result, as we have seen from even firms of national stature, can be disastrous. As a result, facility owners and utility companies should avoid being mesmerized by glossy corporate images and "no risk" guarantees.

While the essence of energy efficiency improvements in our inventory of existing buildings is inexorably founded in equipment and systems technology and engineering applications, the main driving force in the current resurgence of energy retrofit, the utility companies, appear to take an approach that is significantly more academic than practical. Com-

Table 13-10. Assessment of Implementation Resources.

Resource	In-House Technical Expertise Required	Owner's Staff Time Required	Financial Risk	Design Control	Construction Control	Equipment Standardization
ESCO	Low	Low	Low	Low	Low	Low
Design-Build Contractors	medium	medium	medium	med	medium	medium
Engineering Consultants	Low	Low	Low	very high	high	high
Vendors	med-high	medium	med-high	medium	Low	high
In-House Staff	very high	high	high	high	very high	high

pounding this situation is the interesting, though seldom-observed fact that it is possible for the utility company to actually get paid twice for DSM—if they get paid an incentive to implement energy retrofit projects and simultaneously receive the energy sales revenue that results from a non-functioning project. This is perhaps the foundation of PUC's concerns regarding the credibility of energy audits and the estimates of energy savings embodied therein (there is currently a greatly heightened emphasis on audits of energy savings estimates in the State of California), as well as concerns regarding the long-term persistence of energy conservation measures once they have been implemented.

The challenge facing the various state PUCs and utility companies, if they are to maintain their credibility, is to implement programs with both technical and administrative rigor and produce energy savings that can be counted on. The challenge facing experienced energy engineers is to make their presence felt in the process, for the good of all concerned.

Chapter 14

Lighting System Upgrade: A Comprehensive Management Approach

Craig DiLouise

I. INTRODUCTION

The pressing need to produce goods and services competitively in the emerging global economy has led U.S. industry to invest in modernization of structurally sound existing facilities. According to National Lighting Bureau, some two out of three such buildings in the U.S. are at least 25 years old. These buildings are typically equipped with electrical systems which do not offer the high performance and efficiency available from today's technologies. Retrofit of these systems (lighting, motors and drives, or transformers) presents an opportunity to permanently reduce electrical operating costs and thereby make the investing corporation more competitive.

Of all building systems, electric lighting often presents the greatest opportunity. Older lighting systems are typically inefficient, overlighted and, in many cases, ineffective by today's technological standpoint. Proper specification, installation and maintenance of new equipment can reduce energy consumption aggregately by more than 50%, while providing a quality environment for the occupants of the lighted space. Measurable energy savings can achieve paybacks of two to three years and an internal rate of return that will in many cases satisfy corporate investment requirements.

Energy managers must be cautious when planning a retrofit so as not to neglect the lighting system's original purpose—to serve people. Lighting must be effective as well as efficient. In this chapter, we will discuss retrofit as a component of a discipline called lighting management. Lighting management is the strategic management of resources to ensure that the lighting system consistently provides appropriate lighting at the lowest operating and maintenance cost. This comprehensive approach requires that the energy manager: a) provide an effective lighting environment appropriate to the activities performed in the space, b) retrofit the existing lighting system wherever viable to reduce energy consumption, and, c) properly maintain the system to optimize retrofit opportunities and to ensure effective long-term performance.

II. PROVIDING AN EFFECTIVE LIGHTING ENVIRONMENT

Quality lighting is the first priority of a successful lighting management program because it can increase worker productivity. According to Building Owners and Managers Association (BOMA), the total annual cost to operate one square foot of typical office space is about $180. Of this, $150 is devoted to wages, salaries and benefits while approximately $1.00 is allocated to the electric, material and labor costs associated with lighting. Therefore, a lighting efficiency project which reduces energy consumption by 50% can save approximately $0.50 per square foot, while a simple 1% increase in worker productivity can gain a $1.50-per-square-foot profit.

A. Quantity of Light

The process of vision requires visible light and a functioning eye. The human eye is extremely adaptive, able to adjust to a wide range of light levels (illuminances, expressed in footcandles). This range includes some 10,000 footcandles on a sunny day to about 0.01 footcandle under full moonlight (refer to Figure 14-1).

Reading this book, however, is much easier under sunlight than

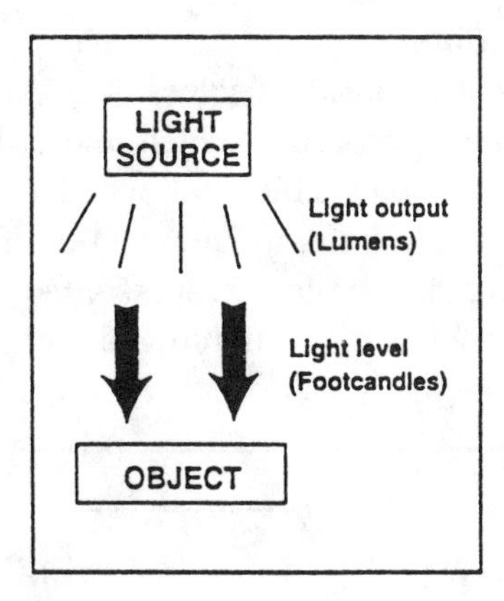

Figure 14-1. Relationship Between Light Output and Light Level.

moonlight. Conversely, 10,000 footcandles is an excessive amount of light to successfully read printed text. Successful task performance is accompanied by a light level range which is appropriate to the complexity, difficulty and importance of the task.

The difficult assignment of assessing proper light levels for given tasks has been undertaken by Illuminating Engineering Society of North America (IESNA), which recommends a minimum-maximum range for some 600 tasks and spaces from drafting to dance halls and discotheques.

Proper light levels are produced by correctly specifying light system output (luminous flux, expressed in lumens) using methods commonly employed by the lighting management and design industries (manufacturer-produced software is also available). Overall, ensuring an appropriate quantity of light can result in fewer errors, accidents, headaches and occurrences of eyestrain and fatigue.

B. Quality Lighting

Additionally, proper lighting quality should be provided. Glare perceived directly from an overhead fixture or on the task itself can impair or even disable performance of a task. Consider the difficulty of reading a glossy magazine as it reflects glare from an overhead fixture. The entire page can become white and the magazine must be tilted to be properly read. The same applies to assembly line operations, drafting and other tasks.

Proper control of glare can increase productivity and produce other advantages in terms of increased morale and reduced rejects, absenteeism, accidents and occurrences of headaches.

C. Specifying Proper Color

Lighting quality also considers color. Objects do not inherently contain color, only chemical properties which allow them to reflect certain colors once they are struck by visible light, which contains all of the colors of the visible light spectrum (refer to Figure 14-2). If a given beam of light is deficient in a color that is chemically represented in the object, the object will appear a different shade or a different color entirely.

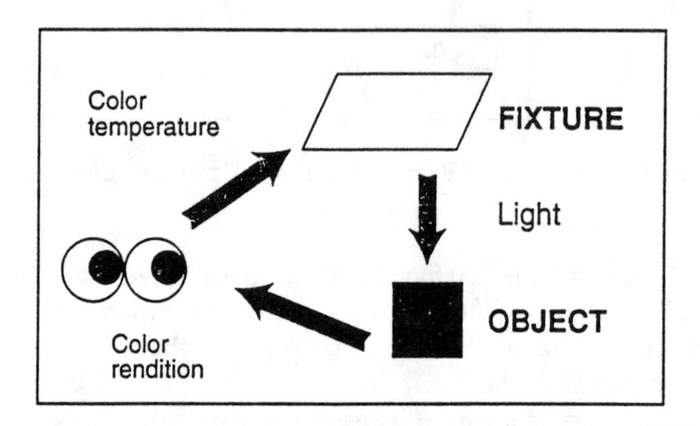

Figure 14-2. Relationship Between Color Temperature and Color Rendition.

Traditional fluorescent F40 cool white lamps render colors less vividly than new fluorescent trichromatic (also called "tri-phosphor") tubes such as F32T8 or F40T10 lamps in addition to some F40T12 lamps. New trichromatic lamps render colors more vividly and show more subtle variations in color shades, the result of a second coat of three phosphors added to the composition of the lamp (refer to Figure 14-3). Trichromatic lamps are typically in the 75-85 Color Rendering Index (CRI) range, while traditional fluorescent lamps are rated at a 62 CRI by most manufacturers.

The result of higher color rendering can be improved morale, which may impact productivity. Additionally, improved color rendering can enrich the retail environment and its merchandise, which may affect sales. Good color rendering can also affect the image of the corporation's environment to visitors. The application requirements will dictate the advantage of using trichromatic lamps.

D. Selling to Senior Management

Energy management projects are sometimes bypassed by senior management although they comply with corporate investment require-

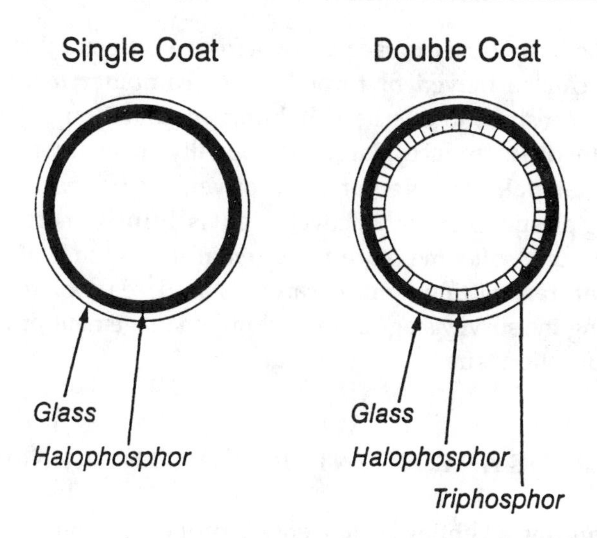

Figure 14-3. Fluorescent Dual-Coat Technology.

ments. This is because the corporation has limited resources to invest internally, and often new process technologies and other capital improvements are chosen due to the emphasis on increasing productivity over investing in reduced operating costs. Therefore, demonstrating to senior management that a lighting management project will both reduce operating costs and potentially increase productivity demonstrates a greater sensitivity to corporate priorities, and therefore may stand a greater chance of being approved by senior management.

E. Quality Improvement Options

Improving the quality of a lighting system is generally more art than science. To gain the full benefits of quality lighting, it is often desirable to seek the consultation of an experienced and competent lighting professional. Numerous publications, periodicals, software programs and other sources are also available from a variety of sources including trade associations, manufacturers and the Federal government.

The appropriate quantity of light should always be specified. Ranges of light levels for generic tasks and specification procedures can be located in the IESNA's *Lighting Handbook* in addition to other sources. Additionally, steps should be taken such as specifying a planned maintenance program to ensure that minimum appropriate light levels are consistently provided.

Generally, glare can be minimized by either moving the light source or the task itself. A light fixture can be moved to a more appropriate location; the individual can face the task from a different direction; or the task itself can be moved or turned. Some computer monitors, whose screens are sensitive to overhead lighting, mounted on adjustable arms and able to tilt to reduce glare. Additionally, new fluorescent fixture shielding material, such as a parabolic louver, can be specified to replace the existing plastic lens or other cover. Light is diffused more strongly as it exits the fixture, softening glare but often at the expense of the fixture's efficiency at transmitting light. Awareness of glare problems can be surfaced during the survey stage of the lighting management program via an occupant questionnaire.

III. PROVIDING AN EFFICIENT LIGHTING ENVIRONMENT

Retrofitting a lighting system entails proper planning of resources to replace components in the existing system (or entire fixtures) in order to improve efficiency. There are two ways to reduce energy consumption, derived from the formula for energy:

$$\text{Energy} = \text{Power} \times \text{Time}$$

or

$$\text{Kilowatt-hours (kWh)} = \text{Watts} \times \text{Operating Hours} + 1{,}000$$

The first method of reducing energy consumption is to decrease power requirements (the wattage required by the equipment to operate). The second method is to reduce the amount of time the system is operated (its operating hours over a given period of time, typically a year).

A. Reduce Power Requirements

Reducing the power requirements of the lighting system presents two economic benefits. First, it reduces the energy consumption of the system, expressed in savings in kWh. Second, it reduces the total demand of the lighting system, expressed in kilowatts (kW), for which the utility may levy a monthly demand charge based on total load requirements.

We will review the strategies for reducing energy use and load by examining a typical lighting system. Suppose we are responsible for upgrading a space lighted by rows of four-lamp fluorescent light fixtures containing traditional F40 cool white lamps powered by traditional magnetic ballasts. Each fixture requires 192 watts (American National Standards Institute wattage). The on-off control is a standard light switch.

Technology options to replace or modify this system include those shown in Table 14-1. The first option represents the current system and establishes the standard reference point for interpreting this table. The third column reveals system light output in lumens after ballast factor is assessed and then relates itself as a percentage against the lumen output of the current system. Input wattage is handled similarly. The last column shows the energy savings resulting from the current system's replacement with the proposed option should the system operate 4,000 hours per year at a commercial kWh rate of $0.10.

Table 14-1. Lamp-Ballast Retrofit Options, Four-Lamp Fluorescent Rapid Start System.

Lamps	Ballast(s)	Light Output (lumens)	Input Wattage	Annual Energy Savings Per Fixture
4 F40T12	2 Trad. Magnetic	11590 (100%)	192 (100%)	$ 0.00
4 F40T12	2 T12 Efficient Mag.	11590 (100%)	172 (90%)	$ 8.00
4 F40T12	2 T12 Electronic	10736 (93%)	142 (74%)	$20.00
4 F40T12	2 T12 Hybrid	9982 (86%)	142 (74%)	$20.00
4 F40T12	2 F.L.O. T12 Hybrid	11590 (100%)	160 (83%)	$12.80
4 F40T12 Trichrom.	2 T12 Efficient Mag.	12540 (108%)	172 (90%)	$ 8.00
4 F40T12 Trichrom.	2 T12 Electronic	11616 (100%)	142 (74%)	$20.00
4 F40T12 Trichrom.	2 F.L.O. T12 Hybrid	12540 (108%)	160 (83%)	$12.80
4 F40T12/ES	2 T12 Efficient Mag.	11100 (96%)	144 (75%)	$19.20
4 F40T12/ES	2 T12 Electronic	9435 (81%)	118 (62%)	$31.20
4 F40T12/ES	2 T12 Hybrid	8547 (74%)	116 (60%)	$30.40
4 F40T12/ES	2 F.L.O. T12 Hybrid	9768 (84%)	132 (69%)	$24.00
4 F40T12/ES Trichrom.	2 T12 Efficient Mag.	10296 (89%)	144 (75%)	$19.20
4 F40T12/ES Trichrom.	2 T12 Electronic	9945 (86%)	118 (62%)	$31.20
4 F40T12/ES Trichrom.	2 T12 Hybrid	9009 (78%)	116 (60%)	$30.40
4 F40T12/ES Trichrom.	2 F.L.O. T12 Hybrid	10296 (89%)	132 (69%)	$24.00
4 F32T8 Trichromatic	2 T8 Magnetic	11020 (95%)	142 (74%)	$20.00
4 F32T8 Trichromatic	2 T8 Hybrid	10092 (87%)	124 (65%)	$27.20
4 F32T8 Trichromatic	1 T8 Electronic	10788 (93%)	112 (58%)	$32.00
4 F40T10 Trichrom.	2 T12 Efficient Mag.	14060 (121%)	172 (90%)	$ 8.00
4 F40T10 Trichrom.	2 T12 Electronic	13024 (112%)	142 (74%)	$20.00

NOTE: The "Traditional Magnetic" ballast in the current system is no longer manufactured under Federal law enacted in 1988, which set minimum ballast efficiency standards for the majority of ballasts in use. Many fluorescent lighting systems still contain these older ballasts, however, and therefore it is in this table for comparison purposes.

NOTE: Annual energy savings are based on a commercial kWh rate of $0.10 and 4,000 annual operating hours.

NOTE: The data in this table are an average representative sample of lumen output, ballast factors and system wattages gathered from sources across the industry; actual data may differ depending upon manufacturer; these data are provided for comparison purposes only.

Note that other options are available; only lamp-ballast system options are presented in this table. While reducing power requirements is desirable, retrofit options should be linked to providing appropriate light levels. Three light-level-based options are presented which can reduce power requirements:

1. Reduce Light Levels.

Older buildings are typically overlighted, the result of the popular philosophy in the design community at the time which believed in "more light, better sight." The design community has since revised its recommendations for appropriate light levels which are typically more lean. Additionally, spaces not presently overlighted may become overlighted due to the benefits of a planned maintenance program.

In overlighted areas, equipment can be specified which produces less light (lumen output) but also saves energy. In a hallway, for example, the primary task is orientation, which requires a low light level relative to most task-oriented spaces. This excess light can be converted into energy savings with a variety of equipment such as energy-saving lamps and ballasts, and in some cases, current limiters. Some spaces may present the opportunity to delamp, although for aesthetic reasons this may not be a desirable option.

Example: Suppose a space is lighted to 100 footcandles but we determine that it requires 65 footcandles. Planned maintenance will increase light levels 15% in this application, effectively making our light level requirement about 75 footcandles. A sensible retrofit option is to install four F40T12/ES lamps powered by two cathode cut-out ("hybrid") ballasts, a system which produces 74% of the light output as the existing system but consumes 40% less energy (refer to Figure 14-4). This retrofit system and planned maintenance will provide approximately 75 footcandles (disregarding all other light loss factors for the purposes of example) while saving $0.40 for every dollar spent on the electrical operating cost of that fixture.

2. Maintain Light Levels.

It may be important to provide light levels comparable to those supplied by the original design. In this case, equipment can be specified which will produce a comparable light level but reduce energy consumption. A planned maintenance program can play an important supporting role, increasing light levels to the required levels while the retrofit system itself produces less light.

Example: Suppose we install four F32T8 trichromatic lamps pow-

Figure 14-4. A Technician Installing a Cathode Cut-Out (Hybrid) Ballast.

ered by a single electronic ballast in each fixture; this can generate up to an approximate 42% energy savings while providing comparable light levels. Although based on Table 14-1 it would appear that this system produces 7% less light, the energy manager must consider optical factor, which takes into account the increased transmission of light from the fixture due to thinner lamp diameter (which can result in up to an estimated 5% increase in light level). Trichromatic lamps also possess superior lumen maintenance characteristics, which allow them to sustain a higher light output over time. All lamps are susceptible to lumen depreciation, which is the gradual deterioration of light output as the lamp operates over time.

Another example of such a retrofit is replacing a 60-watt incandescent light bulb with a 13-watt compact fluorescent lamp-ballast unit, which can result in approximately 80% energy savings while producing comparable light output.

3. Increase light levels.

At times, it may be desirable to increase light levels in order to delamp to ultimately produce specified light levels. Increased light levels can be achieved by installing a specular reflector with F40T10 fluorescent lamps, allowing the removal of two lamps and one ballast from a four-lamp fixture. In this case, both the reflector and the lamp increase light levels. A reflector is a highly reflective piece of aluminum or other material designed to more efficiently transmit light directed at the "house-side," or ceiling of the fixture, down into the task area. The F40T10 lamp produces higher lumen output than traditional F40 cool white fluorescents due to its trichromatic properties (refer to Table 14-1). Some reflectors are designed to provide the appearance of a fully lamped fixture.

Note that reflector installations, although they appear satisfying in that they produce approximately 50% energy savings if a four-lamp fixture is delamped to two lamps, do not always produce comparable light levels. Additionally, reflectors must be aggressively cleaned to maximize their reflective properties, as dirt and dust accumulation characteristic of any workplace will deteriorate their ability to reflect light intended for the workplane. A final reason for concern is that should one F40T10 or other lamp in a reflector-retrofitted fixture fail, the other will also fail, or operate poorly, as the ballast operates these lamps on the circuit in series. Therefore, when one lamp fails, the entire fixture will go out, which affects uniformity of light provided by the overall lighting system and reduces light levels. Generally, reflectorized fixtures must be aggressively maintained—cleaning and spot/group relamping—to be fully effective as light sources.

Increasing light levels can also be achieved via a planned maintenance program and/or repainting room surfaces to a more reflective color.

B. Reduce Operating Hours

To calculate the approximate operating hours of the system, the following formula may be used:

OPERATING HOURS =
Hours System is Operated per Day
× Days Operated per Week
× Weeks Operated per Year

The supporting data are gathered during the survey stage of the retrofit via observation, questionnaire or monitoring equipment. During the survey, the energy manager should check for excess use of the lighting system—during lunch hours, periods during the workday, after hours—

whenever lighting is left operated when the space is vacant. Typical spaces where this occurs include offices, storage rooms, lavatories, closets, kitchen areas and after hours, general work spaces. Excess use of the lighting system presents a retrofit opportunity with installation of control devices.

Control devices shave off the excess lighting time and subsequently, excess energy use. The available technologies for accomplishing this retrofit goal include occupancy sensors and energy management systems.

F. Occupancy Sensors

They replace the on-off function of the wall switch, rely on ultrasonic or passive infrared technology to detect occupancy in a space by sensing motion. They may be wall- or ceiling-mounted depending on the requirements of the application (refer to Figure 14-5). Ultrasonic sensors utilize inaudible sound waves to detect motion, while passive infrared sensors detect abrupt changes in heat in the space. When motion is detected, the

Figure 14-5. A Technician Installing an Occupancy Sensor.

occupancy sensor activates the lighting system; when the sensor detects no motion for a specified period of time, the lighting system is turned off.

A disadvantage of early occupancy sensors was that they would turn off the lights while people were motionless for a period of time, and while people were masked from the sensor by a partition such as a lavatory stall door. Additionally, a room's lights were apt to activate when a person passed the room in an outside hallway. Such problems have been addressed with more-sensitive motion technology, by adjusting the timer to deactivate the lights after longer periods, and by partially masking the sensor's lens so that only motion in the monitored space would be detected.

Because the mortality rate of fluorescent lamps is linked to how often the lamp is started, questions have arisen concerning occupancy sensor's affect on lamp life. Occupancy sensors can reduce lamp life, but the cost of purchasing and installing a new lamp must be weighed against the energy savings. The result depends on the characteristics of the application, but the energy savings (up to 35% depending on the application) generally outweigh the added cost of lamp replacement. If a space is frequently trafficked, it may not make sense to install an occupancy sensor, or the period before turning the system off should be lengthened.

2. Energy Management Systems

They are centrally located systems which are programmed by the user to activate and deactivate lighting systems in various spaces on preset schedules over an entire year. If a given space anywhere in the building is not used after normal working hours, for example, the energy management system will deactivate the system except for those days the cleaning crew will be working in the area. Energy management systems can also dim lighting systems which are compatible with a dimming function.

Besides specifying control devices, the energy manager can attempt behavioral modifications such as an education program which encourages (and requires) employees to turn off lights when leaving a space.

IV. PROPER MAINTENANCE OF SYSTEM

A successful lighting upgrade, whether the primary goal is to produce a more effective or a more efficient lighting system, does not end once the new equipment is installed. It is instead an on-going process that incorporates continuous improvement and an aggressive planned maintenance program.

Planned maintenance is the practice of organizing resources to maintain the fixtures on a scheduled periodic basis. The tactics include group relamping, cleaning, troubleshooting and inspection. The interval(s) for performing these activities is determined by the characteristics of the application and economics. The benefits include optimized use of labor, resources and energy. The primary benefit, however, is a lighting system that consistently provides minimum proper light levels for those who need and use light in the space. All lighting systems are subject to depreciation and, therefore, experience light loss; the extent of this loss is "recoverable," meaning it can be controlled via planned maintenance.

A. Planned Maintenance Tactics

1. Group Relamping

This is the practice of replacing all operating lamps in a system *en masse* at periodic intervals (typically with occasional spot relamping in between), as opposed to the common practice of continuous spot relamping, which is the replacement of lamps as they individually fail.

Fluorescent lamps are given a rated service life by manufacturers in their catalogs. The service life expresses the interval at which half of a large group of lamps are expected to fail. The service life rating therefore is an average. Fluorescent lamps fail in small numbers along a mortality curve until about 70% of rated life, then the failure rate accelerates. At 70% of rated life, 10% of the lamps are expected to fail. Between 70% and 100% of rated life, another 40% are expected to fail (refer to Figure 14-6). The remainder will typically fail soon after. Group relamping is typically most viable at 70% of life; at this time, the lamps are replaced with assembly-line efficiency before large-scale burnouts occur. As the nominal labor cost of spot relamping a burnout is considered 5 to 10 times higher than the cost of group relamping one lamp, significant labor cost savings can be realized both as a dollar figure and in terms of hours of lost opportunity.

Fluorescent lamps also experience lamp lumen depreciation, a recoverable light loss factor that describes the gradual loss of lumen output as a lamp ages (refer to Figure 14-7). At rated life, some 20% of the lamp's lumen output is lost, while the operating wattage remains the same. Group relamping ensures that the lamps are replaced at an interval that minimizes this waste.

The benefits of group relamping therefore are to save opportunity costs in labor, minimize energy waste, and ensure a consistently lighted, effective system. Generally, group relamping is considered economically viable if the labor cost of spot-replacing one lamp, minus the cost of group-replacing it, is greater than the new cost of the lamp.

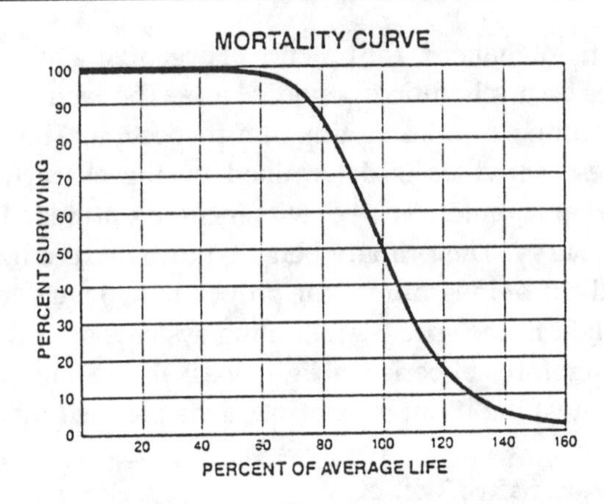

Figure 14-6. Typical Mortality Rate of Fluorescent Lamps.

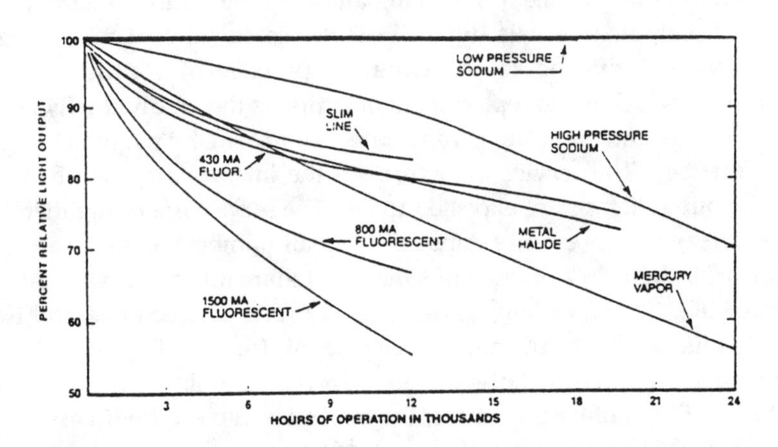

Figure 14-7. Lamp Lumen Depreciation of Standard Lamps.

2. Cleaning

This is the practice of washing the fixture interior, shielding material such as a prismatic lens, and in some cases, the lamps themselves. All lighting systems experience fixture (luminaire) dirt depreciation, which can impede the transmission of light from the fixture via absorption by dirt and dust (refer to Figure 14-8). Cleaning operations are typically combined with group relamping for greater economy, although dirtier environments such as industrial facilities or facilities with poor ventilation may benefit from more frequent cleaning intervals. Additionally, specular reflectors also benefit from more frequent cleaning as the lumen output of the fixture depends more highly on the reflectance of the fixture surfaces.

Additionally, room surfaces should be cleaned and/or repainted

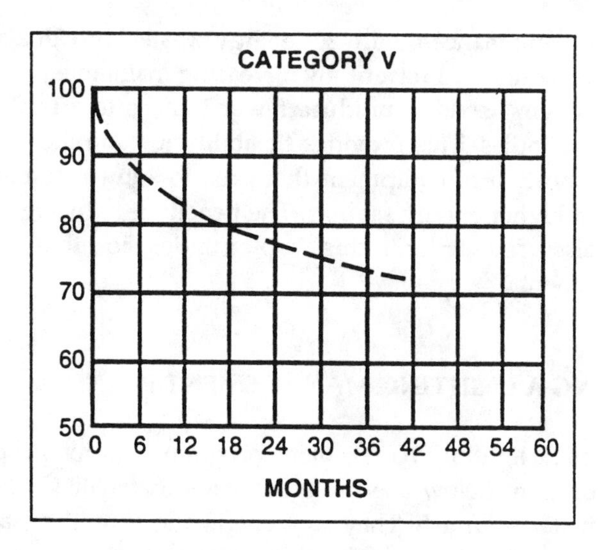

Figure 14-8. Typical Fixture Dirt Depreciation for a Fluorescent Four-Lamp Lensed IESNA Maintenance Category V Fixture.

periodically to ensure maximum reflectance, although typically at greater intervals. Dirt and dust on the walls, or a dark color, will absorb and waste light. Room surfaces should be cleaned aggressively in indirect lighting environments where these surfaces are relied upon to reflect light from a fixture onto the workplane. The benefits of cleaning include higher maintained light levels and an appearance of good housekeeping.

3. Troubleshooting and Inspection

This should occur at intervals to correct problems and replace defective components; these services can be economically combined with a group relamping/cleaning operation.

B. Planned Maintenance Benefits

Lighting management calls for the strategic management of the lighting system to ensure an effective lighting environment for the lowest operating and maintenance cost. Planned maintenance ensures a more effective lighting environment by increasing maintained light levels. It ensures consistently lighted fixtures, which can enhance corporate image and housekeeping to visitors and employees alike. In addition, planned maintenance reduces the maintenance cost via effective management of labor and other resources. Finally, there is the often overlooked benefit of reduced operating costs.

Planned maintenance minimizes energy waste, but can also play a strong supporting role in a retrofit. By increasing maintained light levels, the lighting system needs to produce fewer lumens to affect the same number of footcandles. This provides flexibility in equipment selection, allowing specification of equipment that produces lower lumen output for subsequent higher energy savings. The benefit of reducing operating costs is also discussed earlier in this chapter under the subtitle, "Reduce power requirements."

V. PREPARING A LIGHTING MANAGEMENT PLAN

An effective lighting management program requires proper planning and execution. Below are the basic steps that will see a lighting upgrade from start to finish. They are general and should be tailored to the application:

A. Prioritize Benefits

The first step to a successful upgrade is to determine corporate priorities concerning the purpose of the upgrade. Priorities include energy savings, productivity, retail sales, beautification and/or reduced air cooling requirements.

B. Write a Lighting Energy Policy

Energy efficiency is often a top priority, and should be institutionalized within the corporation via a lighting energy policy. This policy can be expanded to include other building systems such as power generation, distribution and other end-use equipment such as electric motors and drives. The lighting energy policy should be prepared as a lighting management plan, and can be used as a permanent reminder and reference for present and future managers at the corporation.

C. Gather Education, Training and Resources

Education and training materials are available from numerous organizations such as the Federal government, trade associations, professional societies and manufacturers. These materials include training textbooks, handbooks, computer software, survey tools and other helpful resources.

D. Prioritize Use of Labor

The energy manager can become sufficiently knowledgeable to conduct a survey, analysis and equipment specification. However, it may be desirable to contract a lighting management company, energy service company or consultant to assist with these steps. Other sources of assistance, often free, include the local electric utility, lighting designers and lighting manufacturers and their distributors.

Additionally, installation, disposal and maintenance services will be required; these can be performed in-house or contracted to a lighting management company or electrical contractor. In-house personnel may lack the expertise and equipment to perform fully effective installation and maintenance operations. The cost of training and equipping in-house personnel must be weighed against the cost of contracted services.

E. Conduct Building Survey

No analysis and equipment specification should be performed without full information about the system being operated and the operating environment. The characteristics of the lighting system need to be assessed, including conditions, energy consumption and operating hours. The lighting quality needs of the occupants need to be assessed via observation or an occupant questionnaire. Additionally, other information should be gathered such as the dimensions and characteristics of each lighted space, the number of fixtures, the local external climate, dirt conditions, the local electric utility rate and other factors that will affect equipment decisions, opportunities and economic analysis. Conducting a proper survey will help avoid costly future mistakes and ensure that the lighting management plan is fully sensitive to each individual application within the facility.

F. Address Lighting Quality Issues

Using current IESNA recommendations, prescribe light levels appropriate to each application. Additionally, examine lighting quality problems such as glare and prescribe curative action.

G. Research Financing Opportunities

Investigate financing options such as utility incentives, performance contracts, leasing options and other types of agreements in addition to conventional financing from local financial institutions.

Numerous electric utilities in high-energy-demand regions of the country, such as the east and west coasts, are embarking on demand side management (DSM), the practice of controlling energy demand rather than building new generating supply. DSM programs often take the form of a direct cash rebate based on purchase of qualifying technologies, reductions in demand for power, or reductions in energy consumption.

Energy service companies (ESCO) and some other organizations offer performance contracts, also called shared savings agreements, as a financing mechanism (refer to Figure 14-9). In a performance contract, the ESCO pays for the entire cost of the lighting upgrade, then splits the resultant energy savings with the customer. The customer benefits from no immediate cash outlay while experiencing immediate cash flow. Typically, when no utility rebate is involved, the ESCO requires the majority of the energy savings up front, and as the typical 10- to 15-year contract progresses, the split becomes more even. Ownership of the equipment is retained by the ESCO.

H. Establish Lighting Management Options

Determine options to upgrade the present lighting system to meet corporate priorities established in Step A. These options will feature energy-efficient and quality-sensitive components and should incorporate planned maintenance of the system. Other lighting management options should be established, such as a lighting schedule, an educational program for employees, and movement of workstations or fixtures to reduce glare.

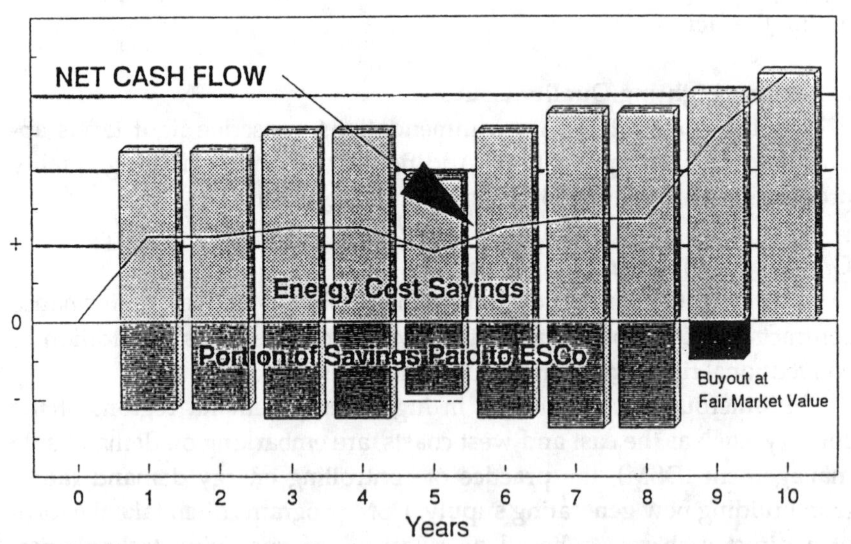

Figure 14-9. Shared Savings Agreement (Performance Contract).

I. Test for Lighting Quality

All retrofit options determined to produce energy savings should be tested for lighting quality. Options which do not produce sufficient lumen output for the benefit of the occupants should be eliminated.

J. Test for Economic Return

Each retrofit option must be tested for economic viability against the corporation's internal rate of return requirement. Calculate energy savings (which may include net heating, ventilation and air cooling savings), payback periods and internal rates of return. Other economic benefits, such as anticipated increases in productivity or sales, or labor savings via planned maintenance, can be incorporated to make options appear more attractive to senior management.

K. Assess Environmental Impact

Since the first Earth Day in 1970, both consumers and businesses have been aware of the environmental benefits of energy conservation. Deferred generation of electricity at power plants results in reduced consumption of valuable natural resources such as oil, coal and natural gas. It also prevents the emission of potentially harmful gases such as carbon dioxide, sulphur dioxide, nitrogen oxides and other gases and particulates. Carbon dioxide is believed by a significant portion of the scientific community to be related to global warming, the so-called "Greenhouse Effect." Sulphur dioxide is linked to acid rain, and nitrogen oxides to ozone pollution, or smog.

For years, the business community regarded energy conservation and environmentalism as synonymous with depressed economic growth. The proliferation and subsequent advancement and reliability of energy-efficient technologies in the 1980s, however, has presented businesses with a new choice: energy efficiency. This allows the corporation to upgrade a lighting system and benefit both its own bottom line and the environment.

In 1990, one of nine goals of the National Earth Day Committee stated, "A transition to renewable energy resources and increased residential and industrial energy efficiency." A Gallup poll revealed that three out of four Americans consider themselves "environmentalists." Numerous other studies and polls of American consumers in the early 1990s indicated that environmental protection should be regarded as a national (and corporate) priority.

At this step of the upgrade, the energy manager should calculate the environmental benefits to provide the marketing department with valuable public relations material. This will help the energy manager ultimately sell senior management on the lighting upgrade proposal by posing a possible significant public relations benefit.

Additionally, the energy manager can recommend to senior management participation in the U.S. Environmental Protection Agency's Green Lights Program. Green Lights Program encourages large U.S. corporations to upgrade lighting systems in 90% of all qualifying building space and implement upgrades within five years. Green Lights Program provides assistance in the form of education materials and a computerized survey/analysis program called the Decision Support System. As of April 1993, 428 corporations of America's Fortune 1000 have agreed to become Partners and implement Green Lights goals. Combined, Green Lights Partners—and Allies (manufacturers, lighting management companies and electric utilities), who also agree to comply with Green Lights goals—represent some 2.9 billion square feet, or 3% of all U.S. industrial/commercial building space.

L. Conduct Trial Installation

After successful proposal to senior management, the most viable lighting management options can be implemented in trial installations to test its effectiveness and measure energy savings directly with monitoring equipment. This allows the energy manager flexibility to adjust the lighting upgrade for more effective full installation, or choose another option as more viable in the application.

M. Full Installation

The energy manager will oversee full installation of new lighting components and other changes to the lighted environment.

N. Equipment Disposal

Old lamps, ballasts and other equipment should be disposed of in accordance with the state and Federal guidelines. In some states, fluorescent lamps are considered hazardous waste and must be treated accordingly when disposed of in significant quantities. PCB-containing ballasts must be also disposed of appropriately or the "hazardous waste generator" may face Superfund liability. The proliferation of lighting retrofits has produced demand for and subsequent founding of lamp and ballast recyclers. Lamp and ballast recycling is generally both an economical, and the most environmentally sensitive, disposal option.

O. Maintain the System

The lighting system should be properly maintained after installation to ensure consistent proper light levels and to optimize the value of the system.

P. Monitor Results

Rather than rely purely on engineering projections, the energy manager may wish to measure actual energy savings via monitoring equipment. This can be particularly advantageous in performance contracts with ESCOs or in shared savings agreements or other DSM programs with the local utility.

Q. Market Environmental Commitment

The marketing department can utilize the positive environmental effect of energy efficiency in public relations and other communications to both the consumer and investment markets.

VI. A COMPREHENSIVE APPROACH

Lighting management offers a comprehensive approach to upgrading a lighting system integrating all possible considerations. A lighting system is not purely a vehicle of energy use to produce light; it is a critical component to a corporation's productivity and image. Therefore, upgrades that can result in productivity and image enhancements should be granted equal weight to those which generate energy savings. Additionally, when upgrading to reduce energy consumption, the energy manager should consider all aspects, opportunities and pitfalls in a well-organized planning process.

Section 4
Case Studies of Retrofit Projects

CHAPTER 15

AN ENERGY CONSERVATION CASE STUDY

Francis M. Gallo, P.E.

I. INTRODUCTION

In the design and operation of any building, the consumption of energy is an important consideration. Though a building may have been designed with excellent energy conserving features, the operating staff holds the key to the successful administration of the facility. Some energy efficient buildings never operate as designed because of improper commissioning, lack of staff training, poor maintenance, and/or lack of management commitment. Sound management principles in combination with a solid technical understanding of the building systems is requisite for operating efficiently. Careful measurements and record keeping provide critical feedback on the success or failure of the energy program. This case study not only illustrates some of the excellent energy conservation features that can be designed into a facility, but demonstrates how a determined operating staff can find ways to realize the full energy saving potential of any building.

II. BUILDING DESCRIPTION

Corporate Center, completed in 1982, is located in Danbury, Connecticut. The building's unique shape is a result of the architect's integration of the facility into the surrounding environment (refer to Figure 15-1). The building was constructed on a 53-acre crescent shaped parcel surrounded by wetlands. The site has 2.5 miles of roadway transversing 674 acres. The building is elevated on 1,008 concrete columns so that both the roof and the underside of the first floor are exposed to the environment. This allowed for minimum excavation since the mechanical and service areas were constructed at grade level.

The 1.3-million ft^2, four story office building is 1,360 ft long and 675 ft wide, and is compartmentalized into nine large and six small office sections or pods (refer to Figure 15-2). The building also contains a main

Figure 15-1. Union Carbide Corporate Center, Danbury, Connecticut.

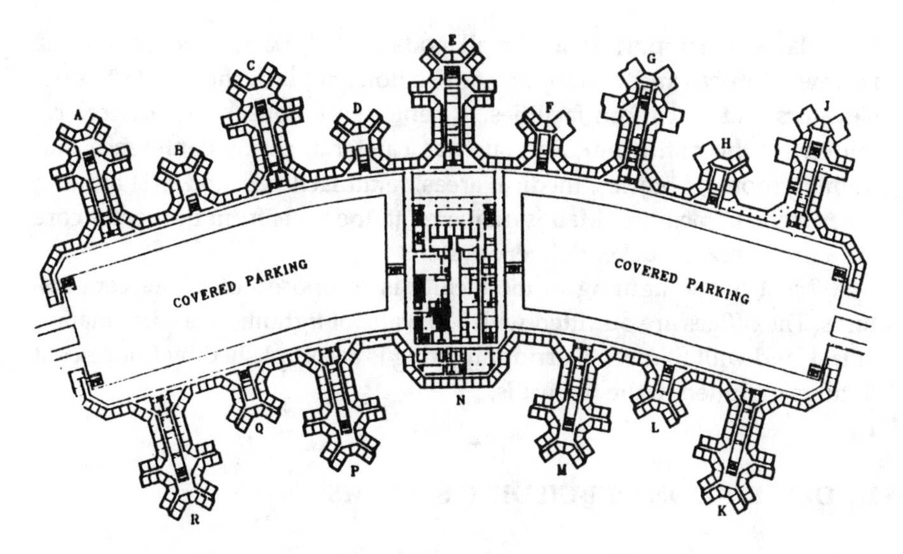

Figure 15-2. Floor Plan of Corporate Center.

core area that houses common services such as the cafeteria. A 900,000-ft^2 covered garage is surrounded by the office sections. The garages are separated from the office sections by a 30-foot clearance. The occupants have only to drive onto the floor in which they work and park outside in their office sections. Individually inclined ramps provide access to each garage level. Enclosed bridge walkways provide entry to the pods from the garages.

The structural columns, floors and roofs of both the building and the garages are reinforced concrete. The buildings exterior is composed of insulated aluminum panels and one inch thick double glazed windows. Translucent insulated fiberglass panels run the length of the two corridors that face the garages. The pods have a built-up roof while the garages have a single-ply ballasted system.

The building can accommodate 3,500 occupants. There are 2,350 private offices in the building. Each office is the same size, 182 ft^2 and has two fixed drywalls and one file wall. The fourth wall is the building's windowed perimeter. The drywalls afforded economy and acoustical segregation while the file wall accomplished utility and privacy. A single office size was selected to minimize the rearrangement costs and provide all office occupants with a windowed room.

There are nine large office sections and six smaller ones. The layout of a pod is fairly typical throughout the entire facility. Large pods are approximately 80,000 ft^2 and contain 45 to 54 perimeter offices, 18 interior support staff stations, conference rooms, rest rooms, copy rooms and miscellaneous support areas. Small pods are 33,000 ft^2 and contain 22 perimeter offices and 10 support staff stations but lack the support areas, elevators and rest room facilities. A center core houses the services required by the entire building such as cafeteria, major conference and training rooms, libraries, medical areas, retail facilities, and first aid and other medical facilities. Mail is delivered to the pods from the center core via a mechanized bucket delivery system.

Most of the lighting in the facility is composed of fluorescent fixtures. The offices are outfitted with 530 watts of lighting in a combination of task and wall fixtures. Corridor lighting is accomplished by fluorescent fixtures mounted in the file walls.

III. DESCRIPTION OF BUILDING SYSTEMS

A. Central Plant

The central chilling plant consists of three 720-ton Trane Centrifugal R-11 chillers and one 350 ton Dunnham-Bush R-22 vertical screw machine. Two of the three centrifugal chillers have double bundle condensers for heat recovery. The screw machine also has a heat recovery bundle. The centrifugal chillers are capable of dropping chilled-water temperature from 56°F to 42°F with a rise in condenser-water temperature of 86°F to 95°F. The full load ratings of the centrifugal chillers are 0.67 kW/ton (0.69 kW/ton in heat recovery mode). The Trane chillers are mainly used for summer-time cooling and winter weekday energy recovery. The Dunnham-Bush chiller is mainly used in the winter as the primary heat recovery machine on the weekends. A three-cell, counterflow mechanical draft cooling tower is located at grade level several hundred feet from the building.

The central heating plant consists of two 250-Bhp firetube and one 145-Bhp watertube Cleaver Brooks boilers. The boilers are dual fuel types, utilizing either #2 fuel oil or natural gas. Two boiler converters are provided to raise hot-water temperature from 12°F to 14°F on cold winter days when heat recovered from the chiller bundles is insufficient.

B. Tank Farm

The major feature to conserve energy at Corporate Center is the heat recovery system. A chilled- and hot-water storage system consisting of 10 180,000-gallon underground concrete tanks serves the building (refer to Figure 15-3). The tank farm is located under the south garage. Water used to heat and cool the building is charged during off-peak times and stored

Figure 15-3. Corporate Center Central Plant Schematics.

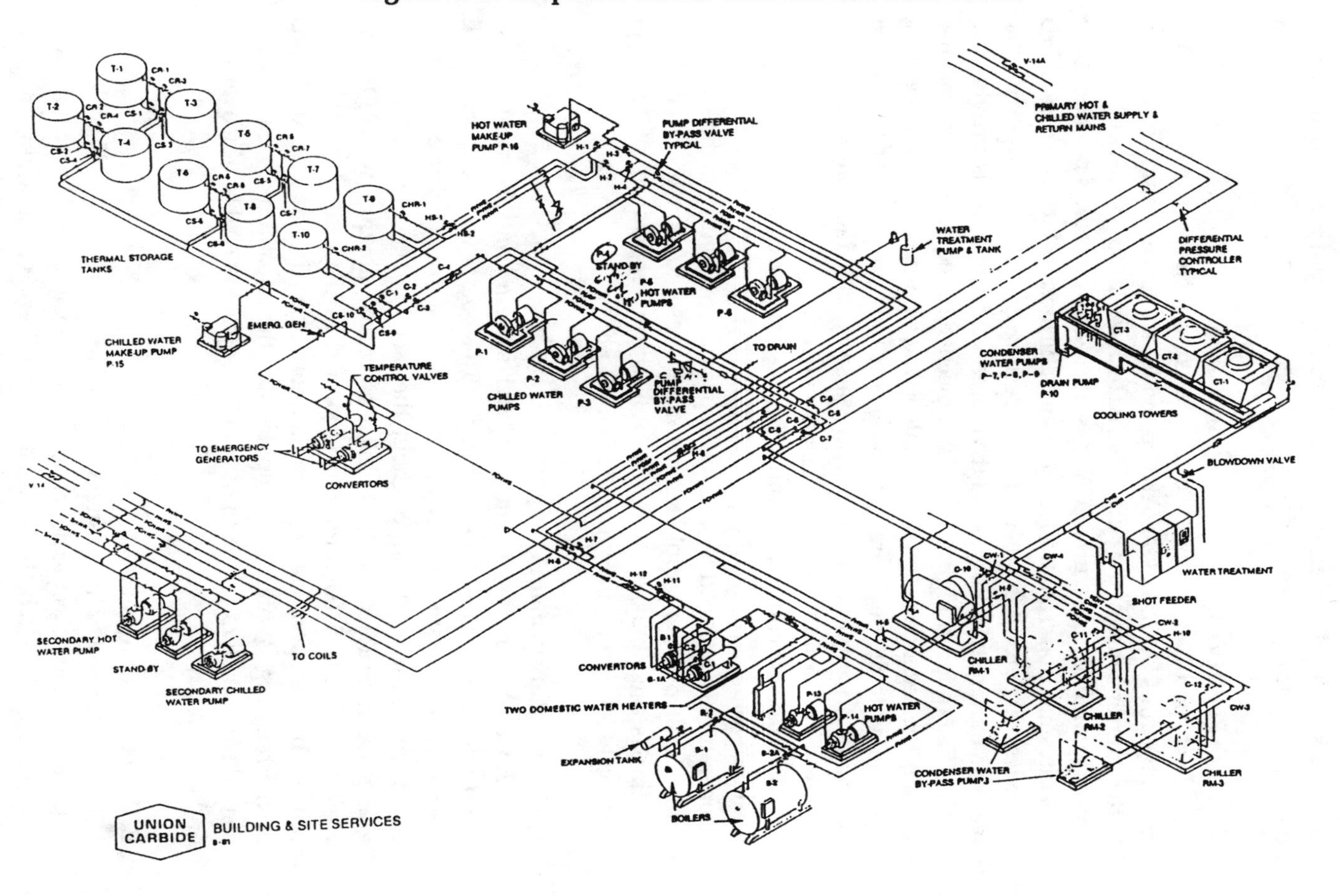

in tanks for use during the day. This enables the building operators to take advantage of lower off-peak electric rates (11:00 pm and 7:00 am weekdays and all day on weekends).

C. Office Pod HVAC

Each pod has its own mechanical equipment room with two air-handling units (AHUs). One AHU provides air conditioning and heating to the interior zones of the pods. This AHU has a 10% fresh air make-up. It serves a variable-air-volume (VAV) system with four zones per floor (two zones in small pods) controlled by pneumatic thermostats. A second AHU provides 100 percent fresh air for each perimeter office. Twenty-five cubic feet per minute (cfm) of constant volume ventilation air is provided to each office (refer to Figure 15-4). The mechanical equipment rooms also contain the secondary hot- and chilled-water pumps providing heating and cooling to the perimeter fan coils (refer to Figures 15-5 and 15-6). An electric steam humidification system is also provided to maintain at least 20% relative humidity (RH) on design winter days.

Each office contains a four-pipe fan coil air-conditioning system with individual temperature control (refer to Figure 15-7). Chilled-water temperature is designed for a 42°F-supply and 56°F-return in the summer. Winter chilled-water supply temperature is reset to a maximum of 48°F. Heating is accomplished with a supply water temperature of 105°F returning at 95°F (refer to Figures 15-7 and 15-8).

D. Energy Management Computer System

An energy management computer system controls many of the building systems. The Honeywell Delta 5100 computer also monitors all life-safety and fire systems. An optimum start/stop algorithm controls the operation of the AHUs throughout the facility. The duty-cycling programs were disabled in the interest of indoor air quality (IAQ). The Honeywell Delta 5100 computer is located in the central plant and is utilized by the operating engineers to monitor environmental conditions in the building. While the computer has the capability to control the central plant chillers, it is used mainly to control the AHUs, office fan coil, and hydronic systems.

E. Power Distribution

Primary power to the site consists of two separate 13,800- volt feeders. They supply power to the primary switchgear located in central plant. Power is supplied to both the east and west sides of the building from the central plant. The small pods contain high-voltage substations that step

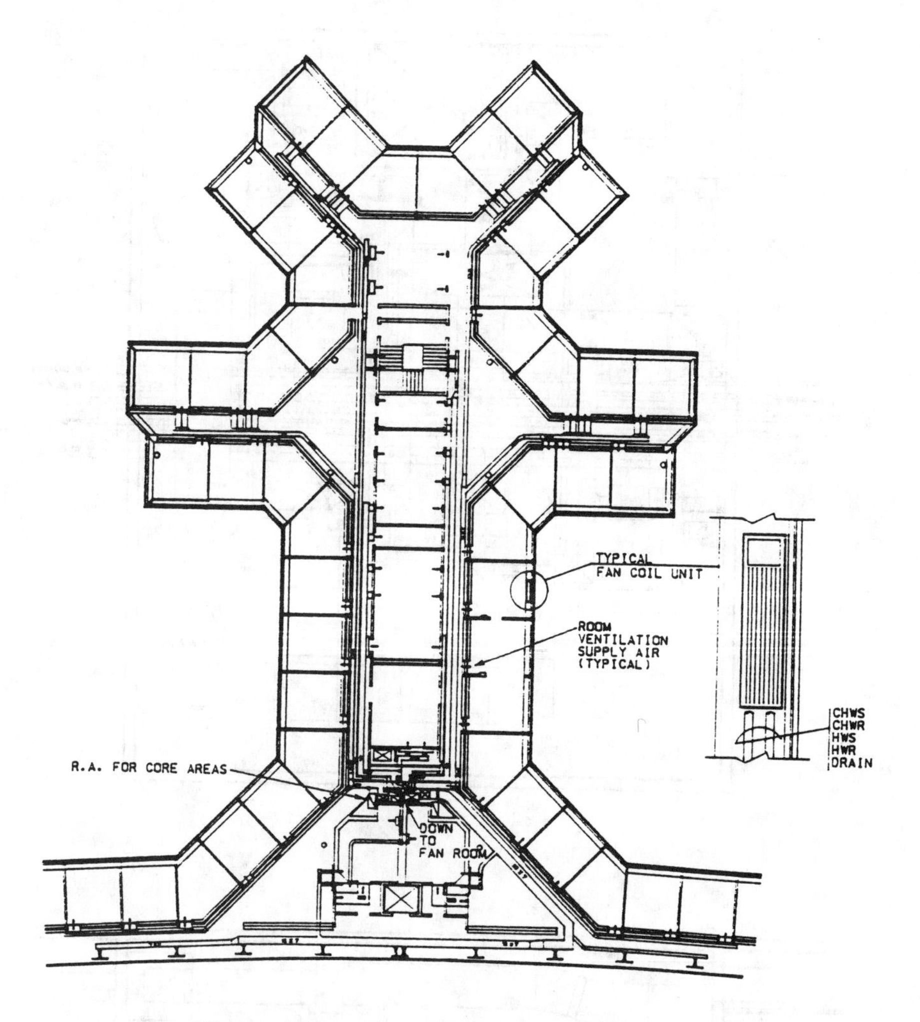

Figure 15-4. HVAC System for an Office

down the 13,800-volt primary power to 480 volts and lower it as required and distribute the power to the adjacent large pod. The plant has two diesel emergency generators that are capable of generating 850 kW each. The emergency generators power critical loads such as elevators; emergency lights; telecommunication equipment; and critical heating, ventilating and air-conditioning (HVAC) loads during commercial power interruptions.

F. Other Energy Saving Features

One energy disadvantage of a low-rise building is the large roof to

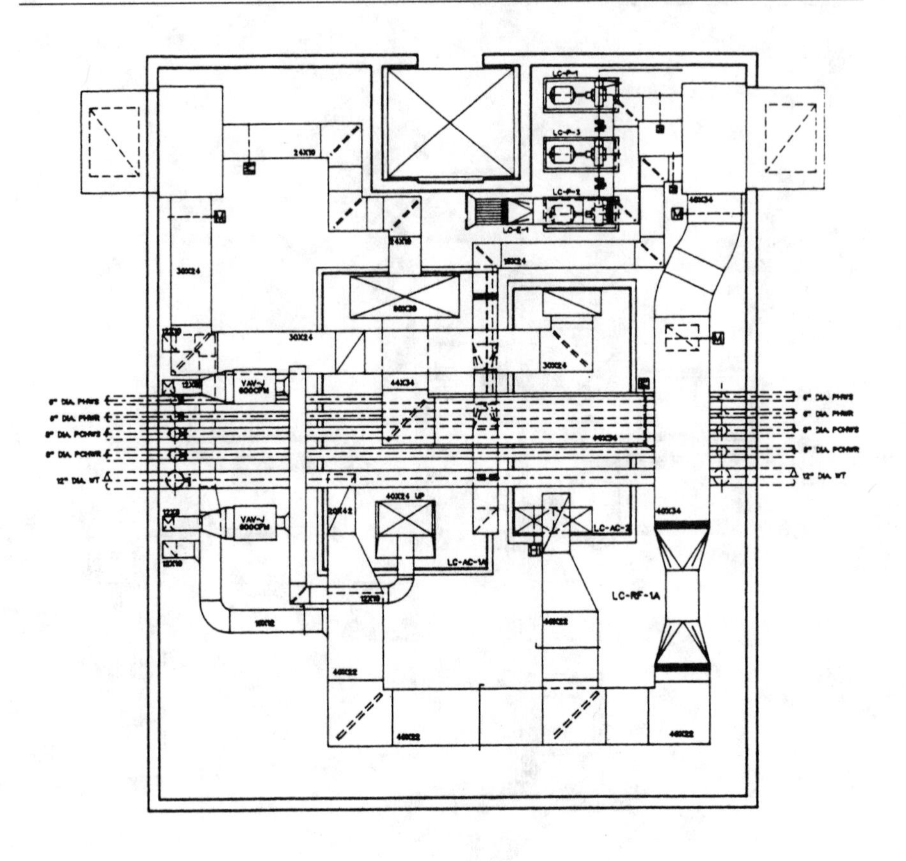

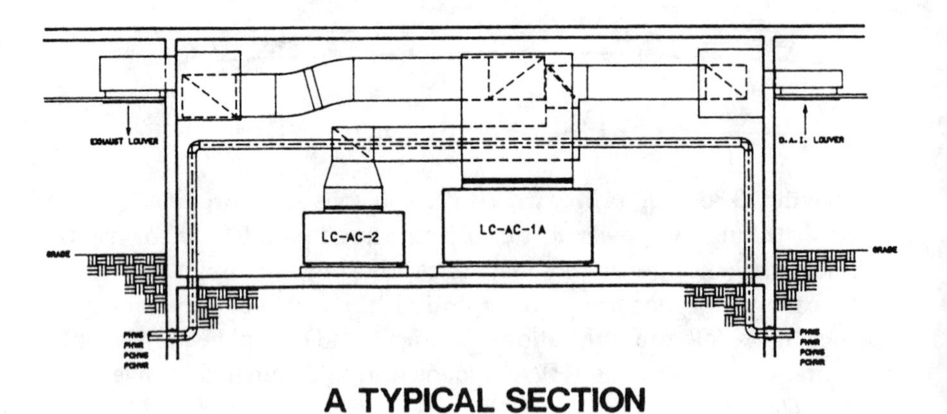

A TYPICAL SECTION

Figure 15-5. Mechanical Equipment Room Plan, Large Component.

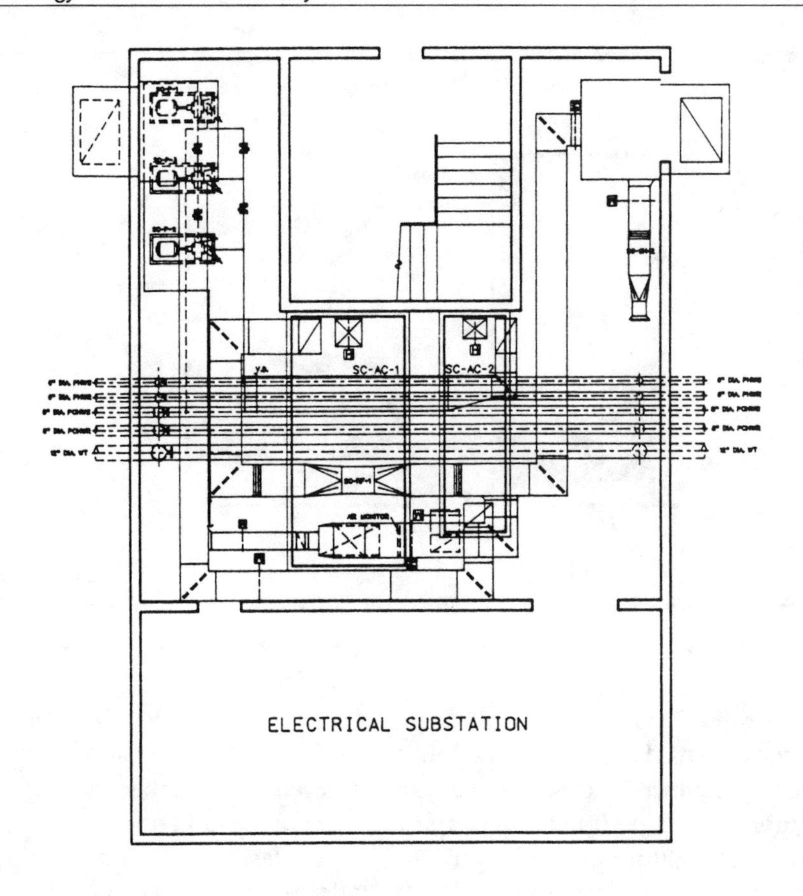

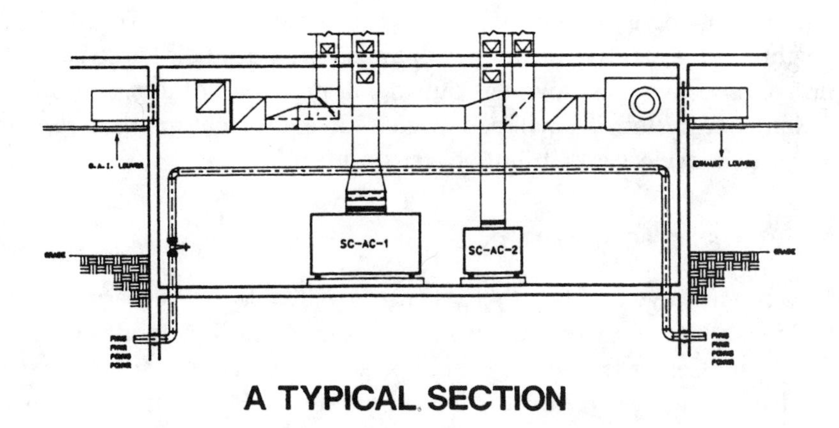

Figure 15-6. Mechanical Equipment Room Plan, Small Component.

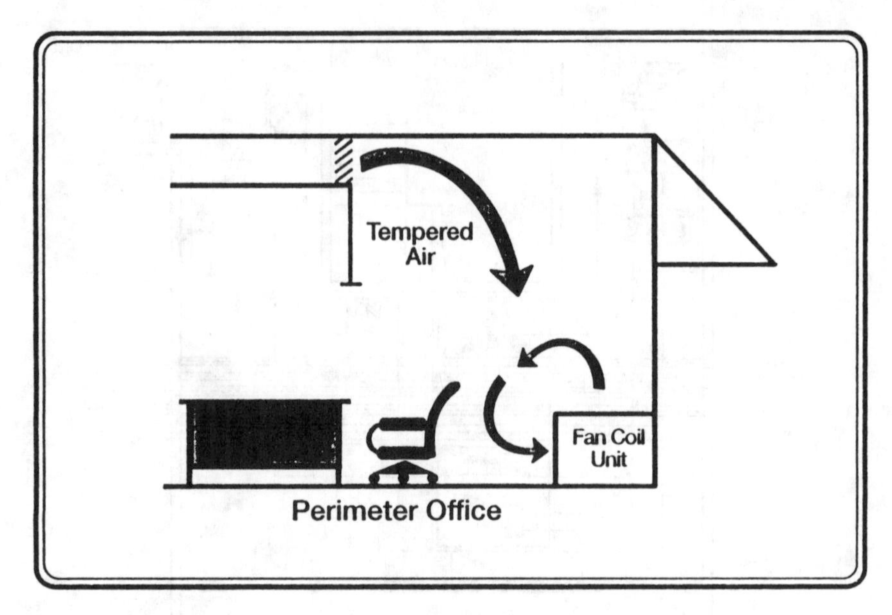

Figure 15-7. A Four-Pipe Fan Coil Air Conditioning System

floor ratio. This leads to an increase in energy transfer to and from the environment. However, some low-rise building characteristics compensate for the increased heat transfer such as a reduction in the stack effect, a larger floor-to-perimeter ratio, and increased availability of natural light for office lighting. Low-rise buildings can be designed with smaller central fan systems that require lower static pressures. The use of more efficient fan coil units in each office takes care of the major HVAC load.

Other energy conserving features consist of a low floor-to-floor height of nine feet five inches, the elimination of dead air space between floors, reflective glass awnings outside all perimeter offices, and a heat reclamation system in small pod mechanical equipment rooms to recover the heat from electric distribution transformers.

IV. MODES OF OPERATION

In any large building there is energy available for recovery, storage and reuse. Heat is generated by electrical equipment, lights and people. At Corporate Center this waste heat is picked up by HVAC system and is used in the winter to heat the building (refer to Figure 15-9). The heat recovery system has several modes of operation. There were originally three summer modes and six winter modes. A combination of these modes was used in spring and fall.

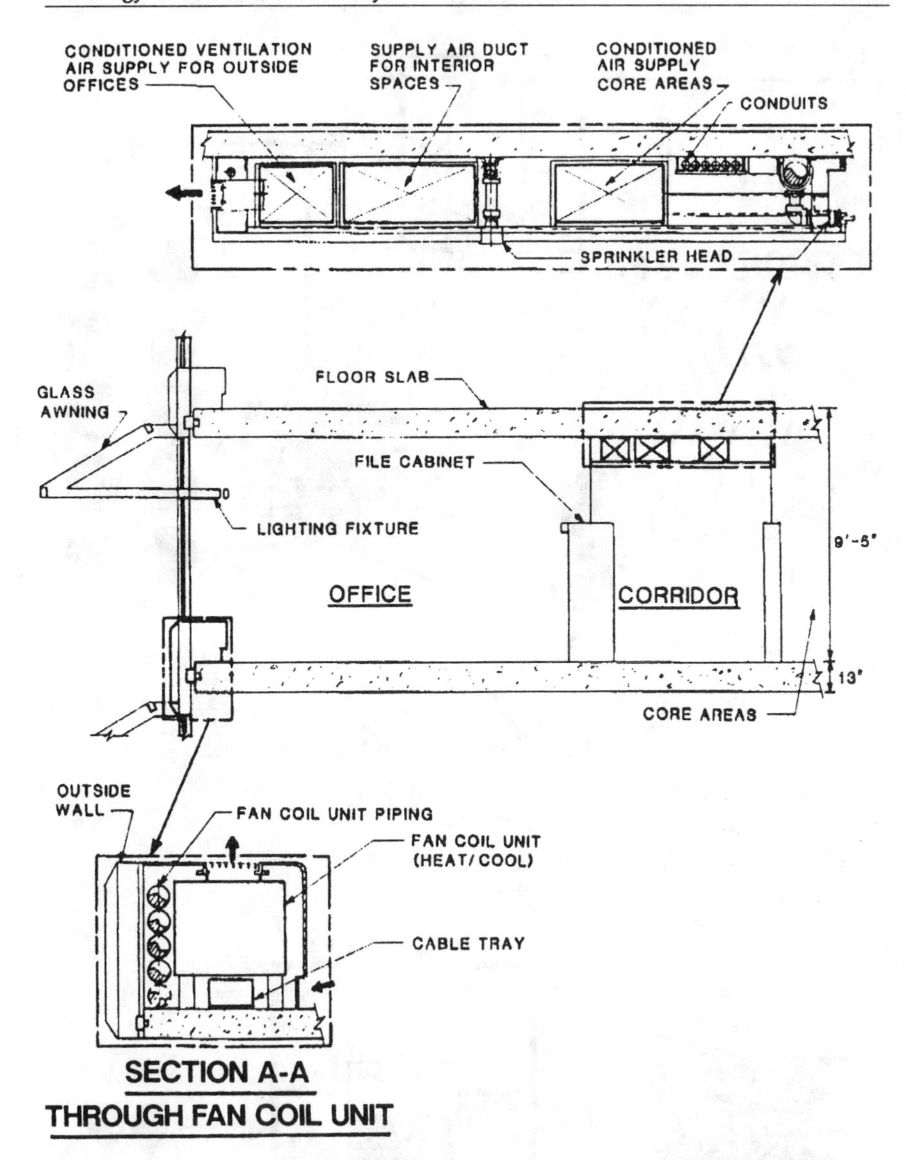

Figure 15-8. A Typical Office and Corridor at Corporate Center.

A. Original Summer-Time Operation

Mode 1: Summer Morning Start-Up

The refrigeration machines are used to cool the building. Chilled water stored in the tank farm is saved for on-peak times (refer to Figure 15-10).

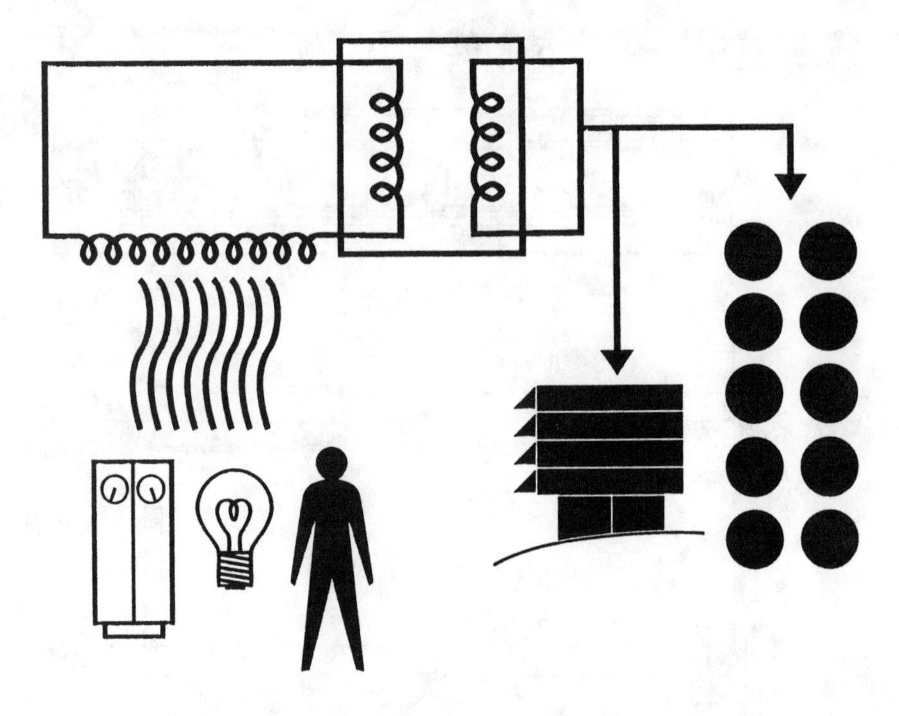

Figure 5-9. Re-Use of Waste Heat During Winter.

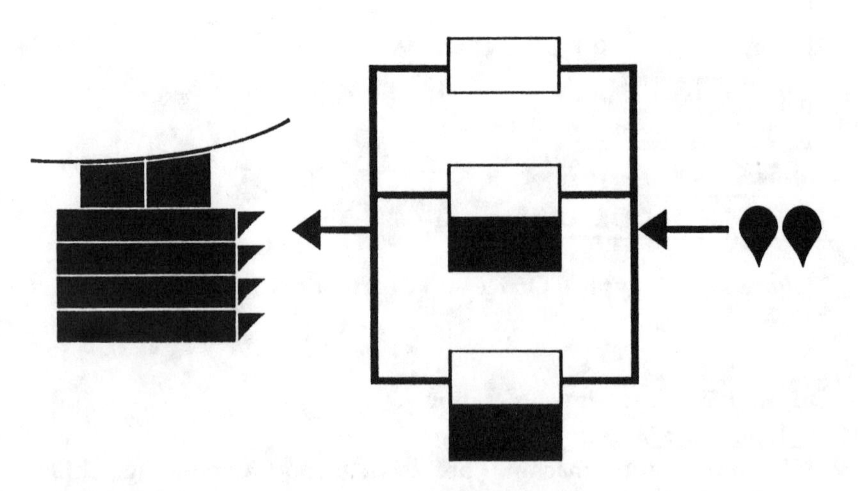

Figure 15-10. Summer Morning Start-Up (Mode 1).

Mode 2: Summer Daytime Operation

Water chilled the night before is circulated through the building. Nine chilled-water tanks are full of 42°F chilled water and one tank is empty to receive the return. Chilled water return is at 56°F and is stored until evening when it is rechilled to 42°F (refer to Figure 15-11).

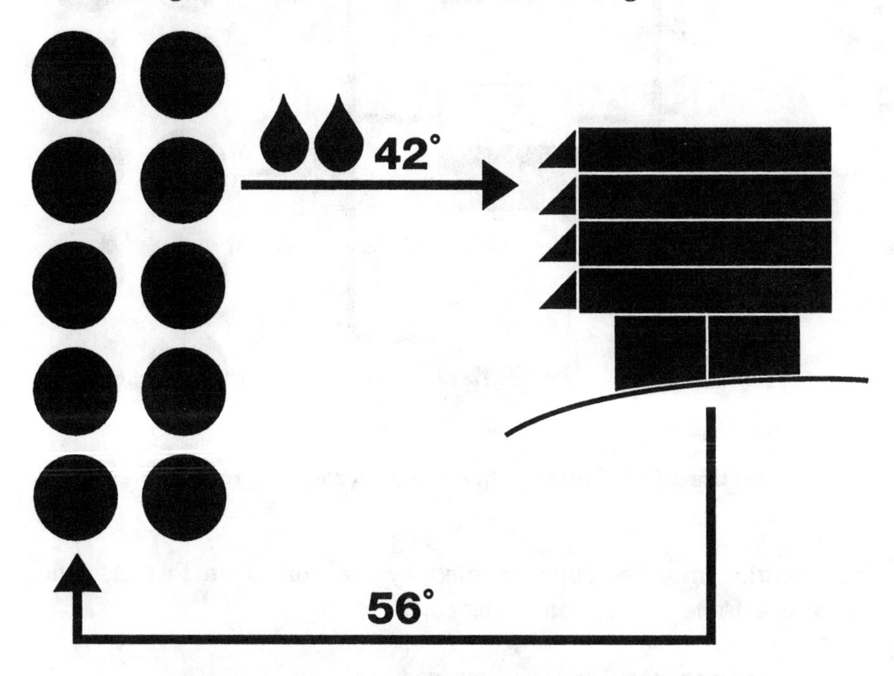

Figure 15-11. Summer Daytime Operation (Mode 2).

Mode 3: Summer Evening Operation

The refrigeration machines are used to chill the water in nine tanks back to 42°F for the next-day operation. After-hours cooling can be provided by utilizing the chilled water left in the piping system. If more after-hours cooling is required, the chilling process is stopped and chillers are used. However, at some point charging the tanks must resume (refer to Figure 15-12).

B. Original Winter-Time Operation

During all winter modes two tanks (one empty and one full) were used for heating and eight tanks (one empty and seven full) were used for cooling. Tanks were charged the night before for the next-day operation. The central plant can heat the building to recover energy from the building interior with outside temperatures down to 35°F. At lower tempera-

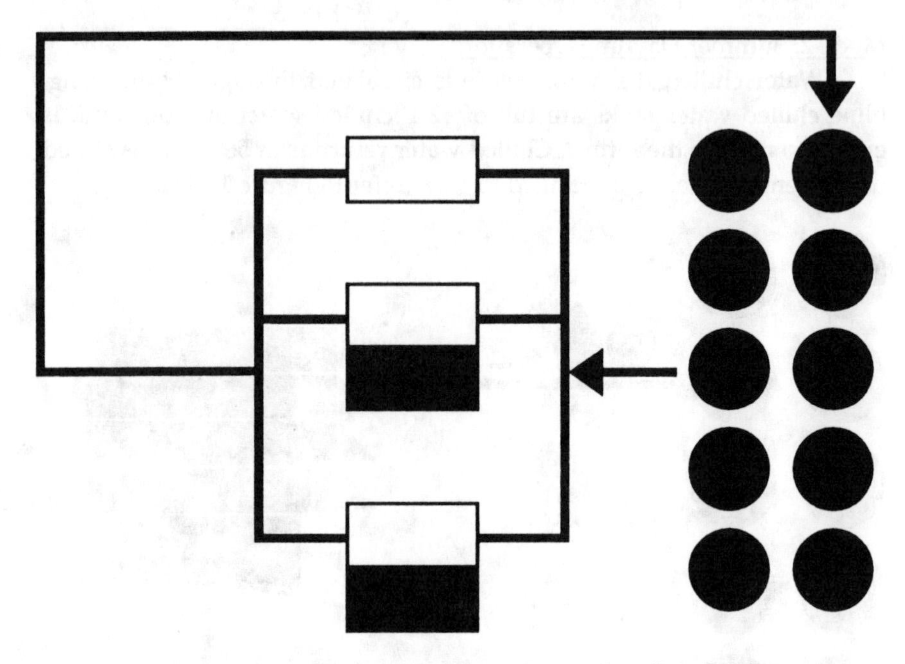

Figure 15-12. Summer Evening Operation (Mode 3).

tures, heating must be supplemented by the boilers. In this case the chillers are used in series with boiler converters.

Mode 4: Morning Start-Up of Unoccupied Building

Water at 105°F from hot water tanks is circulated to building and returned at 95°F to empty hot water tank (refer to Figure 15-13).

Mode 5: Morning Heating After Building Is Occupied

The refrigeration machines are operated to simultaneously cool interior zones while transferring the rejected heat to the perimeter offices (refer to Figure 15-14).

Mode 6: Daytime Winter Heating

Water at 105°F from hot water storage tank is circulated through the building. The boiler can increase the hot-water temperature above 105°F if required. The interior of the building is cooled by 48°F water from the chilled water storage tank (refer to Figure 15-15).

Mode 7: Evening Operation

This mode provides night heating and charges the tanks for the next-

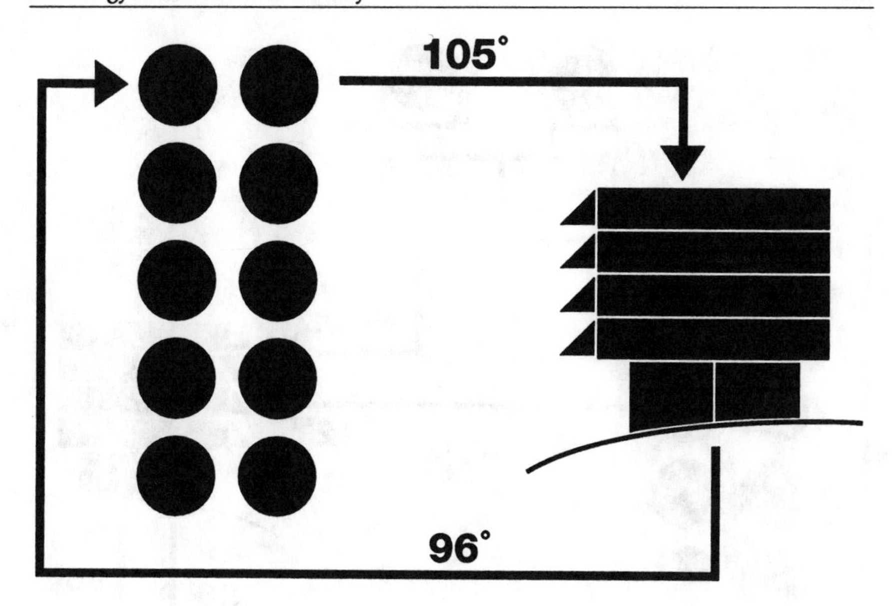

Figure 15-13. Morning Start-Up of Unoccupied Building (Mode 4).

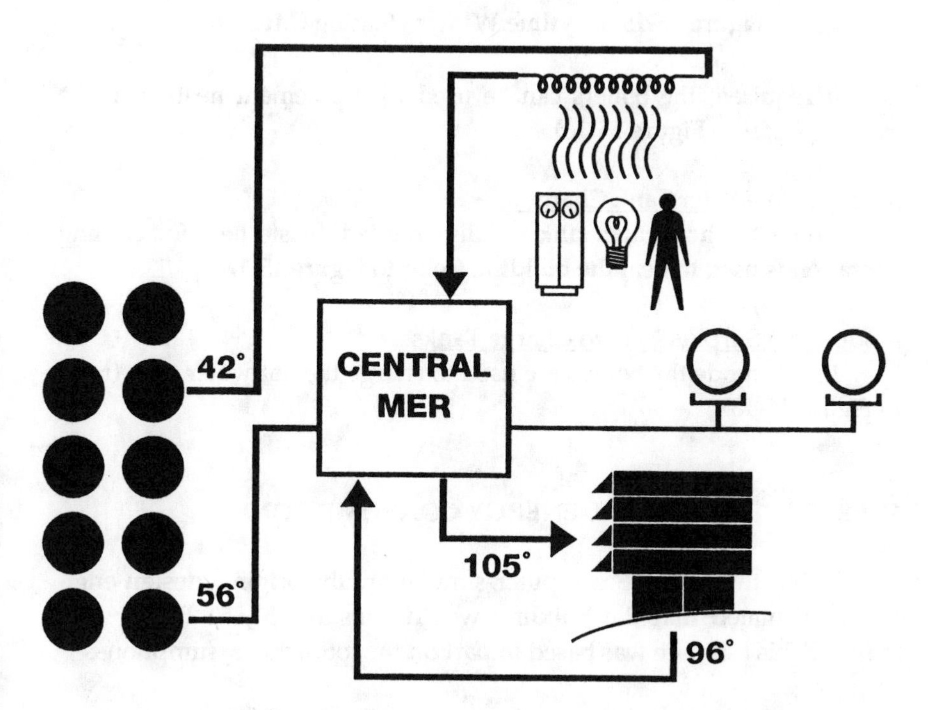

Figure 15-14. Morning Heating After Building is Occupied (Mode 5).

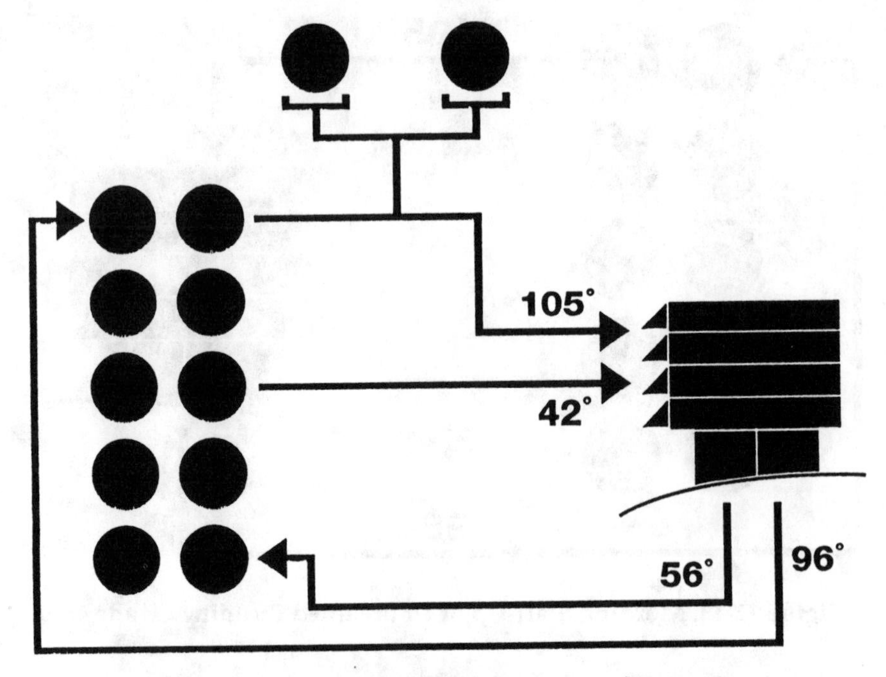

Figure 15-15. Daytime Winter Heating (Mode 6).

day. If required, the boilers can be used to supplement heating in the evening (refer to Figure 15-16).

Mode 8: Hot-Water Tank Charging

When the hot water tank is fully charged, waste heat for chilling operation is used to heat the building (refer to Figure 15-17).

Mode 9: No Surplus Heat to Charge Tanks

In this mode the boilers are used to charge the hot-water tank (refer to Figure 15-18).

V. PREDICTED ANNUAL ENERGY CONSUMPTION

Using the AXCESS computer simulation, the original design engineers calculated that the building would consume 51,851 Btu/ft^2 per annum. This estimate was based in part on the following assumptions:

- Maintaining indoor space conditions at 75°F db, 50% RH in the summer and 68°F db, 30% RH in the winter.

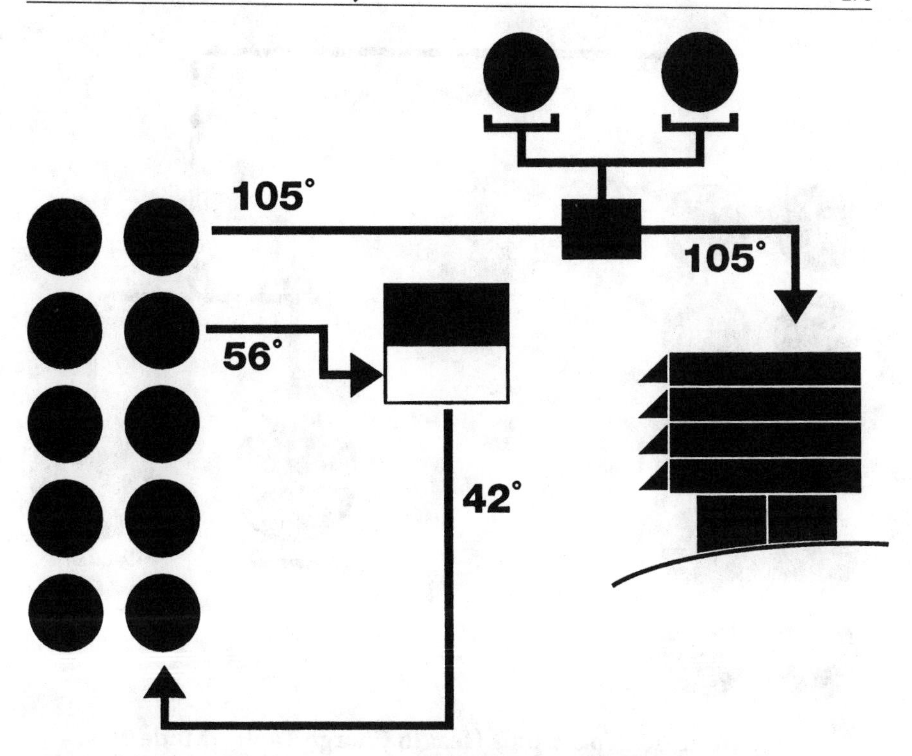

Figure 15-16. Evening Operation (Mode 7).

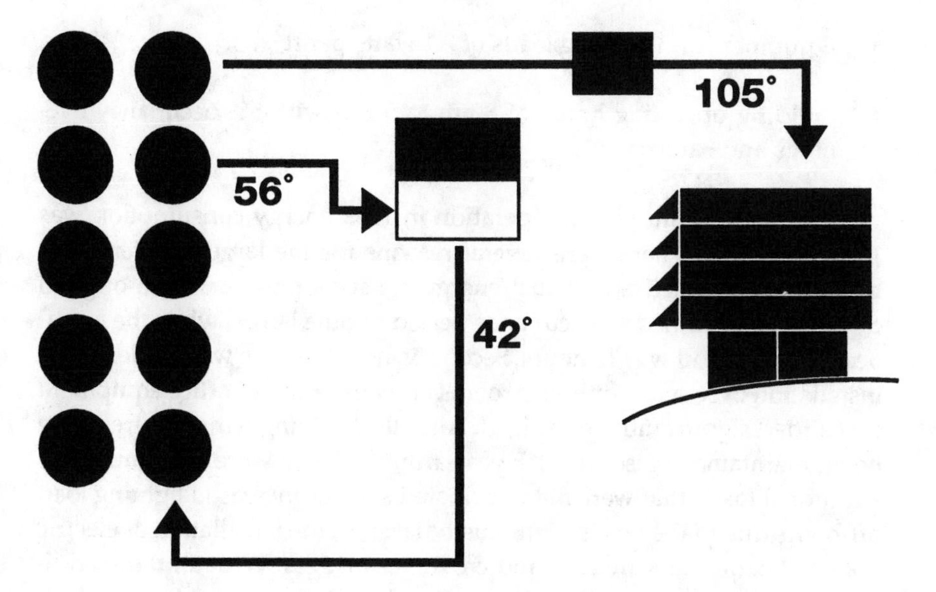

Figure 15-17. Hot-Water Tank Charging (Mode 8).

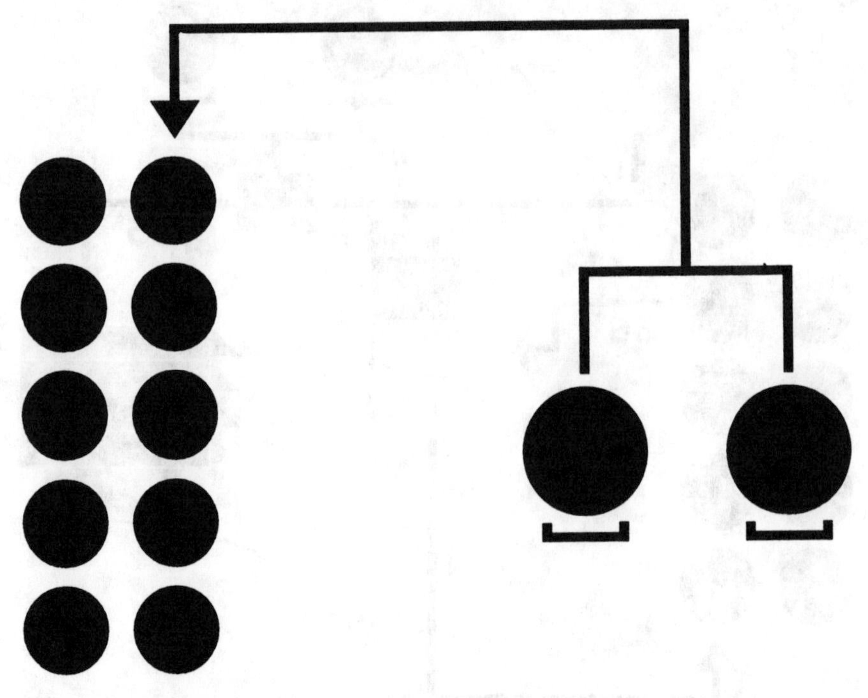

Figure 15-18. No Surplus Heat to Charge Tanks (Mode 9).

- Equipment and lighting loads of 2.1 watts per ft^2.
- Building operating hours of 8 am to 5 pm with 5% occupancy evenings and Saturdays.

In the first full year of operation in 1983 energy consumption was 131,659 Btu/ft^2. There were several reasons for the large difference between the predicted and actual energy consumption. First, the original calculations assumed an occupancy period of nine hours, while the actual occupancy period was 13 hours. Secondly, no allowance was made for the installation of several computer rooms and heavy use of office equipment that added significantly to the load. Also, the building temperatures were being maintained closer to 72°F year-around. There were also numerous additional loads that were not anticipated such as increased lighting load off-hours due to the needs of the custodial staff, the installation of electric booster heating in stairwells and crossover bridges. With all these additional loads factored in, the new predicted annual consumption was calculated by the building operating staff to be 96,270 Btu/ft^2 per year.

VI. MEASURING ENERGY CONSUMPTION

In managing energy at a facility it is important to understand the existing capabilities of the system, set realistic goals, and provide a feedback system to measure progress toward the goal. An energy measurement system must be provided to monitor energy consumption over time. A common unit of energy consumption can be used such as Btu/ft^2. In many facilities there are a number of energy sources used. At Corporate Center there are three energy sources: electricity, #2 fuel oil and natural gas. Each of these energy sources can be readily converted to Btus, divided the gross area of the building and charted. Figure 15-19 shows a plot of the Btu/ft^2 consumed in 1992. Monthly utility bills provided an excellent source of energy consumption history. These bills were adjusted to a 30.4-day per month billing period so that constant month-to-month comparisons can be made.

Once a realistic goal was determined, a method of measuring actual energy consumption and adjusting for weather and vacant space had to be developed. Weather conditions do affect the heat transfer across the building shell and infiltration from the environment. The heating and

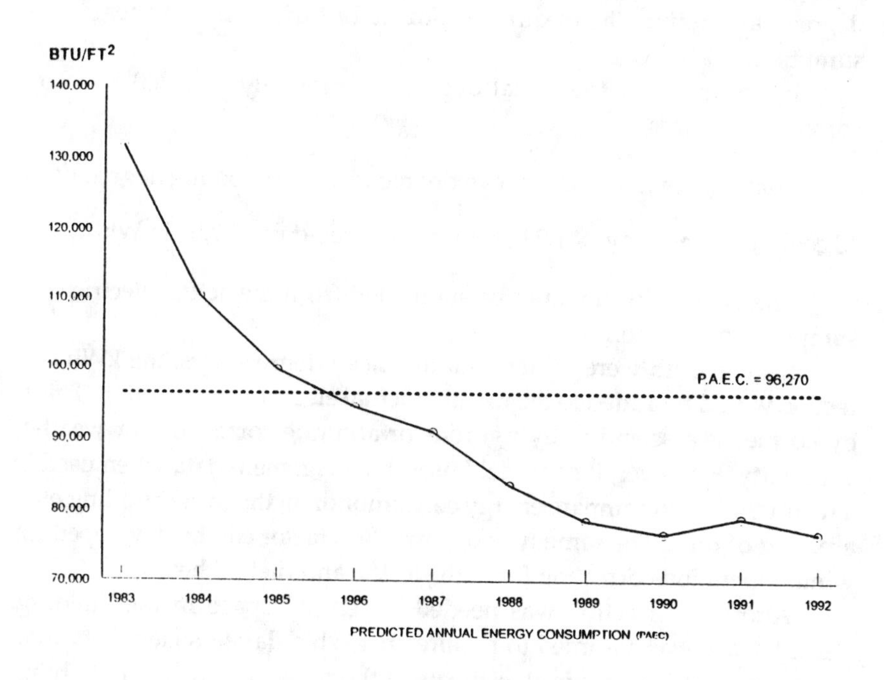

Figure 15-19. Total Energy Consumption (1983 through 1992).

cooling degree-day data for the Danbury, Connecticut area was obtained from National Weather Service. The data combined with some assumptions concerning the variability of energy consumption because of weather, allowed for a correction in the monthly reports. The assumptions made normalize the energy consumption to the monthly average degree-day conditions for the Danbury, Connecticut area.

During a cool summer day, the building base cooling load was estimated to be 8,900 kWh consisting of operating one chiller and three chilled-water pumps on low-speed for 12 hours. Typically on a hot summer day, three chillers and three pumps on high-speed would be required using 24,600 kWh. The difference between the two was assumed to represent the variability of the chilling load due to weather. That difference is multiplied by the working days per month and divided by the mean cooling degree-day data for July (282 degree-days). Thus, the correction factor for Corporate Center is 1,226 kWh per degree-day assuming 22 working days a month. This calculation normalizes the chilling load to a constant weather condition, and gives a correction factor in kWh per cooling degree-day. If the actual cooling degree-day total for the month is greater than the mean, then the kWh correction (as determined by the product of the correction factor, multiplied by the difference in actual degree days, minus the mean) is subtracted from the actual energy consumption.

For example, if the actual degree-days for July was 300, then the correction factor is:

300 actual degree-days – 282 mean degree-days = 18 cooling degree-days.

18 cooling degree-days × 1,226 kWh per degree-day = 22,068 kWh.

The 22,0068 kWh would be subtracted from the actual electric consumption for the month.

If the month were cooler than the mean degree-days, the kWh correction would be added to the month's actual electric consumption. This is by no means a scientifically rigorous or accurate correction for weather variability. However, it does yield useful management data when used in a relative sense to compare energy consumption in the same building over a period of time. The same type of correction factor can be developed for winter operation correcting for both electric and fuel oil heating.

Another correction was needed for empty space in the building. Vacant space was assumed to require six kWh a day less electricity than an occupied office. This is an estimate of the reduction mainly in lighting and office equipment load. HVAC loads were assumed not to be that materially different for an isolated empty office as compared to an occupied one. With these two factors actual monthly energy consumption can be normalized to full occupancy and average degree-day conditions for the month (refer to Figure 15-20).

VII. ENERGY GOAL SETTING

The revised predicted annual consumption became the initial energy goal for the facility. With the 1983 energy consumption at 131,659 Btu/ft^2, there was a 35,389 Btu/ft^2 energy savings opportunity. The first step taken was to operate the chilling and heating plant thermal storage system as intended. During the first couple of years the boilers were used for most winter heating. The heat bundles of the chillers were not fully utilized as intended until 1986. A complete hydronic system rebalance was performed. This corrected problems with insufficient pressure at some fan coil units and central fans. It also corrected a negative differential pressure between the supply and return at the outlying pods. Optimum start/stop of the HVAC system was fully implemented and monitored. By the end of 1986 the revised predicted annual energy consumption (PAEC) was achieved.

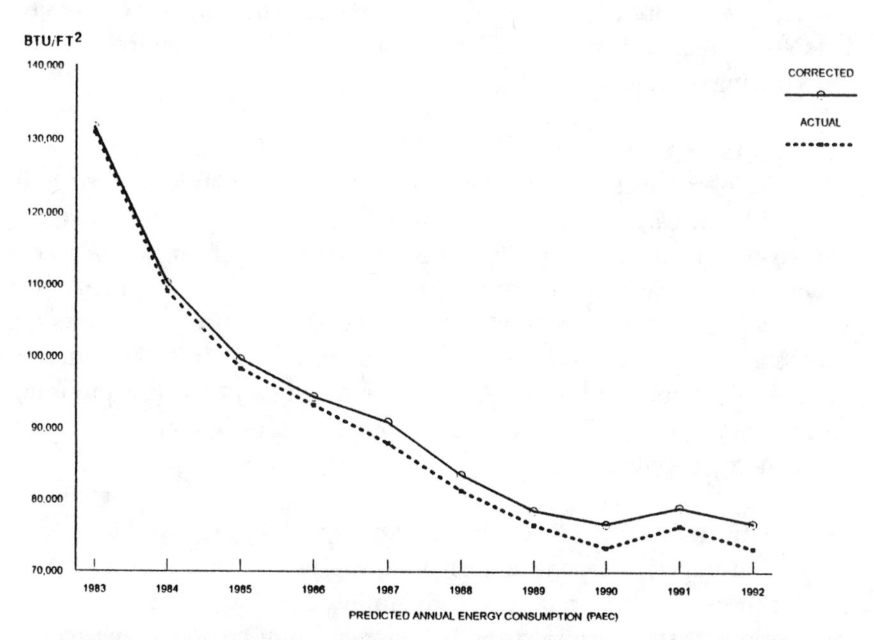

Figure 15-20. Total Energy Consumption (1983 through 1992).

Once the PAEC was achieved the operating staff set new goals yearly to reduce energy consumption still further while enhancing occupant comfort and IAQ. Some of the energy projects and actions taken were:

A. Installation of Computerized Lighting Control Throughout the Building

Corridor lighting was originally manually controlled. This resulted in many lights left on 24 hours a day. During 1986 the lighting panels were rewired and connected to the energy management computer. A time-of-day program was utilized to turn off the lighting after the office cleaning crew was expected to be finished.

B. Installation of Stand-Alone Computer Room Air Conditioning Units

Many computer rooms were located on office floors and cooled using the pod AHUs. This resulted in running the entire pod central fan units 24 hours a day, 7 days a week in order to maintain temperature in a computer room. Stand-alone air-conditioning units were installed in the computer rooms. A secondary loop was constructed similar to that used for the office fan coil system to take advantage of the inherent energy efficiency and control stability of these systems. The majority of the savings was achieved from the ability to shutdown the pod central fan systems during off hours.

C. Utilizing Office Fan Coils as a Heat Sink in Winter

There are times in the winter when the hot water tanks become fully charged in the early morning but there still is a need to continue charging chilled-water tanks. Normally, the operating engineer would have switched from the heat recovery mode to the cooling tower mode. A plan was developed and implemented that utilized the office fan coils as the heat rejection device. This allowed us to reject heat into the building instead of the cooling tower in the early morning and use less pumping and fan horsepower. This preheats the building for occupancy operation at no energy penalty.

D. Instituted "OWL" Program (An Acronym for Out with Lights)

Encouraged occupants to turn off the lights in their offices if they are out for more than five minutes. A five-minute period was selected after calculations that determined the break-even point between energy savings and decreased lamp life was three and half minutes.

E. Aggressive Measures Were Taken to Reduce Lighting, Heating and Air Conditioning in Unoccupied Areas

Unoccupied space was consolidated into contiguous blocks of space to leave whole floors empty. Temperatures in unoccupied areas were then reset to 58°F. HVAC units were only run during off-peak hours to ventilate the space and maintain temperature. In order to reduce sun load in the summer, all window blinds were closed. All lighting was turned off except for emergency lighting.

F. Installation of Energy Saving Lighting

When the corridor lighting system was due for a re-lamping, 34-watt fluorescent spec 34 tubes were used versus the original 40-watt warm white lamps. Since many of the lamps were in for several years they had decreased lumen output. Re-lamping increased lighting levels and color rendition while reducing energy consumption. The re-lamping was performed under a rebate program sponsored by the local utility company so that lamp costs were significantly reduced. Center core was re-lamped with electronic ballasts and T-8 lamps.

G. Revised Operating Modes

Originally there were nine operating modes of the heating and chilling plant. Some modes were never fully utilized. Through experimentation the original modes were modified and the resultant modified modes were more economical and/or energy efficient. For example, a variant of mode 2 was developed that utilized water from a tank that was only moderately warm (e.g., 54°F). This tank would be from an early morning mode 2 operation. The tank would be run through the chiller instead of a closed-loop operation in the late afternoon when all 42°F tanks had been expended but there was still a need for full building cooling. Winter-time tank operation has been modified in that both hot-water tanks are always full. Hot water is recirculated within the tank instead of from a full to an empty tank. This had two main advantages: (a) doubling the hot water capacity that can be reclaimed and stored, and (b) a savings of 180,000 gallons of water and associated water treatment chemicals each year since it was necessary to prepare for winter operation two tanks (one chilled and one hot) that had to be empty.

H. Improved Differential Water Temperatures

One historical problem that was corrected was that on design summer days supply-air temperature from the AHUs were 4°F above the design temperature. This was due to inadequate passes in the chilled-water coils. This affected central plant operations in that chilled-water return temperature was 3°F to 4°F below the design temperature. An inexpensive solution was to connect the return side of the chilled-water coil with the supply side of the hot water side in the AHU. This added two more passes and the design supply temperature was achieved. The central plant was then able to operate at the design conditions for greater efficiency (refer to Figure 15-21).

VII. ENERGY CONSERVATION PROGRAM RESULTS

The energy conservation efforts described above have resulted in a 40% reduction in energy consumption between 1983 and 1992. However, this was not accomplished at the expense of occupancy, comfort or IAQ. A total number of temperature complaints has shown a steady decrease over the past six years (refer to Figure 15-22). Additionally, measurements of key IAQ parameters have shown an improvement.

The success of the program has been due to not only the excellent energy conserving features of the building but also the team approach used by the operating staff. The management, line personnel and building occupants shared a common goal in managing energy. A goal oriented operating philosophy, along with the willingness to experiment and explore alternative ways of saving energy, contributed to this achievement.

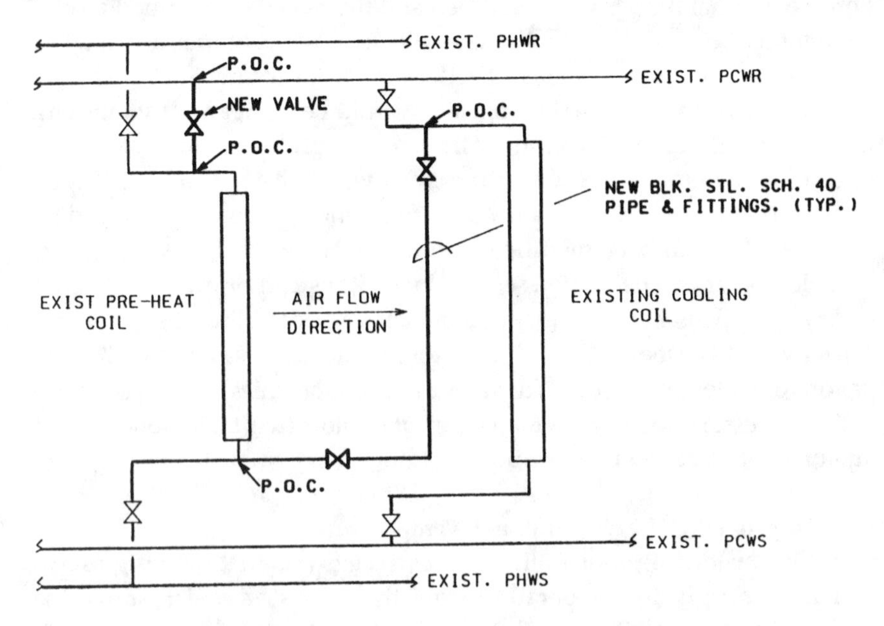

Figure 15-21. New AC-2 Supercool Piping.

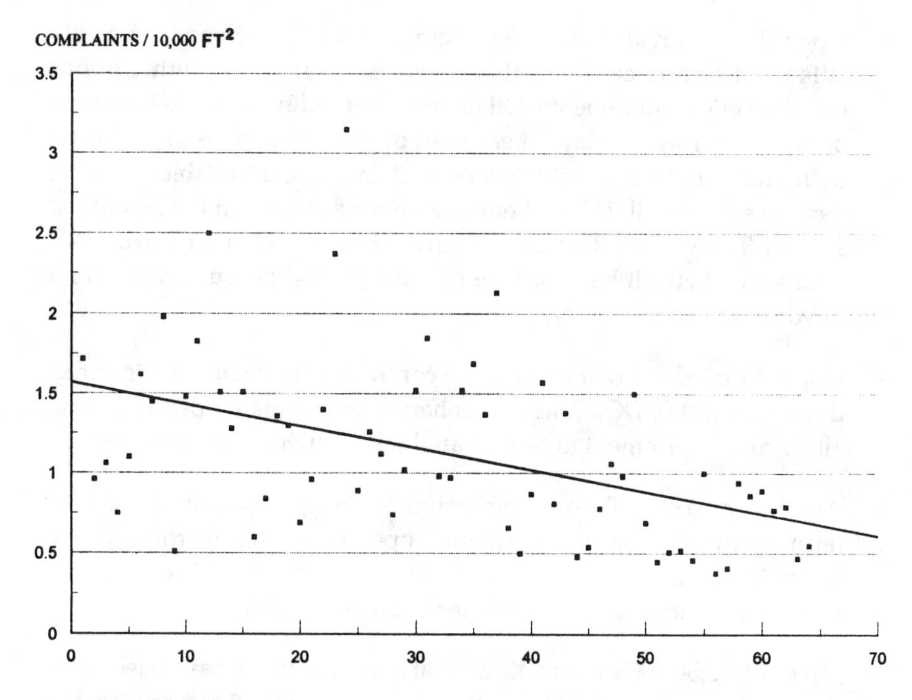

Figure 15-22. Total Temperature Complaints at Corporate Center.

IX. FUTURE PROJECTS

A recently completed energy analysis survey sponsored by the local utility company identified several opportunities for energy enhancement. These are:

- Upgrading all corridor fluorescent lighting and magnetic ballasts to a T-8-lamp and high-frequency electronic ballast combination. Most of the lighting in Corporate Center would qualify for this retrofit. The payback period is estimated to be less than six years with an electric company rebate buying it down to less than three years.

- Installing premium efficiency pump and fan motors wherever appropriate. Most of the motors in Corporate Center have efficiencies between 72% and 94%. Motors with substantial run times would qualify for replacement with more efficient motors in the 84% to 95% efficiency range. The simple payback period was calculated to be 8.6 years for replacement of 115 motors. Electric utility buydown reduces the payback period to four years.

- Upgrading of an air-handling system with variable-frequency-drives (VFDs). There is an AHU that serves several zones with 24-hour conditioning requirements. Other zones served by the AHU are only occupied during the day. There is an opportunity to install VFDs on both the supply and return fans and install controllable isolation dampers. This will isolate the non-24-hour zones from the rest of the system. The VFDs will allow the fans to be ramped down during off-hours. This retrofit has a simple payback period of four years after a buydown from the utility.

- Upgrading central plant tank farm controls from pneumatic to direct digital control (DDC). This will enhance central plant operations and allow most equipment to be operated with much more precision.

- Another project will be to increase the storage capacity of the tank farm by installation of a horizontal opening in the overflow pipes. When installed in all of the 10 tanks, storage capacity will be increased by the equivalent of one tank (refer to Figure 15-23).

In total, 2,250,000 kWh of additional energy savings have been identified in the energy audit. This is equivalent to another 5,868 Btu/ft^2 per year in reduced energy consumption.

X. MEASURES CONSIDERED BUT NOT PRESENTLY ECONOMICAL

There were several measures that the energy audit examined but proved at the present time not to be economical at Corporate Center. Detailed below are some of the options considered. Of course, all of the following options may be economical and entirely appropriate at other facilities.

A. Installation of VFDs on Central Plant Chilled- and Hot-Water Pumps

While a variable-flow pumping system would improve system operation, the installation costs could not be justified on energy savings alone. The existing pumps have a 2-speed capability. This feature combined with the available total number of pumps (four chilled- and three hot-water) yields a fairly flexible pumping combination that can be used to operate economically under almost all load conditions. The payback periods for conversion to VFDs were around 18 years.

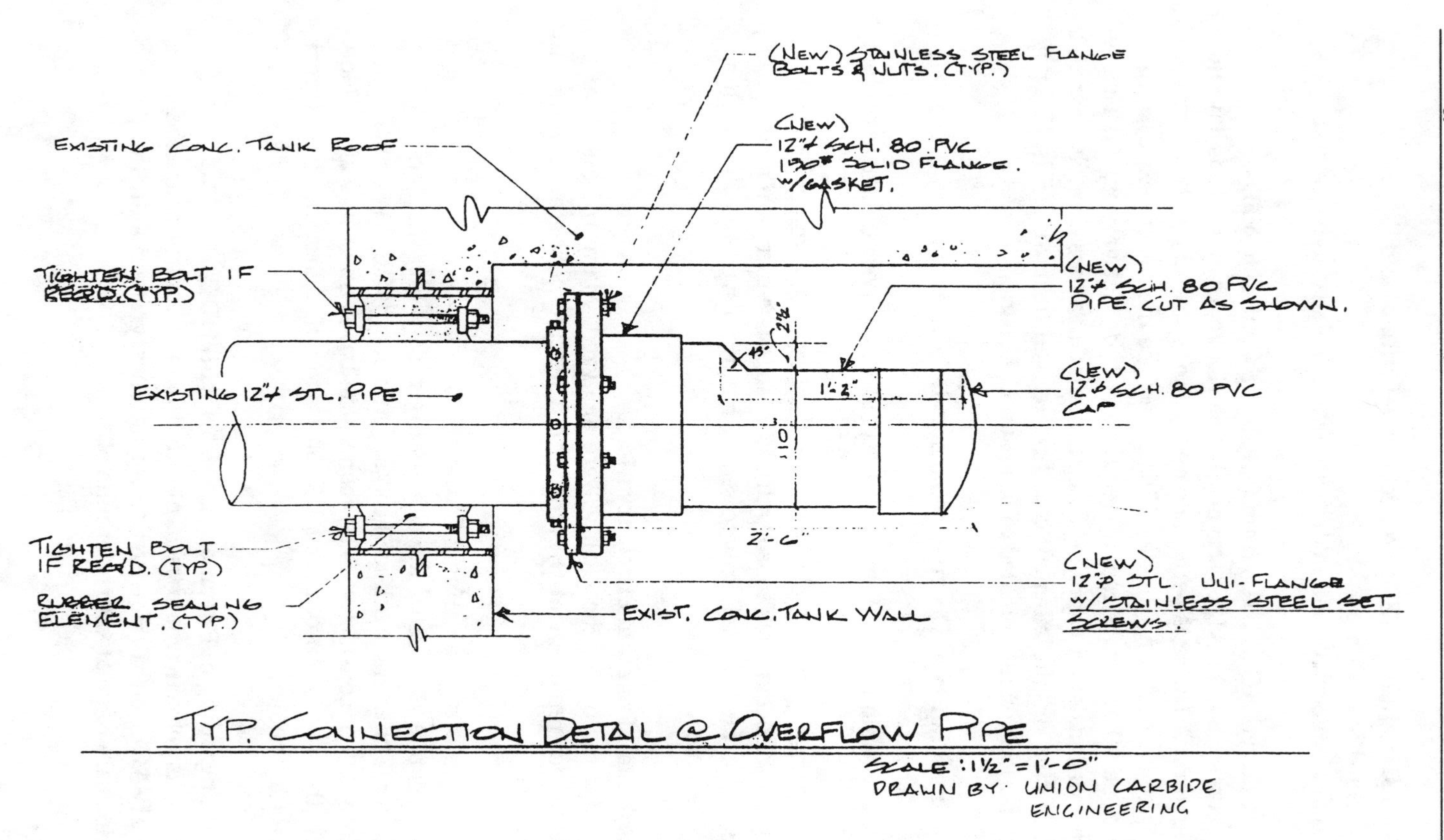

Figure 15-23. Proposed Change to Overflow Pipes.

B. Installation of VFDs on Secondary Chilled- and Hot-Water Pumps

The payback period in this case was calculated to be over 11 years.

C. Convert AC-1 AHUs from Inlet Vane Control to VFDs

Presently, the AHUs serving the interior zones of the pods have inlet vane control. The tests performed on the fans showed that the system ran fairly constant during the day. This is because the system serves an interior zone where space cooling requirements do not vary much. This results in a constant low profile. Since the fans run almost entirely over 80% of capacity very little benefit could be derived from the VFDs. The payback periods were estimated to be around 21 years.

D. Installation of Occupancy Sensors in Offices and Conference Rooms

The unique architecture and layout of Corporate Center presents challenges in the use of occupancy sensors. There are 2,350 offices with five separate light controls. All of the light switches are not in a suitable location for occupancy sensor retrofit. Also, since there are no drop ceilings in the facility, ceiling type sensor mounting would be difficult and uneconomical.

E. Install Premium Efficiency Motors

All motors in the facility were analyzed. Many motors were not selected for replacement with higher efficiency types because of the low runtime.

F. Upgrading Centrifugal Chillers

Despite the age of the chillers (over 12 years) they are relatively efficient. The average full-load cooling efficiency is 0.67 kW/ton. Replacement was not considered to be economical based on energy alone. However, the chlorofluorocarbons (CFCs) issue will determine the ultimate fate of the chillers.

G. Upgrading of Pneumatic Controls with DDC

This retrofit would certainly improve the control characteristics of the HVAC system. However, from an energy standpoint the payback periods are calculated to be around 60 years.

XI. RECOMMENDATIONS BEYOND DESIGN STAGE

This case illustrates that the attainment of energy efficiency is as much dependent on building operation as on the design. Improper commissioning, poor operator training, improper operation of installed energy conserving devices, lack of energy goals and monitoring can and do prevent the attainment of energy conservation. A program to constantly audit operations and encourage experimentation should be developed. Engineering studies should be periodically performed to determine if the latest technology in conservation can be applied economically to the facility. Many local utilities offer financial assistance for such studies and the subsequent conservation projects.

A person should be designated as the facility energy manager who would view energy conservation as a full-time project. As with any other project, the energy manager must plan, lead, organize and control energy consumption at a facility. A very important part of any energy conservation effort is to involve both the users of energy in a building and the operating staff in a common goal. For example, the OWL Program has sensitized office workers to the goals of energy conservation. They can contribute to the bottom line of the organization by saving energy in an easy way.

One valuable tool in seeking out opportunities for enhancement is the energy use profile. For example, Corporate Center's energy consumption profile is illustrated in Figure 15-24. The profile reveals that miscellaneous office equipment, lighting and fans account for over half the energy

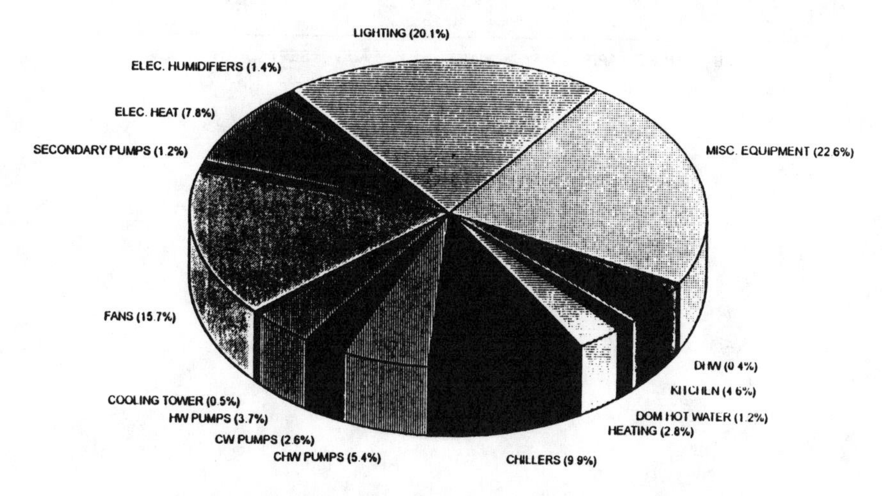

Figure 15-24. Energy Consumption Profile of Corporate Center.

consumption in the building. Equipped with this information, the energy manager can concentrate on the systems that consume the most energy and promise the most benefits from conservation. A method of measurement must be developed and used so that effects of any action taken can be measured. Monthly charts detailing energy consumption are invaluable in spotting trends (refer to Figures 15-25 through 15-30). Perhaps, most importantly, involve your operating staff in all major decisions and empower them to experiment.

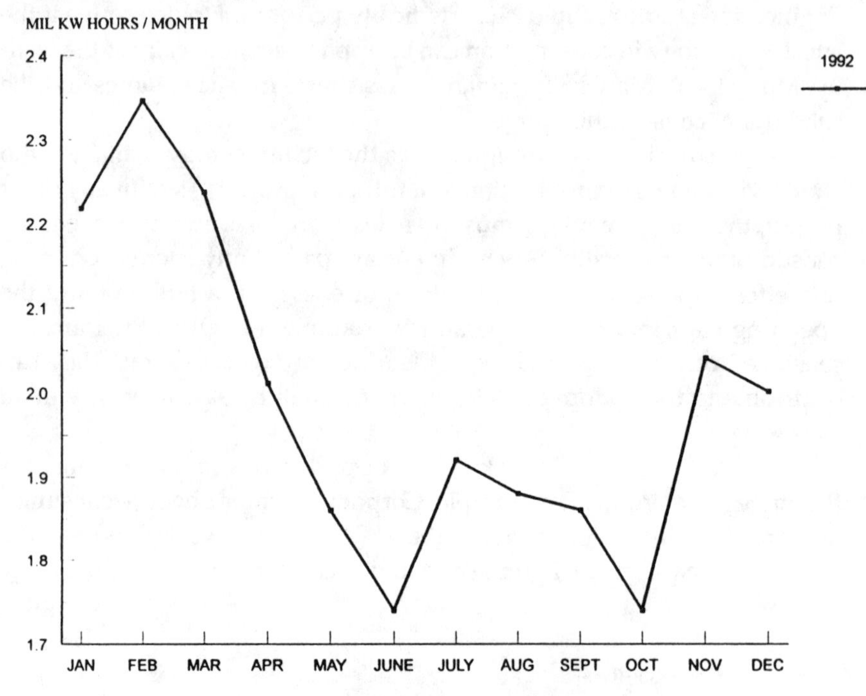

Figure 15-25. Electrical Consumption.

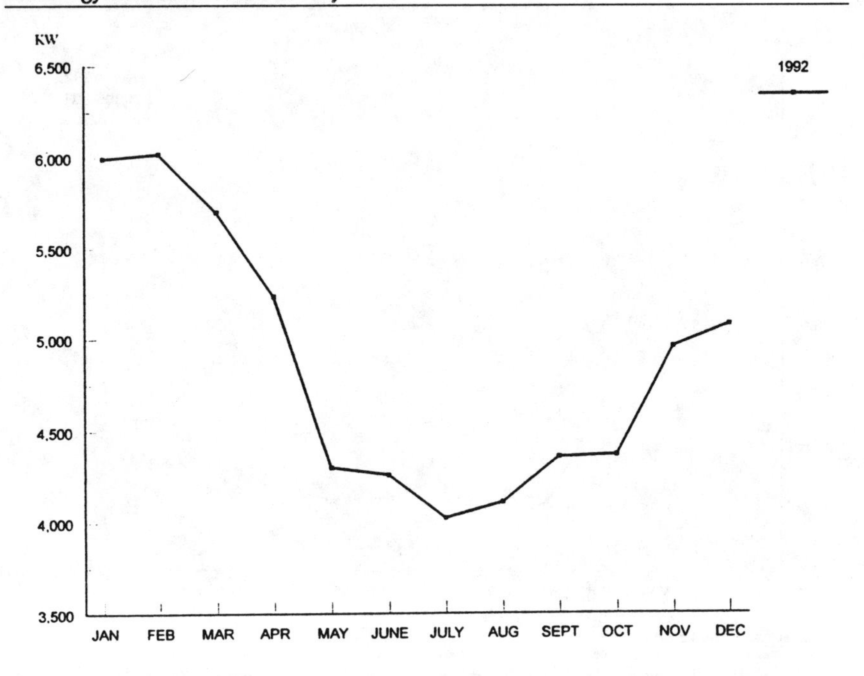

Figure 15-26. Energy Demand.

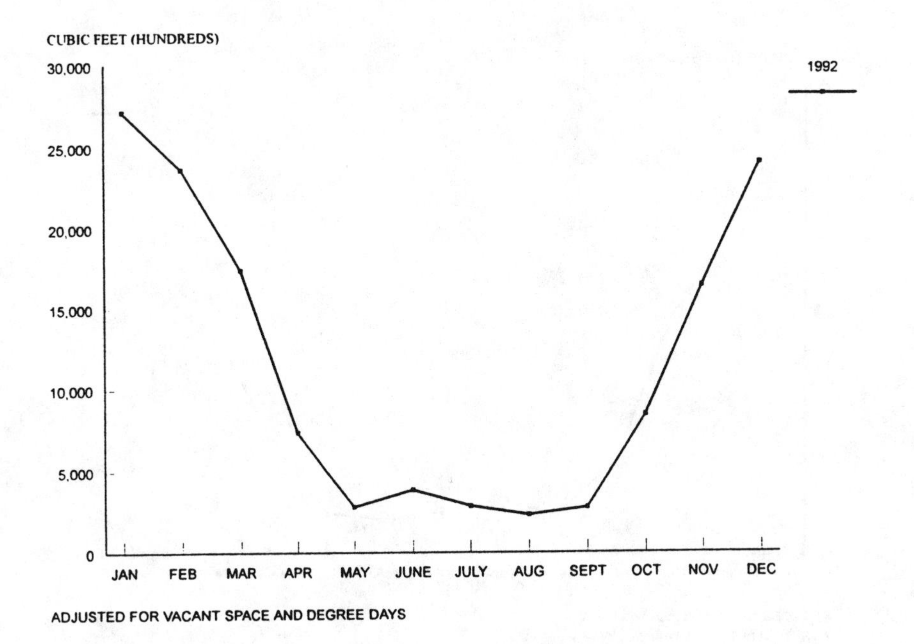

Figure 15-27. Gas Consumption.

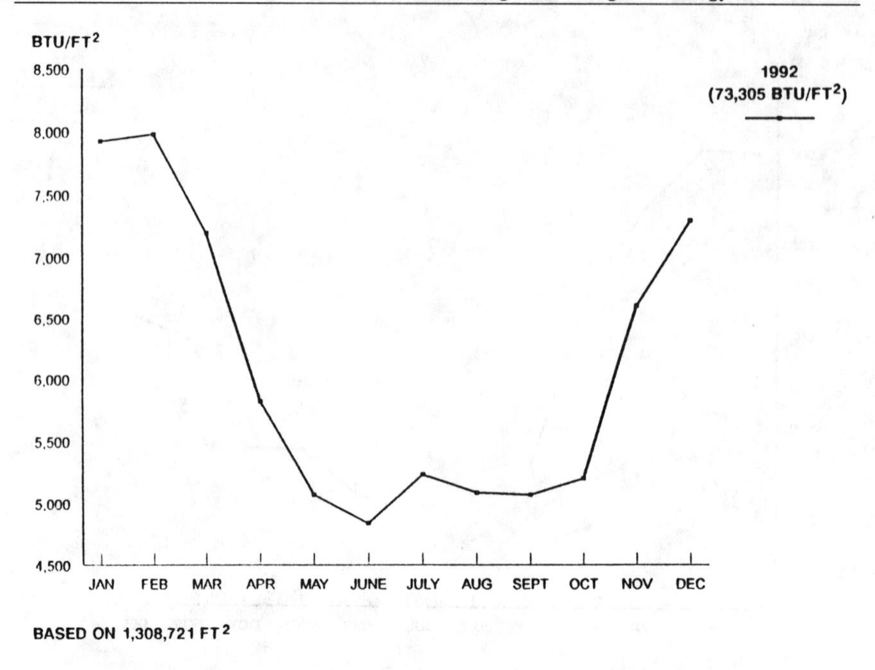

Figure 15-28. Total Energy Consumption (Btu/ft^2).

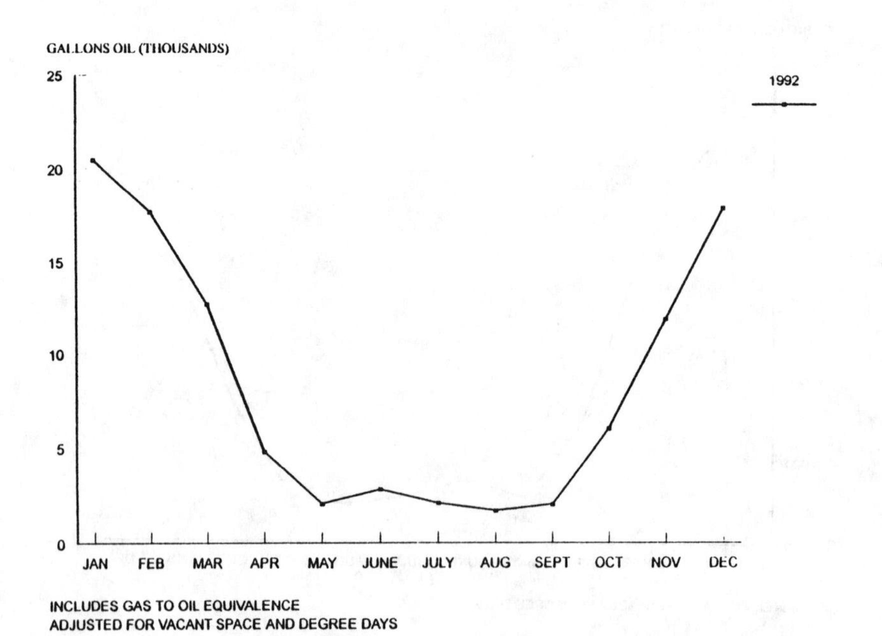

Figure 15-29. Oil Consumption.

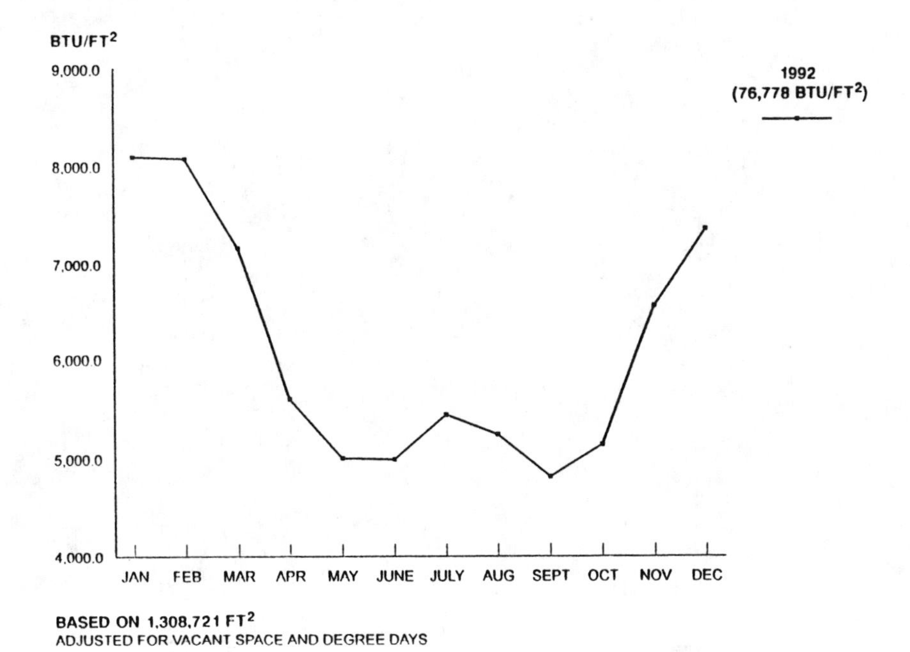

Figure 15-30. Total Energy Consumption (Btu/ft^2).

CHAPTER 16

INSTITUTIONAL REHABILITATION: AN OPPORTUNITY FOR SIGNIFICANT ENERGY SAVINGS

John J. Halas, P.E.

I. INTRODUCTION

The aim in institutional energy conservation is twofold: (a) to establish and prepare a present energy consumption profile, including all categories of energy, and convert all forms of energy to one common base (Btu or calorie) for comparison and evaluation purposes; and (b) to generate actual energy conservation measures, and information about their costs, benefits, and paybacks, resulting in the reduction of energy consumption in the particular institution. To meet these goals we generally undertake the following tasks:

1. Establish and study data relating to energy consumption. In this respect, it is generally helpful to analyze sharp "peaks" in electrical or steam demand curves. Further, establish a priority system based on consumption of energy and conduct an institutional survey from the largest energy user to the smallest in descending order, as time and costs permit. With this in mind, the largest and most beneficial energy savings are identified first, before large sums are spent for relatively small and unimportant projects.

2. Collect and analyze weather data, energy costs, and projected energy cost increases.

3. Establish the energy conservation measures and list them in order of cost savings, energy savings and payback periods.

The following sections describe the techniques for identifying energy conservation measures that grew out of the author's work on behalf of the New York State Energy Office and various other clients of the firm of Pope, Evans & Robbins, Inc. In response to plans of the United States Department of Energy, the New York State Energy Office launched major energy conservation studies directed at several classes of facilities, health care institutions, schools and municipal buildings.

This chapter deals with the results of the health care study. Clearly, one cannot hope to enumerate all energy conservation measures in such a short format, nor can one cite the numerous individual unique examples that apply to a single institution. These latter are usually the result of a thorough survey conducted by a skilled professional engineer. Rather, the general discussions here can provide an important base required for typical energy conservation reports.

II. GENERAL CHARACTERISTICS AND OVERALL RESULTS OF THE SURVEY

It is well known from past energy audits that, compared to commercial, educational and speculative type structures, hospitals are major users of steam and/or hot water as well as large users of heating and, in most cases, year-round air conditioning. Further, to comply with a unique set of rules and regulations embodied in the Federal Hill-Burton guides, they are users of large amounts of outside air in "once-thru" systems and have relatively minor diurnal load variations.

A recent joint United States Department of Housing and Urban Development USDOE study concluded that the mean annual energy consumption of heating, cooling and lighting for hospitals is almost three times that of office buildings or universities.

In recent years, teams of engineers visited 14 hospitals and 10 nursing homes (selected randomly by computer analysis) throughout New York State to establish a data base for common energy conservation measures as well as energy reduction by the institutions themselves. The hospitals surveyed were identified by number and analyzed according to their energy consumption per unit area.

In performing such surveys, care must be taken to remember that no ECM, no matter how beneficial, can interfere with the critical nature of the care provided by the facility. Audit efforts should always be aimed at institutional, HVAC, non-process-energy-consuming systems. Most ECMs identified during the survey resulted in a 5-year payback or less. It is generally not cost-effective for most hospitals to consider longer-payback ECMs except in exceptional circumstances, such as with solar projects. For the ECMs discussed here, energy savings were based on state-of-the-art information available from weather charts and manufacturers. Generally, no fuel escalation costs were considered, and no life cycle studies were performed. In most cases, it was found that a 30% or better reduction in energy consumption can be achieved by considering only the largest and most obvious measures (refer to Figure 16-1 for details).

The study indicated that an investment of $47,000,000 applied to the ECMs for the 234 voluntary nonprofit hospitals in New York State would result in a 6×10^{12} Btu/yr (6.33×10^{12} kJ/yr) savings, or the equivalent of about 43 million gallons (162,970 m^3) of oil per year. Results for the nursing homes are shown in Figure 16-2. The savings are somewhat smaller; for a $5,900,000 investment, a 0.9×10^{12} Btu/yr (9.495×10^{11} kJ/yr), or 6.5 million gallons (24,635 m^3) of oil per year energy reduction is anticipated.

III. CENTRAL PLANT TYPE RETROFITS

The following central plant type retrofits have been found in most institutions:

A. Boiler combustion efficiency increase
B. Oil burner replacement
C. Waterside—Scale buildup removal
D. Fireside—Soot buildup removal
E. Reduction of energy losses due to domestic hot water leaks
F. Reduction of energy losses due to malfunctioning steam traps
G. Preheating combustion air for higher efficiency
H. Hot water storage tank insulation for reduced heat loss
I. Insulating hot pipes for reduction of heat loss

The following examples provide somewhat simplified illustration of these ECMs. It is hoped that the reader may enlarge and further develop these examples to fit particular applications.

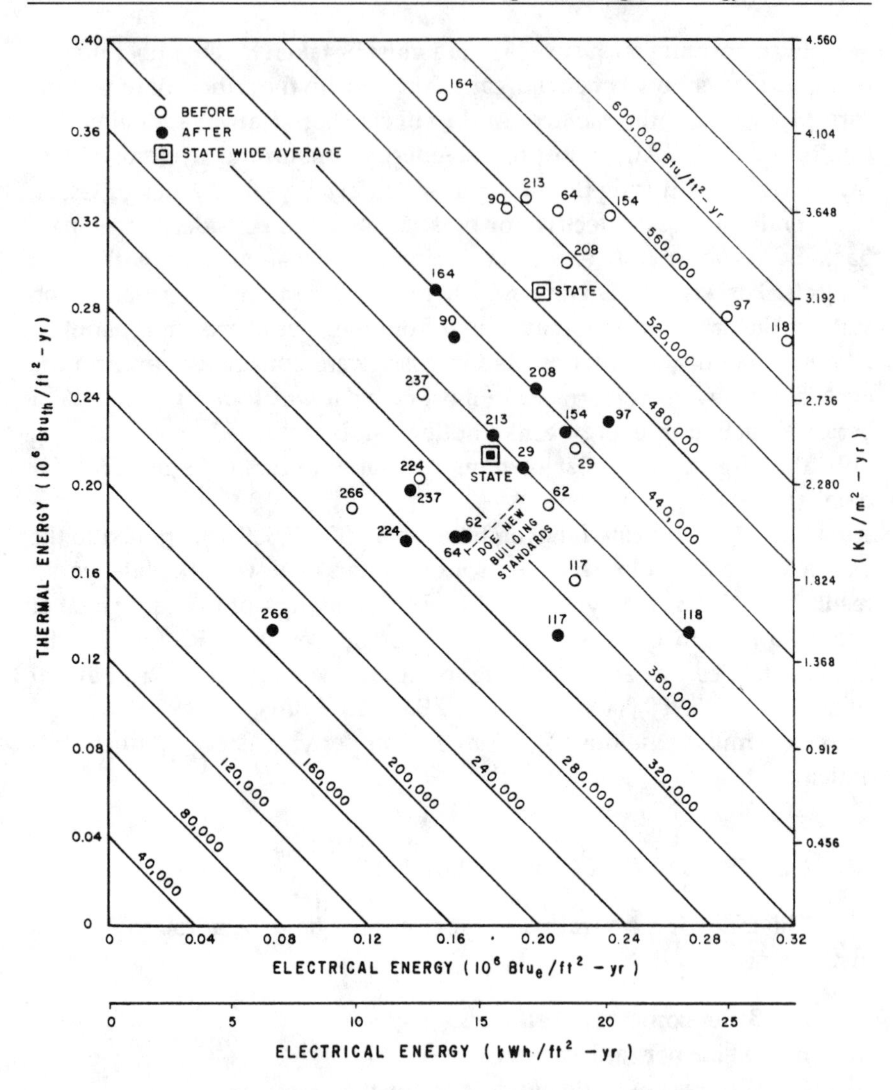

Figure 16-1. Hospital Annual Energy Consumption per Unit Area. (Reproduced by permission of Pope, Evans and Robbins, Inc.)

A. Boiler Maintenance and Operation Improvements

It is important to remember that general boiler plant retrofits usually result in large energy savings. The boiler room or plant, as the primary consumer of thermal energy in most hospitals, deserves special consideration for energy conservation. It is imperative that all boilers operate at maximum efficiency to derive maximum energy from the fuel. For smaller

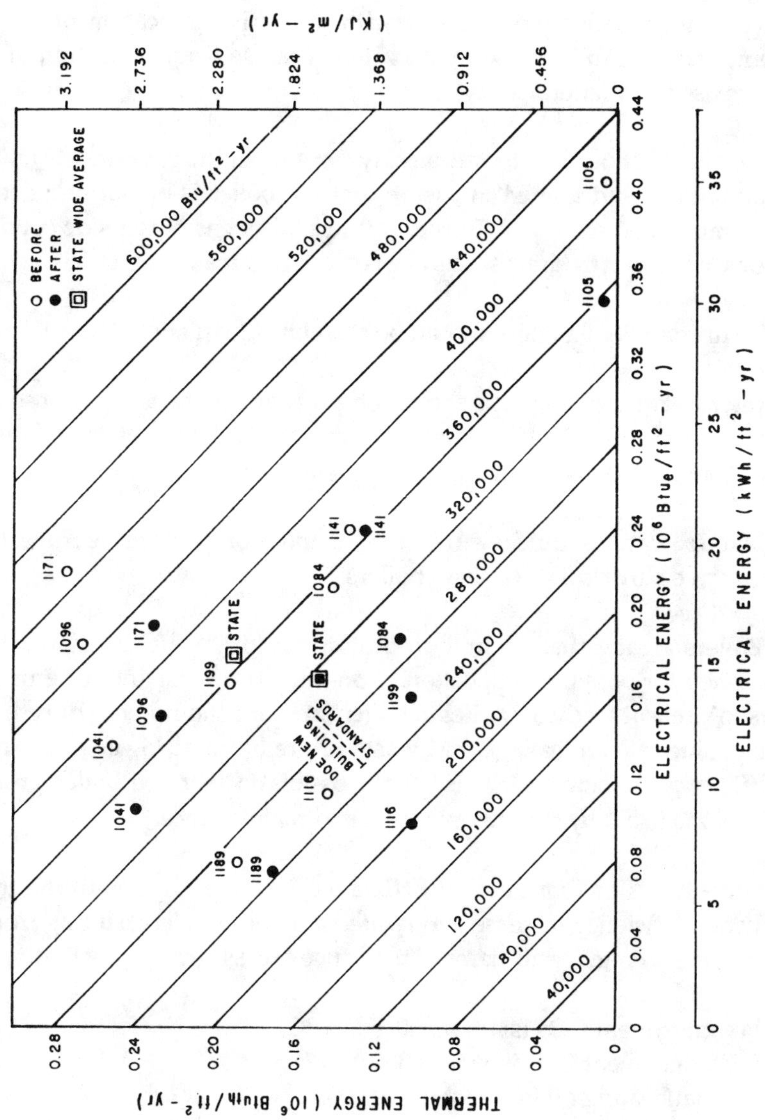

Figure 16-2. Nursing Homes Annual Energy Consumption per Unit Area. (Reproduced by permission of Pope, Evans and Robbins, Inc.)

boilers, of less than 30 million Btuh input (8.79×10^6W), it is recommended that:

1. A permanent stack gas thermometer should be located in the breeching. This provides the operator with a positive indication of stack temperature. An increase in stack temperature is an indication of a decrease in boiler efficiency.

2. Stack gases should be tested weekly for temperature, draft, CO_2, O_2 and smoke, and logged to assure that the boiler is operating at the recommended efficiency. Burner settings, dirty gas, water side conditions, or improper draft should be corrected to maximize efficiency.

3. The fireside of all boilers should be cleaned at least annually.

4. Water treatment programs should be made adequate, to prevent scaling of the waterside. If scaling is present, the boiler waterside should be cleaned. A daily log of water treatment tests should be maintained.

5. Makeup water should be metered to monitor possible leakage or excessive blowdown on steam boilers.

Boilers greater than 30 million Btuh input (8.79×10^6W) should be investigated for possible application of on-line O_2 control and/or economizers to preheat feedwater. Steam boiler pressure should be reduced, if possible, since steam leakage and waste are reduced at lower pressure, and efficiency is increased. It is often possible to increase boiler gross efficiency 2% to 5% by careful operation and maintenance.

Example #1. Boiler Combustion Efficiency Increase. To determine the combustion efficiency increase and related annual savings resulting from adjusting the secondary air damper to reduce excess air, assume that:

1. Preadjustment CO_2 is 7% at 500°F (260°C).
2. Postadjustment CO_2 is 11% at 350°F (176.67°C).
3. Annual heating oil fuel cost is $40,000.

Procedure:

1. Determine the pre-adjustment combustion efficiency and post-adjustment combustion efficiency as follows: Enter the graph in

Figure 16-3 at 7% CO_2 and proceed vertically to intersect the 500°F (260°C) curve; then proceed horizontally to determine efficiency, 74%. Proceed similarly for post-adjustment efficiency, 84%.

2. Determine increase in efficiency:
   (84%—74%)/74% = 13.5%

3. Calculate annual savings:
   Annual savings = 13.5% × $40,000 = $5,400

4. Note that for natural gas the corresponding increase in efficiency is the same, while the savings are somewhat smaller owing to the lower price of the fuel (refer to Figure 16-4).

B. Oil Burner Replacement

Example #2. An old, inefficient burner with 60% efficiency is replaced by a 75% efficient new burner. If the burner consumes 25,000 gallons (94.7501) of oil per year, determine the annual energy saved by this replacement.

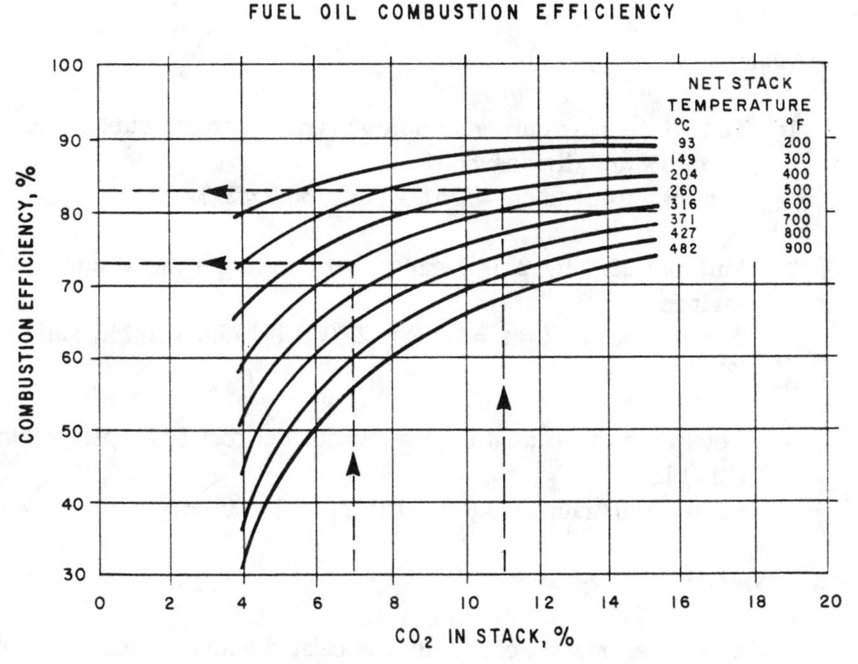

Figure 16-3. Boiler Combustion Efficiency Increase. (Reproduced by permission of Pope, Evans and Robbins, Inc.)

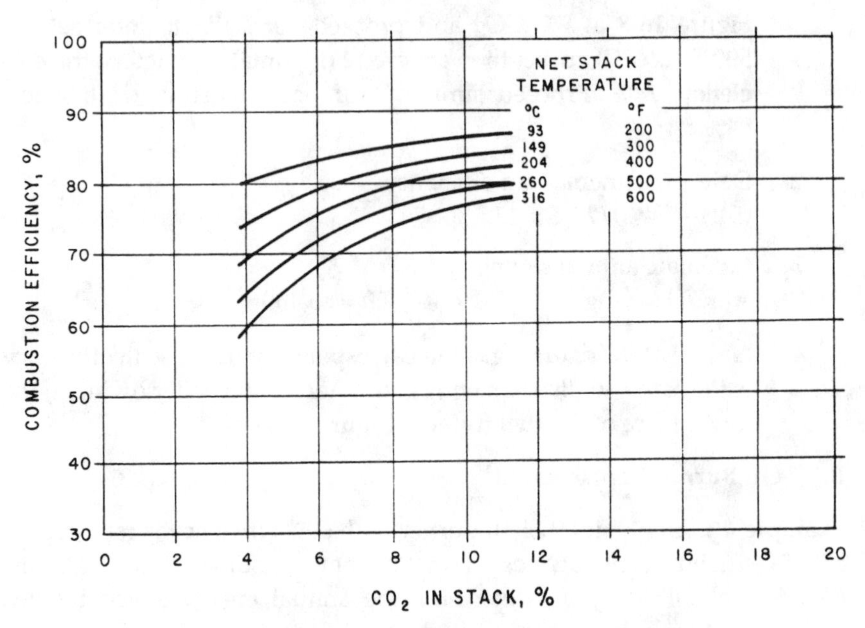

Figure 16-4. Natural Gas Combustion Efficiency.
(Reproduced by permission of Pope, Evans and Robbins, Inc.)

Procedure:

1. To find the new burner's improvement in percent, subtract 60% from 75% and divide by 60%.
   Percent improvement = (75% – 60%)/60% = 25%

2. Multiply 25% by 25,000 gallons to determine the annual fuel savings:
   Annual fuel savings = 25% × 25,000 (94,750) = 6,250 gallons (23,690 L) of oil

3. Determine the annual savings when fuel cost $0.80 per gallon ($0.21 L):
   6,250 × 0.80/gallon (23,690 × 0.211/L) = $5000/yr.

C. Waterside Scale Buildup

Example #3. Determine the energy loss associated with a scale buildup of 1/32 inch (0.79 mm) on the waterside of a boiler. Annual fuel cost is $40,000.

Procedure:

1. Enter the Figure 16-5 graph at 1/32 inch (0.79 mm) and proceed vertically to intersect the "normal scale" curve; then proceed horizontally to determine gross energy loss: 2%.

2. Calculate annual savings by multiplying energy lost by annual fuel cost:
   Annual savings = 2% × $40,000 = $800

D. Fireside Soot Buildup

Example #4. Determine the associated fuel losses resulting from 1/32 inch (0.79 mm) buildup on the fireside of an oil-fired boiler. The normal measured net stack temperature, when the boiler is clean, is 350°F (176.67°C), and CO_2 is 10%. Annual fuel cost is $40,000.

Procedure:

1. Enter the Figure 16-6 graph at 1/32 inch (0.79 mm); proceed vertically to intersect the curve; and then proceed horizontally to determine the increase in flue gas temperature, 80°F (26.67°C).

2. Determine the increase in flue gas temperature by adding the clean temperature, 350°F (176.67°C), to the increase in temperature due to the 1/32 inch (0.79 mm) soot buildup, 80°F (26.67°C). Flue gas temperature = 430°F (211.11°C).

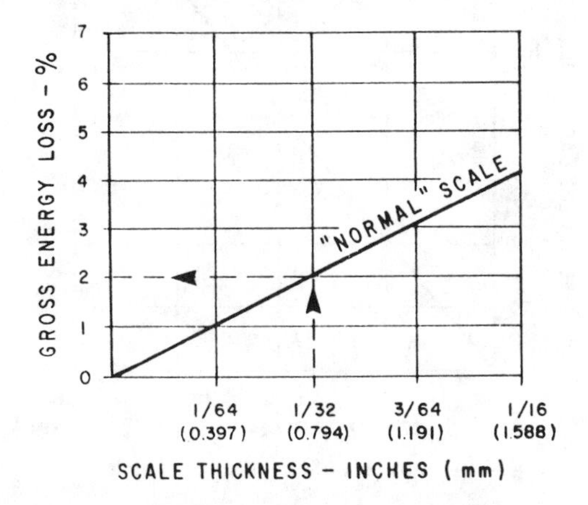

Figure 16-5. Waterside Scale Buildup.
(Reproduced by permission of Pope, Evans and Robbins, Inc.)

3. Determine the combustion efficiency for the dirty and clean boilers by referring to Figure 16-3. Enter the graph at 10% CO_2 and proceed vertically until intersection with the 430°F (211.11°C) curve (dirty boiler); then proceed horizontally to determine the combustion efficiency, 81.5%. Perform the same procedure for the clean boiler except use the 350°F (176.67°C) curve, 83.5%.

4. Determine the percent energy loss in efficiency by subtracting:
   % loss = Clean efficiency – Dirty efficiency
   % loss = 83.5% – 81.5% = 2%

5. Calculate annual savings by operating a clean boiler:
   Annual savings = 2% × $40,000 = $800

E. Heat Loss Due to Domestic Hot Water Leak

Example #5. A domestic hot water system is leaking hot water at the rate of 5 gal/hr (5.25 ml/s). If #6 oil is used as fuel for the hot water system, determine the annual energy savings that will be achieved from stopping this leakage.

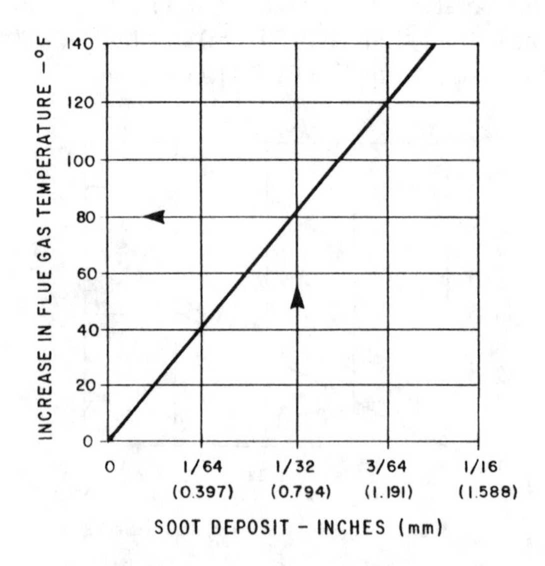

Figure 16-6. Fireside Soot Buildup.
(Reproduced by permission of Pope, Evans and Robbins, Inc.)

Procedure.

1. Enter the Figure 16-7 graph at 5 gal/hr (5.25 ml/s) of hot water and proceed vertically upward to intersect the curve; then follow horizontally to energy lost, 5,000 Btu/hr (1,465 W).

2. Multiply 5,000 Btu/hr (1,465 W) by 365 days and by 24 hours to get annual energy savings:
   Annual energy savings = 5,000 (1,465 W) × 364 × 24 = 44 million Btu (46.42×10^6kJ)

3. Fuel conversion: Divide 44 million Btu (46.42×10^6kJ) by 149,700 (41,766.3 kJ/l) to convert Btu to gallons of oil.

$$\text{Annual Fuel Savings} = \frac{44 \text{ million Btu } (46.42 \times 10^6 \text{kJ})}{149{,}700\ (41{,}766.3 \text{ kJ/l}) \times .75 \text{ (Boiler efficiency)}} = 392 \text{ gallons } (1485.68 \text{ l}) \text{ of \#6 oil}$$

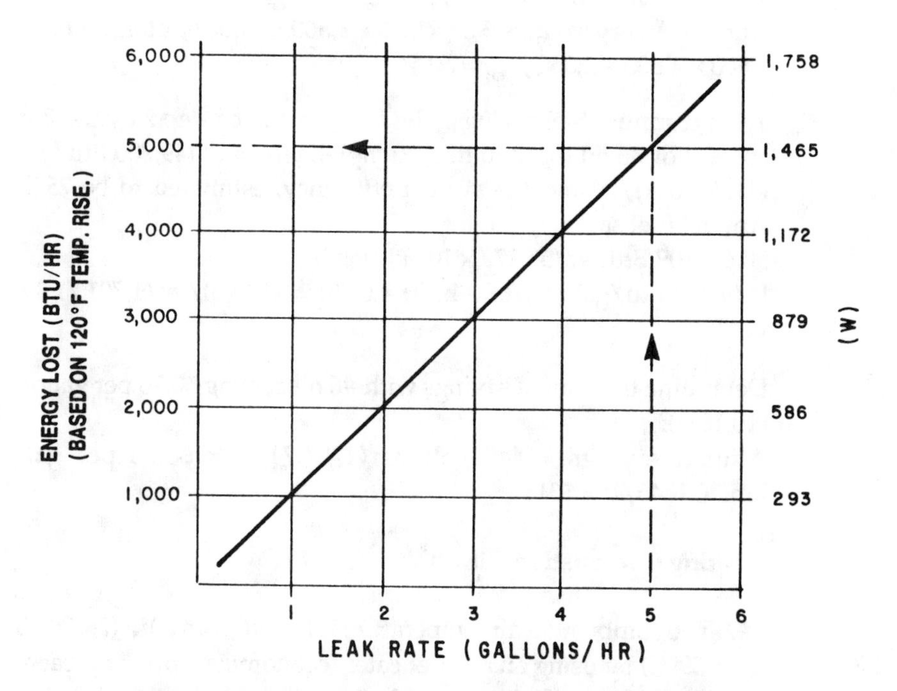

Figure 16-7. Heat Loss Due to Domestic Water Leak. Hot Water Temperature Rise 100°F (38°C); System Efficiency 83%. (Reproduced by permission of Pope, Evans and Robbins, Inc.)

4. Annual savings when #6 oil costs $.80 per gallon ($0.21/l):
Annual savings = 392 × $.80 (1485.68 × 0.21) = $314

F. Energy Losses Due to Malfunctioning Steam Traps

Example #6. Determine the total annual energy saved by repairing a steam trap in a building with an orifice size of 0.20 inch (49.8 Pa) for a IS-psi (10,335 kPa) steam system. The heating fuel used is #6 fuel oil, and steam is generated 30% of the total plant operating time of 4,800 hours.

Procedure:

1. Enter the Figure 16-8 graph at 0.20 steam trap orifice size and proceed horizontally to intersect the 15-psig (103.35 k Pa) curve; then proceed vertically to the steam loss, 35 lb/hr (4.41 g/gs).

2. To find annual energy savings, multiply 35 lb/hr (4.41 g/s) by 30% and the number of plant operating hours, 4,800, and by the steam enthalpy of 1,000 Btu/lb (2330 kJ/kg):
Annual energy savings = 35 × 0.30 × 4,800 × 1,000 (4.41 g/s × 0.3 × 3,600 × 4,800 × 2,330 J/g) = 50.4

3. To find annual fuel savings, divide the result of step 2 by the Btu content of #6 oil (used in this example), which is 149,700 Btu/gal (41,766.3 kJ/l), and the system efficiency, estimated to be 75%.
Annual fuel savings:
50.4 × 10^6 Btu/yr (53.172 × 10^6 kJ/yr)/
[149,700 Btu/gal (41,766.3 kJ/l) × 0.75] = 449 gal/yr (1,701.71 1/ yr)

4. Determine the annual savings with #6 oil costing $0.70 per gallon ($0.18/l):
Annual savings = 449 gallons (1701.71 l) × $0.70 per gallon($0.1845/l) = $314

G. Preheating Combustion Air

Example #7. The combustion air temperature is raised from 80°F (26.67°C) to 300°F (148.89°C) by using an air preheater (economizer) prior to reaching the #2 fuel oil burner. If the annual boiler fuel consumption is 25,000 gallons (94,750 1), determine the achieved savings by raising the combustion air temperature.

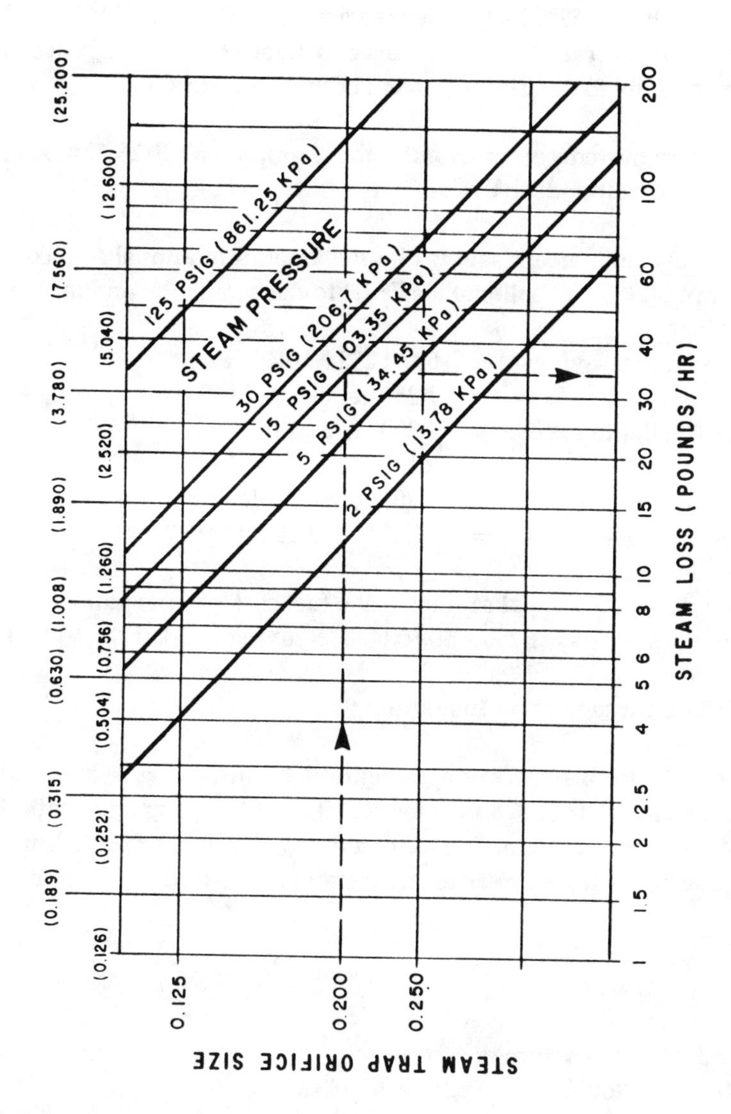

Figure 16-8. Energy Losses Due to Malfunctioning Steam Traps. (Reproduced by permission of Pope, Evans and Robbins, Inc.)

Procedure:

1. Refer to the Figure 16-9 graph and determine the fuel savings for the 300°F (148.89°C) air. Enter the graph at 300°F (148.89°C) and proceed vertically until the curve is intersected; then proceed horizontally to determine percent of fuel saved (5%).

2. The baseline temperature for the above graph is 80°F (26.67°C) air, which corresponds to 0%.

3. Multiply the percent savings by the amount of annual fuel consumption (25,000 gallons) (94,7501) to determine the annual fuel savings:
   Annual fuel savings = 5% × 25,000 (94,7501)
   = 1,250 gallons (4,737.5 l)
   For #6 oil, Btu saved = 1,250 (4,737.5 l)
   × 149,700 (41,766.3 kJ/l)
   ×.75 (System efficiency)
   = 140,344,000 Btu (148,062,920 kJ)

4. Determine the annual savings when fuel cost $.80 per gallon:
   Annual savings = 1,250 × $0.80 (4737.5 savings × $0.211) = $1,000

H. Hot Water Storage Tank Insulation

Example #8. Determine the savings incurred by insulating a bare hot water storage tank, 6 ft (1.828 m) long × 3 ft (0.914 m) diameter with 3 inches (76.2 mm) of insulation. The water temperature is 140°F (60°C), and the average surrounding ambient air temperature is 80°F (26.67°C). Fuel is #2 oil.

Procedure:

1. Find the temperature difference:
   140°F—80°(60°C—26.67°C) = 60°F (33.33°C).

2. Find, from the Figure 16-10 graph, the heat loss for the 3-inch insulated tank and the uninsulated tank by entering the graph at 60°F (33.33°C) and proceeding vertically until the appropriate curves are intersected; then proceed horizontally to determine values.

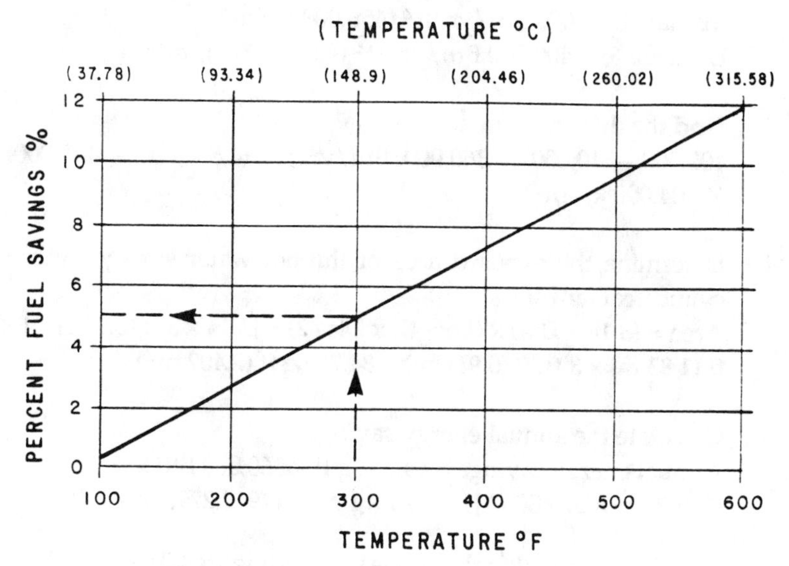

Figure 16-9. Preheating Combustion Air.
(Reproduced by permission of Pope, Evans and Robbins, Inc.)

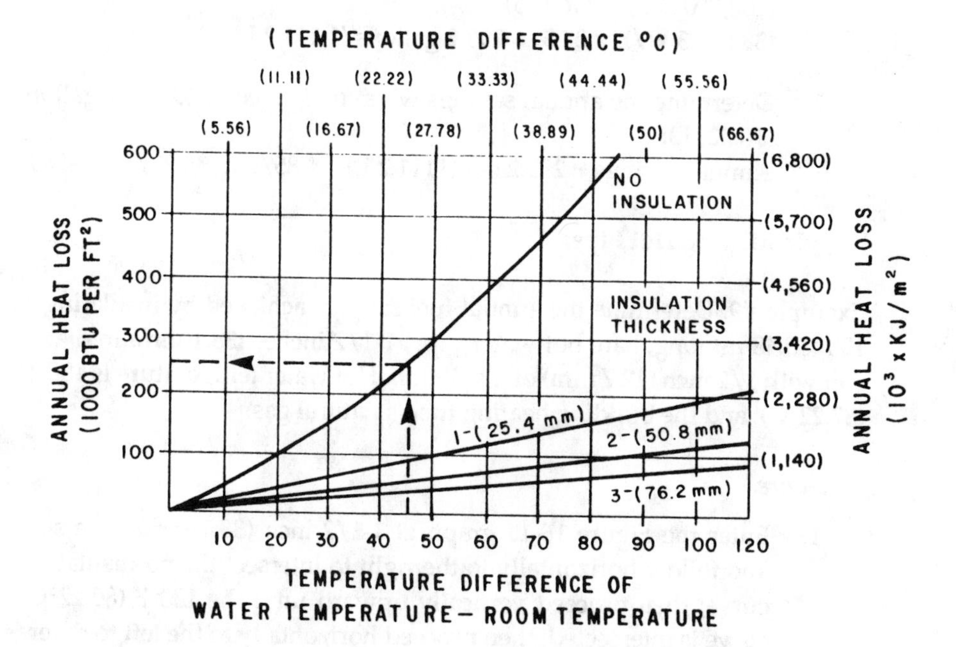

Figure 16-10. Hot Water Storage Tank Insulation.
(Reproduced by permission of Pope, Evans and Robbins, Inc.)

Insulated: 40,000 Btu/yr/ft^2 (456,000 kJ/m^2)
Uninsulated: 400,000 Btu/yr/ft^2 (4,560,000 kJ/m^2)

3. Find the difference in heat loss:
400,000 – 40,000 = 360,000 Btu/yr/ft^2 (4,560,000 – 456,000) = (4,104,000 kJ/m^2)

4. Determine the exposed area of the hot water storage tank. For cylindrical tanks:
Area = (3.14 × Dia) × (Length × Dia/2) = [3.14 × 3 ft (0.914 m)] × [6 ft (1 83 m) × 3 ft/2 (0.914 m) = 84.78 sq ft (2.402 m^2)

5. Calculate the annual energy savings:
Annual energy savings = 84.78 sq ft × 360,000 Btu
(2 402 m^2 × 379 800 kJ) = 30.5 × 10^6 Btu (912,279.6 kJ)

6. Determine annual fuel savings by dividing step 3 by 138,700 Btu/gal (38,697.3 kJ/L) (fuel Conversion number for #2 oil and by the estimated system efficiency of 75%:
Annual fuel savings = 30.5 × 10^6 Btu/yr (32.178 × 10^6 kJ/yr)/
(138,700 Btu/gal × 0.75)
(38,697.3 kJ/L × 0.75) = 293.2 gallons (1,111.2 L)

7. Determine the annual savings when fuel oil costs $.85 per gallon ($0.22/L):
Annual savings = 293.2 gal (1111.2 L) × $.85/gal ($0.22/L) = $249

I. Insulating Hot Pipes

Example #9. Determine the annual fuel savings achieved by insulating a 100 (30.48 m) long, bare hot water pipe, 1-1/2 inches (38.1 mm) in diameter with 1/2 inch (12.7 mm) of insulation. Hot water temperature is 180°F (82.22°C), and the building heating fuel is natural gas.

Procedure.

1. Enter the Figure 16-11 graph at 1-1/2-inch (38.1-mm) pipe size and follow horizontally to the right to intersect the no-insulation curve; then proceed vertically upward until the 180°F (82.22°C) curve is intersected; then proceed horizontally to the left to determine the heat loss for each 10 feet (3.048 m) of pipe (13 × 10^6 Btu/yr) (13.715 × 10^6 kJ/yr).

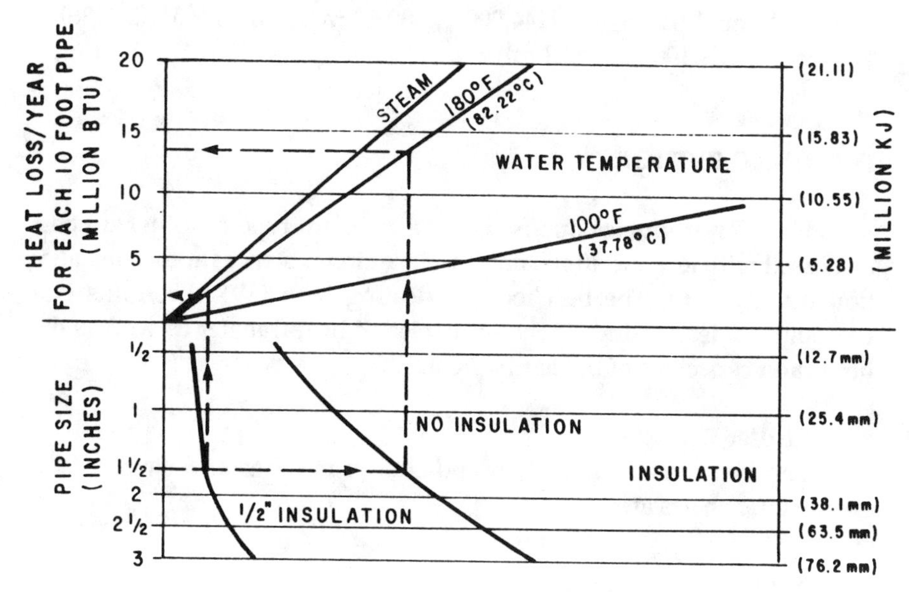

Figure 16-11. Insulating Hot Pipes.
(Reproduced by permission of Pope, Evans and Robbins, Inc.)

2. Enter the graph at 1-1/2-inch (38.1-mm) pipe size and follow to the right to intersect the 1-1/2 inch (12.7-mm) insulation curve; then proceed vertically upward until the 180°F (82.22°C) curve is intersected; then proceed horizontally to the left to determine the heat loss for each 10 feet (3.048 m) of pipe (2.1×10^6 Btu/yr (2.216×10^6 kJ/yr).

3. Determine the annual energy savings by subtracting Step 2 from Step 1 and multiplying by 10, the number of lengths of 10 feet that equal 100 feet:
Annual energy savings = [13 million (13.715×10^6 kJ) – 2.1 million (2.216×10^6 kJ)] × 10
= 10^9 million Btu(114.990×10^6kJ)

4. Determine the annual fuel savings by dividing Step 3 by the fuel conversion for natural gas, 1,030 Btu/cu ft (38.419 kJ/m^3).
Annual fuel savings = (114.99×10^6kJ/38,419kJ/m^3 × 0.75 (System Efficiency)
= 144,900 cu ft (4100.67 m^3) natural gas

5. Determine annual savings when natural gas costs $.35 per therm ($12.36 per 1,000 m^3):

Annual Savings = [144,900 (4,100.67 m^3)/1,000] ×.35 ($12.36) = $51 per 100 feet (2.83 m)

IV. HVAC RETROFITS

HVAC retrofits (as opposed to central plant type retrofits) are best discussed with few specific examples, since circumstances at each installation usually vary. The best tool for dealing with HVAC retrofits is a thorough understanding of the principles of heat transfer as well as the limits and capacities of the hardware used.

A. Chiller Retrofits

The four most commonly found chillers in terms of their part load energy efficiencies are:

1. Steam absorption units
2. Reciprocating chillers
3. Centrifugal chillers
4. Screw-type chillers

Most chillers generally run at less than 100% loading most of the time. If we compare at part loading the amount of energy input versus output (in terms of Btu supplied and removed) the least efficient is the steam absorber, while the most efficient is usually the screw-type chiller. Reciprocating units are generally small in size and somewhat difficult to control in small steps for part load operation. In addition, they present no energy savings if equipped with hot-gas bypass-type capacity controllers.

For new water-cooled units, the trend is to use "double bundle" condensers where one bundle may be used for domestic hot water generation, terminal air, reheat, or other purposes. Older machines may benefit from the addition of a heat-exchanger in the condenser water circuit to generate warm water from the condenser waste heat. One must be careful, however, not to increase the condensing temperatures of existing units, since such an increase often consumes more energy than the amount gained in the subsequent heat exchanger. Because the variations of additions and alterations are almost limitless, only three chiller examples with wide potential application are presented here.

Retrofitting an Older Absorption Chiller.

Older absorption chillers, often in service for 20 years or more, should always be retrofitted to increase their efficiency and cut their steam consumption per ton of refrigeration produced. The retrofit generally consists of one or more of the following types of measures:

1. Improve heat transfer between coolant and refrigerant by utilizing "on-line" cleaning systems that keep the heat exchanger surfaces clean.

2. Provide necessary controls to permit operation with lower condenser as well as higher chilled water temperatures. As these liquids approach each other's temperature, the efficiency of the operation is improved.

3. Improvement of the operation by providing more accurate temperature and load sensing devices, thereby reducing excess energy and waste.

The most commonly applied measures are:

1. **Full range economizer valve.** The valve controls the flow of dilute (weak) solution to the concentrator (generator) by means of a solenoid or three-way by-pass valve. If less weak solution is permitted in the concentrator, less steam is required; hence part load efficiency improves.

2. **Uncontrolled condenser water operation.** Older machines usually need their condenser water temperature regulated to around 86°F (30°C) to 90°F (32.22°C) regardless of the availability of cooler condenser water from the cooling tower. The affinity of the brine solution for water is higher if the temperature is lower. Older machines often crystallized if the condenser water was too "cold" and the concentration high. In uncontrolled condenser water operation [down to 55°F (12.78°C) condenser water], a signal is usually generated by the condenser water temperature. The signal is then sent to the steam supply valve to control its opening, and simultaneously the solution in the absorber is diluted to prevent crystallization. The diluted and colder brine solution still retains its affinity for water; therefore chiller operation is not affected. For each 1°F (0.56°C) temperature drop of the condenser water, 1.5% to 2% savings in steam consumption are possible.

3. **Load anticipation and chilled water modulation.** By sensing outside air or return water temperature or a combination of both, the chilled water temperature is allowed to "float" upward until a set temperature difference, say 5°F (2.78°C) to 10°F (5.56°C) is established between supply and return water temperatures. While such a variation of the leaving chilled water temperature effectively eliminates humidity control for most cooling coils, the benefits of energy savings generally outweigh the negative aspects of close humidity control for units serving noncritical hospital areas. Increasing the leaving chilled water temperature of the chiller 1°F (0.56°C) generally results in a 1.5% to 2% decrease in steam consumption or overall efficiency. An added side benefit for most applications is the additional savings of not having to reheat the overcooled supply air.

4. **On-line brush cleaning of condenser and absorber tubes for better heat transfer.** Overall heat-transfer reduction due to layers of materials deposited on tube-wall surfaces makes all chillers become less and less efficient until the steam rate (lb/ton) (kg/kW) becomes high enough to warrant chemical and mechanical cleaning. Because of fouling, secondary effects such as water velocity changes, feed systems irregularity, algae treatment failure, or microscopic particle deposition may speed up the naturally occurring time-dependent fouling. To overcome fouling and to reduce energy and maintenance costs, a small plastic brush is fitted into the tubes and is made to travel the length of the tubes by the flow of the water. A four-way valve is used to reverse the water flow according to a predetermined schedule. The traveling brushes remove the slowly accumulating scale and dirt and restore the badly fouled tubes' heat transfer to the original "clean" condition.

5. **Control modifications for operational changes.** It was found that depending on actual job conditions the addition of a few control components can make overall operation more reliable. Mention is made of the following items for reference only:

   (a) **Steam flow control for optimum efficiency.** By limiting the open position of the steam valve at transient conditions, steam is saved, since the wide-open steam valve wastes steam.

   (b) **Chiller insufficiency compensation.** This prevents chiller operation if any transient conditions due to mechanical malfunctions exist.

A somewhat shortened example of typical calculations and anticipated results is included below.

Upgrading for Uncontrolled Condenser and Chilled Water. Economic analysis for a 426 nominal-ton (1499.52-kW) machine to be upgraded is based on the following actual conditions found during a field survey:

1. Average load on machine of 60%.
2. Condenser water regulated to be about 85°F (29.44°C) during the cooling season.
3. Usage about 3,600 hr/yr.
4. Steam costs of $7.20/1,000 lb ($15.86/1,000 kg).
5. Usage factor taken from manufacturers catalog.
6. Present steam rate: 19.8 lb/ton (255 kg/kW). (Actual, measured at full load).

Estimated present use:

19.8 lbs/ton × 3,600 hr/yr × 0.6 × 426 tons/hr = 18.2 × 10^6 lb/yr
(2.55 kg/kW × 3,600 × 0.6 × 1499.52 kW/hr) = 8.26 × 10^6 kg/yr
Cost: (18.2 × 10^6 lb/yr) × ($7.2 × 10 3/1b) = $131,040/yr
(8.26 × 10^6 kg/yr × $15.86 × 10^{-3}/kg = $131,004/yr)

Estimated future use:
Average condenser water temp. (future uncontrolled) estimated:
72°F down 13°F from 85°F to 72°F
22.22°C down 22.2°C from 29.44°C to 22.2°C

Average chilled water temp. (in future allowed to rise if load is reduced):
46°F, a rise of 4°F from 42°F to 46°F
7.78°C, a rise of 2.2°C from 5.56°C to 7.78°C

Total combined:
4°F + 13°F = 17°F at 1.5%of (2.2°C) + 7.22°C = 9.42°C at 1.5% of

Savings are:
25.5% reduction of steam consumption (estimated)

Usage factor: 0.6 × 0.745 = 0.447
Reduced consumption:
19.8 × 3,600 × 0.447 × 426 = 13.57 × 10^6 lb/yr
2.55 × 3,600 × 0.447 × 1,499.52 = 6.16 × 10^6 kg/yr

Savings:
(1 8.2—1 3 57) = 4.63×10^6 lb/yr, or at \$7.20/1,000 lb
8.26×10^6—6.16×106 = 2.10×10^6 lb/yr, or at \$15.86/1,000 kg

The cost savings is \$33,336/yr.
On-line Brush Cleaning Savings:
Estimated average fouling (from field tests): $f = 0.002$.

Increase in steam consumption per ton due to fouling as per manufacturer's charts is approximately 22%.

Rated consumption: 19.8 lb/ton (2.55 kg/kW).

Increase in steam consumption per ton due to normal fouling is 3%.

Present increase due to excessive fouling: 22% – 3%= 19%.

Consumption at present fouling:
19.8 lb/ton × 1.22 = 24.156 lb/ton
2.55 kg/kW × 1.22 = 3.11 kg/kW

Annual present energy cost: \$131,040/yr.

Estimated annual savings due to brush cleaning:

Energy cost/(1 + % increase in lb/ton)
= 131,040—[131,040/(1 + 0.19)] = \$20,922/yr

At steam costs of \$7.20/1,000 lb (\$15.86/1,000 kg), this represents 2.9×10^6 lb (1.32×10^6 kg) of steam saved per year.

While the above measures are considered individually, the combined results of both would cut steam consumption drastically. It is obvious that if a machine is retrofitted for uncontrolled condenser, chilled water, and other modifications, the brush cleaning will not result in additional savings of 2.9×10^6 lb/yr (1.32×10^6 kg/yr). Savings will be somewhat less, since, owing to other measures, less steam is used to begin with. In such situations, one should start with the already reduced steam consideration a smaller than calculated further drop of steam consumption. It is also important to remember that many old machines will seldom have their steam rates reduced 20% to 25% or consume less than the original "new" rate regardless of the retrofit. This is mainly due to internal design as well as external conditions.

Heat Exchanger Installation. Our third example is based on a hermetic reciprocating chiller operating at present with a COP of 2.99. The unit provides 69 tons (242.88 kW) of cooling (nominal), with a power input of 80.9 kW [chilled water at 42°F (5.56°C)] and recorded condenser water temperatures of 85°F (29.44°C) water entering the 102.3°F (39.05°C) water exiting from the machine. Water temperatures are regulated at the tower, since this older machine cannot safely operate with much colder condenser water. Saturated discharge temperature is set at 120°F (48.89°C), and 127.6 gpm (8.051 l/s) of condenser water is circulated.

Assume a heat-exchanger (double-walled for safety) is installed in the condenser water circuit to provide tempering for the incoming domestic hot water. The heat-exchanger selected would cool 127.6 gpm (8.051 l/s) of water from 102.3°F (39.06°C) to 87.3°F (30.72°C), a 15°F (8.34°C) drop, by preheating the incoming 40°F (4.44°C) domestic hot water to 70°F (21.11°C), a rise of 30°F (16.67°C). Although 63.8 gpm (4.025 l/s) of domestic water could theoretically be heated, because of inherent inefficiencies, only 60 gpm (3.786 l/s) is considered. Also note that the approach temperature of 87.3°F (30.72°C)— 70°F (21.11°C), or 17.3°F (9.61°C), is rather high to reduce the size and first cost of the heat exchanger. It is very likely that water warmer than 70°F (21.11°C) would be available to the domestic hot water heater; this consideration makes these calculations conservative:

The energy gained is:
60 gpm × 8.33 lb/gal × 60 min/hr × 30°F = 899,640 Btu/hr
0.2274 m^3pm × 999.6 kg/m^3 × 16.67°C = 263,595.0 W

For 3,600 full-load hr/yr, this is equivalent to 3,238 × 10^6 Btu (3,416.09 × 10^6 kJ). Or, if oil is used to generate the steam used in the existing domestic hot water heater, the oil saved is:

(3,238 × 10^6 Btu/yr)/(140,000 Btu/gal × 0.8 × 0.9) = 32,123 gal/yr
(3.416 × 10^9 kJ/yr)/(39,060 kJ/I × 0.8 × 0.9) = (121,466 I/yr)

If boiler efficiency is 0.8 and estimated hot water generation efficiency is 0.9, at a rate of $0.75/gal ($0.198/l) the cost savings are $24,092 ($24,050) per year. It is important to note that the above examples are only approximations, generally on the conservative side. For example, no credit has been given to savings brought about by reduced cooling tower operation. The reader is therefore urged to evaluate each situation independently for more accurate results.

B. Energy Transfer Retrofits

Energy transfer retrofits generally consist of devices that transfer energy from one stream of fluid such as air (exhaust) to another, usually incoming outside air. They are best used when the incoming outside air stream is heated and/or cooled without the addition of return air. Such "once-thru" systems are often found in hospitals, serving operating rooms or other critical areas where the addition of return air may result in cross contamination. They may also be applicable for kitchens, where the exhaust air steam is usually too hot, fouled, or both, for reuse. The best application is for systems that use both heating and cooling. In such systems, heat is transferred from the exhaust stream to preheat the incoming outside air in the winter, while in the summer, the heat is transferred from the hot outside air to the cooler exhaust air. Depending on the relative locations of the intake and exhaust ducts there are generally three types of heat transfer systems available for a simple retrofit:

1. Run-around coils
2. Heat pipes
3. Heat wheels

Run-around coils consist of two air-to-water heat exchangers placed in the appropriate air streams. The coils are interconnected by a pipe loop incorporating a pump and a three-way valve for freeze protection. A glycol/water mixture is usually circulated through both coils to transfer energy from one to another. During winter, especially in places where the climate is severe, 30% to 50% of the heat can be easily recovered. The run-around coils can only transfer sensible heat; this limits their usefulness in the summer. The major advantage is that the air streams need not be near each other. It is generally true that long distance between the air streams reduce efficiency, because of the heat loss in transfer by the piping system, but large quantities of air at large temperature gradients make even distant coil systems cost effective.

If the two air streams (exhaust and supply) are close enough, a minor alteration of ductwork may make it possible to use either heat pipes or a heat wheel. Both devices require parallel and opposite flows of air at different temperatures. A heat-pipe device is usually a coil consisting of Freon elements. The elements are concentric tubes containing a liquid (usually refrigerant) that is allowed to evaporate on the warmer side and migrate to the cooler half of the coil. At the cool side, the refrigerant condenses and flows by gravity back to the warmer side of the coil. Summer/winter switchover is relatively simple; only the tilt of the coil needs to be changed, since the relative temperatures of the air streams are reversed. The heat-pipe is inserted in the ductwork, one half in the intake the other in the exhaust. The edges are sealed, and the system transfers up to 70% to 80% of the energy. Energy transfer is still sensible only, and the air streams must be near each other. Both the run-around coil and a properly sealed heat-pipe prevent cross contamination, since no part of the system moves from the contaminated air stream into the fresh air stream.

The heat wheel uses a wheel-type heat transfer unit inserted into the parallel and near air streams, in a configuration somewhat similar to the heat pipe. The wheel, however, rotates, and cross contamination may become a problem if organisms attach themselves to the heat transfer material. The great advantage of the heat wheel is that latent heat transfer is possible, thereby increasing the overall summer/winter efficiency, especially in areas where the summers are long and humid. While selection of the optimum energy-transfer method should be left to the qualified energy auditor, bear in mind that a once-thru system with large temperature gradient, running 24 hours a day, will result in the shortest payback. Most manufacturers make computer-based economical analysis available, eliminating the necessity for complicated and time-consuming manual calculations.

C. Control and Regulatory Retrofits.

Many control and regulatory retrofits may be identified in the course of hospital energy audits. While there are too many to mention, some stand out as applicable to most facilities. Brief descriptions of the three most important ones are presented below in descending order of usage:

Thermostatic Control Valves on Hot Water or Steam Heating Elements: Lack of temperature control is a frequent cause of overheating and excessive energy consumption in low pressure steam and hot water heating systems. Installation of thermostatic control valves allows room-by-room control of temperature, matching the thermal energy consumption to the heat losses in the space. These valves consist of self-contained, nonelectric thermostatic operators and control valve bodies suitable for modulating steam or hot water flow to radiators, convectors, fan coils, etc. Set temperatures may be locked to minimize tampering. Thermostatic temperature dial and sensor may be mounted remotely, easing installation and operation.

Energy savings is usually a function of existing temperature and controlled future temperature. Any ongoing installation of new thermopane windows or added insulation will reduce heat loss in occupied spaces, increasing the tendency toward overheating. Installation of these valves will compensate for this tendency and provide maximum savings. Savings of 10% to 15% are typical in complete hospital installations where overheating due to insufficient control is a problem.

Cycle of Shutdown of Heating, Ventilation and Air-Conditioning Equipment: Many areas in the hospital are either unoccupied, used only in emergencies, or lightly staffed during the off-hours of 7 P.M. to 7 A.M. Significant energy and dollar savings can accrue if heating, ventilating, and air conditioning (HVAC) equipment serving these areas is duty-cycled or shut down during those periods. Cycling and shutdown should be provided, however, for off-hour or emergency operation. Most HVAC equipment is designed to handle peak loads. Since such loads are infrequent, some equipment may be cycled during the day. Overdesign or equipment functional changes may also allow equipment to be cycled within code or other requirements. Two methods are available for such control:

1. Local electromechanical timers can be installed on each piece of equipment. These timers typically have a seven-day cycle and can turn loads on and off several times a day. They are relatively inexpensive. However, they do not have the versatility to cycle equipment frequently and do not perform predictive demand control.

2. A central microprocessor based load and demand controller can be installed to turn equipment on and off. These units are programmable; they provide versatility in cycling, shutdown and demand limiting. This type of unit is connected to each load with low voltage wiring and can monitor, control and override the operation and shutdown of equipment from a central location. A significant improvement in operation, resulting in cost reductions as well, is the central microprocessor generating high frequency signals that utilize the hospital's existing wiring for transmission. The units to be controlled are equipped with relays only that are activated by the signal generated. Since no additional wiring is required, the installed first cost is reduced.

Installation of Enthalpy-Economizer Controls. Many areas of a hospital are ventilated and heated or cooled using a mixture of outside air and recycled return air. When the outside air conditions are similar to the conditions required for delivery from the air-handling unit, it is desirable to have outside air introduced into the building. If the outside air conditions do not closely match the inside requirements, it is desirable to introduce the minimum fresh air allowed by applicable codes. An enthalpy economizer cycle allows outside air and return (inside) air to be mixed in the proper proportions so the least amount of energy need be expended to heat or cool the supply air to the required condition. This controller automatically controls the dampers in the air ducts (based on the enthalpy wet bulb temperature) of the air streams.

V. LIGHTING RETROFITS

The present trend in lighting ECMs is to reduce the overall general illumination and provide levels of light consistent with requirements in task areas; this is known as task lighting. The U.S. Department of Energy had recommended that 50 foot-candles (538 lumen/m^2) be maintained in office or work space task areas, 30 foot-candles (323.0 lumen/m^2) in office non-task areas, and 10 foot-candles (108.0 lumen/m^2) in other nonwork areas such as storage areas, stairs, and corridors. Special work areas are to be treated according to their specific requirements. Where necessary, ceiling-mounted fixtures should be supplemented with desk lamps in office spaces or with portable lamps in assembly spaces.

Each area should be separately analyzed for task lighting. Two rooms identical in size, containing the same number of lighting fixtures and the same number of desks, may have different recommendations. The furniture arrangement must be considered. Ideally, desks should be located together in a task area; file cabinets, tables, book shelves, etc., should be grouped in non-task areas.

Foot-candle readings should be taken at task or work areas at the surrounding non-task areas and in open unallocated spaces. If there are more than 65 foot-candles (700 lumen/m^2) or 70 foot-candles (753 lumen/m^2) in the task areas, or more than 40 foot-candles (431 lumen/m^2) or 50 foot-candles (538 lumen/m^2) in the nontask areas, lamps should be removed and the foot-candle (lumen/m^2) levels lowered to approximately 50 and 30 (538 and 323 lumen/m^2), respectively.

ECMs can be established for lighting by two basic means: (1) reduction in wattage per given area, or (2) reduction in the hours of usage. Wattage reduction can be achieved by "delamping" of fixtures; (1) replacing fluorescent lamps with lower-wattage energy saving lamps; and (2) replacing incandescent lamps with lower-wattage lamps or with self-ballasted screw-in fluorescent lamps where feasible.

Reduction in the hours of usage can be achieved by: (1) manually switching off lights when they are not needed, (2) using manual timers to control lights in small offices, and (3) using photocells to switch lights off when sufficient daylight is available for requirements of the area involved. Noncritical areas should have their levels of illumination reduced in accordance with the DOE recommendations. Of course in a hospital, many areas require exacting visual acuity. The lighting in these areas should be treated according to the specific requirements applicable.

The following example demonstrates one of the most common energy retrofit methods. It also indicates that most lighting retrofits offer short term paybacks.

Example #10. Determine the amount of energy saved by replacing 3,000 watts of incandescent lighting with fluorescent lighting to achieve an equivalent lighting level. The lamps are lit 12 hours per day.

Procedure:

1. Enter the Figure 16-12 graph at 3,000 incandescent lamp wattage and proceed horizontally to the right to intersect the curve; then proceed vertically downward to fluorescent lamp wattage (600 W).

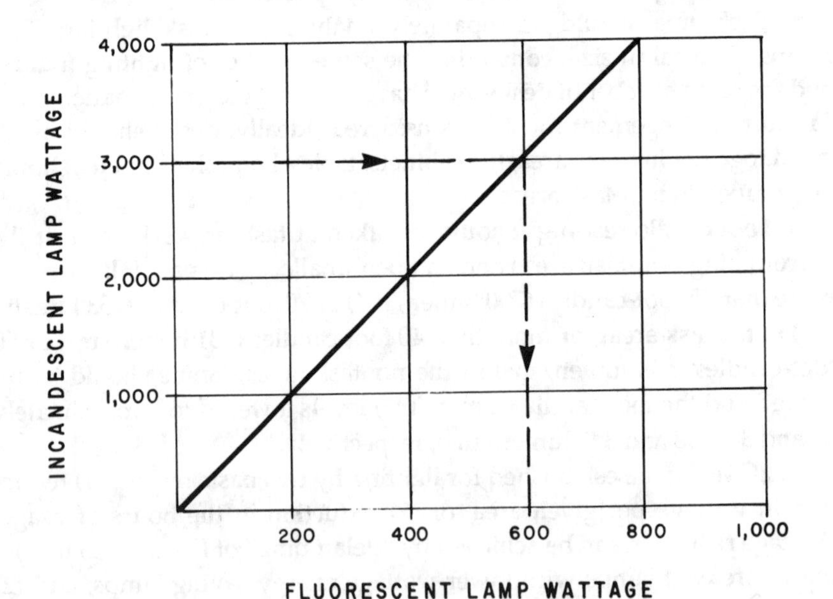

Figure 16-12. Replace Incandescent with Fluorescent Lighting. (Reproduced by permission of Pope, Evans and Robbins, Inc.)

2. Compute the difference of wattage for the two lamps of equivalent light level:
   Incandescent lamp = 3,000 W
   Fluorescent lamp = 600 W
   Difference = 2,400 W
   = 2.4 kW

3. Calculate the annual energy savings by multiplying the difference found above by the number of hours illuminated (12) and days per year.
   Annual energy savings = 2.4 kW × 12 × 365 = 10,512 kWh

4. Determine the annual $ savings when electricity costs $.08/kWh:
   Annual savings = 10,512 kWh × $0.08/kWh = $841

Chapter 17

Schoolhouse Energy Efficiency Demonstration: An Energy Audit Program

Donna L. Rybiski
and
Milton Meckler, P.E.

I. INTRODUCTION

Schools present special needs and special opportunities. Industrial firms, in contrast to schools, have the economic incentive, the capital, and the engineering talent needed for energy management. Although a number of schools employ talented engineers and maintenance directors, few have sufficient staff to undertake new programs. Schools have been unable to pass on to taxpayers the higher energy costs of retrofitting expenditures. The alternative, a reduction in services, is generally considered unacceptable, although higher energy costs have caused a cut in some education programs.

The special opportunity refers to the school's unique ability to influence individuals and other institutions in the community. An earlier study found that schools can reduce energy consumption by at least 25% without major capital expenditures, and by 50% with capital expenditures. Our program was designed to prove the earlier study through demonstration school energy audits, and to communicate the findings to professional educators and patrons in order to encourage support for energy management in schools. Although energy for education in grades kindergarten through high school costs billions of dollars per year, a significant portion of that amount can be saved. Schools with effective energy management programs can influence actions by others throughout the commercial-residential sector, multiplying the benefit of a school program.

The Schoolhouse Energy Efficiency Demonstration (SEED) was developed by Tenneco Inc. and its success can be traced to participation by nine national education groups which agreed to encourage and recommend the SEED program to their members: American Association of School Administrators, National Congress of Parents and Teachers, Association of School Business Officials, National School Boards Association, Education Commission of the States, Council for American Private Education, National Association of Elementary School Principals, National Association of Secondary School Principals, and National Education Association.

Programs were conducted in the public school districts of East Hartford, Connecticut; New Castle County, Delaware; DeKalb County, Georgia; Oak Lawn, Illinois; Lexington, Massachusetts; Livonia, Michigan; Manchester, New Hampshire; Lancaster, New York; Greenhills Forest Park, Ohio; West Irondequoit, New York; Haverford Township, Pennsylvania; Bethel Park, Pennsylvania; Pawtucket, Rhode Island; Newport News, Virginia; Garland, Texas; Roanoke County, Virginia; Houston, Texas; Racine, Wisconsin; as well as a school at Ft. Belvoir, Virginia, operated by Fairfax County, and the Baltimore Friends School.

The SEED program had two components: the technical activity, and the public education or motivational effort. First, the technical activity: We conducted comprehensive energy audits at 20 demonstration schools within a year. Each audit consisted of a fact-finding visit to the school, a pre-audit analysis of building plans and fuel records, a two-day thorough examination of the facility, and the writing of a complete energy study. All phases of the audits were conducted by a team headed by Tenneco's energy conservation consultant, Roger Rasbach of The Woodlands, Texas, an architectural designer. The audits were conducted under the direction of two professional engineers, Dr. Calvin M. Wolff of Houston and Milton Meckler, president of The Meckler Group, Encino, California, working in conjunction with Roger Rasbach Associates, The Woodlands, Texas.

On the second day of the audit, the professional engineer conducted an informal workshop for the host school district's maintenance staff, and others engaged in maintenance and energy management from nearby school districts and state and municipal governments.

Schools selected for the program were required to be representative of the region, so that recommendations could be transferred to the maximum extent. Because nearly 60% of schools in use today were built between 1955 and 1965, most demonstration schools were from that period, and had extensive areas of glass. Most were one- or two-story, with slab-on-grade foundations and uninsulated brick and block walls. The school in Oak Lawn, Illinois, was the only one with wall insulation. Roofs were usually flat with no insulation or limited rigid insulation combined with tar and gravel surfaces. Most buildings were rectangular with double-loaded corridors. One school was three stories, and one was a campus with five classroom buildings. Fifteen were elementary schools, three were junior high schools, one was a high school, and one, the private school, had kindergarten through high school. About half used oil and half used natural gas as the primary fuel, and one had supplemental electric baseboard heating. Five were air-conditioned. Most used either stream or hydronic heating systems, with unit ventilators or natural convection.

All school superintendents involved were aware of the need for energy efficiency. Some had advanced energy management programs, and others were in the early stages of program development. Resources of the districts varied considerably, although all had access to some assistance from state agencies. In almost every district, the key administrators had attended energy workshops. They were acquainted with the issue, and the well-publicized innovative new schools and expensive capital equipment for retrofitting. The need was for practical advice for low-cost, quick-fix solutions with the greatest return. We emphasized building modifications and changes in operating procedures that can be accomplished by the maintenance staff, and financed from the operating budget. Most recommendations have a payback period of two years or less. Major retrofitting (e.g., boilers and roofs) generally was advised only at the time of major repairs or replacement. Schools were advised to specify high-efficiency products when replacing fans, motors, pumps, kitchen appliances, and other energy consumers. Also, a switch to the new high-efficiency fluorescent lamps was recommended on a replacement basis.

The problems encountered most frequently, and the indicated solutions, meet the superintendents' request for guidance on quick-fix actions that provide immediate payback opportunities. For example, massive air movements caused excessive fuel consumption in 90% of the schools in the study. Such movement results from equipment that takes in too much cold fresh air and exhausts too much heated air, infiltration through cracks around doors and windows, open classroom doors and windows, and transmission through the large number of windows. Next, night and weekend temperatures in some cases were little different from daytime temperatures, and in most cases were too high. Fortunately, there are low-cost solutions to these problems, with the exception of heat transmission through windows. The primary finding was that the average school can reduce its energy consumption by between 35% and 40% without major capital expenditures. Since completion of our audits, American Association of School Administrators (AASA) has published a study showing that the nation's schools have reduced energy consumption by nearly 30% and Btus consumed per square foot of school space dropped from 148,113 to 104,445.

This leads to an obvious question: Have schools already achieved the savings we are predicting, or can schools that have cut consumption by nearly one-third now cut consumption another one-third? Although a reduction in energy consumption of more than one-half may sound overly optimistic or unrealistic at first, the potential seems to exist for many schools. The AASA report concerns progress through base year for our study. Also, for the schools in our study with the best existing energy management programs, opportunities to reduce energy consumption in the range of 20% to 40% were recommended.

There are obvious reasons for the exceptional opportunity now facing schools in energy management. Schools constructed in the 1950s and 1960s had to meet the test of minimum first cost, with little consideration for operating costs. Administrators, faced with overcrowded classrooms, were under pressure to build schools as quickly and as cheaply as possible. Energy costs, in real terms, were constant or even declining during the school building boom that responded to the postwar baby boom. In recent years, the well-publicized financial difficulties of schools have created new problems. Schools have been forced to stretch the life of equipment, defer major repairs, and even defer regular maintenance.

Now that the opportunities in energy management have been identified, the task is to encourage support from school patrons. To do so, it is necessary to show energy management as an education issue. Again, AASA data can be used to make the point. The typical U.S. school district, according to AASA, allocates 85% of its budget for fixed costs, primarily compensation. The other 15%, called discretionary, goes for educational materials such as books and other operating costs such as energy. In the last five years, the portion of the discretionary budget that goes to pay energy costs has doubled, from 12.5% to 25%. Because total budgets have not increased proportionately, rising energy costs have replaced classroom materials.

In the SEED program, we have attempted to make this point to persons interested in quality education. Each of our audits received excellent newspaper and television news coverage. Articles, films, and presentations have been used to convey this message.

II. INITIATING THE SURVEY

For the study conducted by AASA, a systematic random sample was drawn from its membership list stratified by district size. The ratio 1:4 for districts over 25,000 enrollment was used and 1:7 for districts under that figure. A total sample of 2,127 was drawn. The sample was matched to the universe of school systems (CIC list) and found to be generally representative of district sizes (refer to Table 17-1.) The sample was also matched to the number of districts in the ten federal energy regions (refer to Table 17-2.)

The survey instrument was developed in consultation with the Department of Energy and with the Educational Research Service which compiled and analyzed the data. The instrument was mailed, and a second mailing was sent out subsequently. The response rate exceeded 50% at deadline. Because of the complexity of the data requested, all responses were cleared by hand, and many verifying phone calls were made. Of the responses received, 629 were deemed usable. The usable response rate was thus 29.57%. While this was statistically sufficient to satisfy the conditions of the study, it is noted that responses by specific cells (i.e., certain fuel types in certain regions) are so low that caution is warranted in interpreting the findings. In certain instances the *n* was too low to treat the data at all. Where this is a concern, *n* is provided in the tables to assist the reader in assessing the data.

An analysis of the usable responses indicated that Region 5 (IL, IN, MI, MN, OH, and WI) was overrepresented and districts under 300 in size were underrepresented (also shown in Table 17-1). The data were not adjusted for climate, but it is noted that the number of degree-days for the study's time frame, school year or calendar year, was significantly higher than for the year upon which comparisons are based.

In other words, the reductions in consumption noted were made in a year when climatic demands on energy were much greater. Indeed, the

Table 17-1. Universe/Sample/Respondent Breakdowns by Enrollment Groups.

	CIC UNIVERSE		SAMPLE			RESPONDENTS	
ENROLLMENT GROUP			SAMPLING RATIO	#	%	#	%
25,000 or more	413	2.8%	1:4	103	4.8%	33	5.2%
10,000 to 24,999	633	4.3	1:7	90	4.2	32	5.1
5,000 to 9,999	1,220	8.4	1:7	174	8.2	58	9.2
3,000 to 4,999	1,491	10.2	1:7	213	10.0	65	10.3
1,000 to 2,999	4,099	28.1	1:7	586	27.6	191	30.4
600 to 999	1,796	12.3	1:7	257	12.1	98	15.6
300 to 599	2,139	14.7	1:7	306	14.4	75	11.9
299 or less	2,784	19.1	1:7	398	18.7	59	9.4
Total	14,575	100.0		2,127	100.0	629*	97.1**

*Response rate was therefore 29.6%.

**No enrollment data provided by 2.9% of respondents.

Table 17-2. Number/Percent of Local Public School Districts in the United States by DOE Energy Region.*

ENROLLMENT RANGE	U.S. TOTAL	DEPARTMENT OF ENERGY: ENERGY REGIONS									
		1	2	3	4	5	6	7	8	9	10
25,000 or more	187	3	8	20	47	26	29	8	7	34	5
10,000 to 24,999	530	37	49	61	78	91	59	19	22	92	22
5,000 to 9,999	1,104	87	135	133	210	221	152	46	18	114	47
2,500 to 4,999	2,067	165	254	257	343	548	171	74	59	134	71
1,000 to 2,999	3,463	253	405	248	331	1,130	368	290	99	206	133
600 to 999	1,864	121	169	18	52	538	312	320	97	140	97
300 to 599	2,323	157	161	7	33	465	507	473	236	158	126
Fewer than 300	4,296	367	141	1	5	373	709	1,189	822	393	296
Total in region	15,834	1,190	1,322	745	1,099	3,392	307	2,419	1,360	1,271	797
Percent of total U.S.	100.0	7.5	8.3	4.7	6.9	21.4	14.1	15.3	8.5	8.0	5.0
Respondents%		5.1	9.2	7.5	6.0	31.3	11.1	11.4	5.9	5.9	6.5

*Source of data: Table 4 (Number of operating local public school systems, by size of school system and state: United States, p. xix, *Education Directory Public School Systems*, by Jeffrey W. Williams and Sallie L. Warf, National Center for Educational Statistics, 1978.

weather bureau indicates it was one of two of the worst years in the last 50. The effects of school closings in the winter of base year due to fuel curtailment were also considered; but since the total number of pupil days lost was less than 1% of the operational demand, it was disregarded in the findings presented. While the number and percentage of electrically cooled schools are given, the schools are not counted in the total figures, as most of these buildings are also heated and thus would be counted twice.

III. FINDINGS

Table 17-3 indicated the median Btu per square foot by total and by the earlier findings of Federal Energy Administration (FEA). It should be noted that while AASA's sample was randomly drawn, FEA's was a fortuitous or problematic sample, thus limiting the opportunity to make inferences. Nevertheless, the FEA national survey provides the only other available data on school energy consumption.

Table 17-3. Median Consumption All Fuels (Btu/sq-ft).

	Prior 2 Years	Base Year	% Changes Prior 2 Years To Base Year
Total	161,312	104,445	35.25
Federal energy region			
1	176,710	114,999	34.92
2	185,999	109,687	41.03
3	166,069	111,241	33.01
4	112,013	75,604	32.50
5	184,948	110,488	40.26
6	132,603	77,907	41.25
7	163,245	117,766	27.86
8	182,640	120,103	34.24
9	131,342	66,523	49.35
10	162,761	85,978	47.17

*Federal Energy Administration's School Fuel Impact Survey.

Consumption figures by region logically follow the respective severity of the climate. Figure 17-1 indicates the states comprising each federal energy region. However, the difference in energy reduction by region cannot be explained by known factors (i.e., cost of fuel, type and availability of fuel, or energy conservation programs at the state or regional level).

The study did not encompass the manner in which the reductions were achieved or the amount of money invested in energy-conserving measures. Knowledge of school energy operations outside the study suggests that savings can be primarily attributed to changes in operations, more energy-conscious maintenance, and implementation of low-cost retrofits. Further consideration of the median Btus per square foot suggests that consumption is also a function of district size. Table 17-4 shows that districts over 5,000 consume 11.6% less than those under 5,000, a figure that suggests the benefits of on-staff expertise. The matter warrants closer scrutiny. If these findings are corroborated, the use of educational service agency resources to assist smaller districts could present a very cost-effective vehicle for training and technical assistance.

Table 17-4. Median Btu/sq ft by District Enrollment.

	DISTRICT ENROLLMENT	
	Less than 4999	5000 or more
n	488	123
Median Btu/sq ft	107,294	94,826

Fuel Types

One of the interesting facets of the findings was an analysis of types of fuel used by the schools, nationally and by regions of the country. No figures were available prior to the AASA study. Natural gas consumption on a national basis was surprisingly higher than anticipated at 55%. Table 17-5 depicts the fuel sources for the reporting buildings and the percentage they represent on a national basis.

The variations in fuel source by region were far greater than expected. Table 17-6, which shows fuel sources by region, reveals differences of considerable importance to decisions relative to energy in the schools. For example, the difference between Region 3's 31% and Region

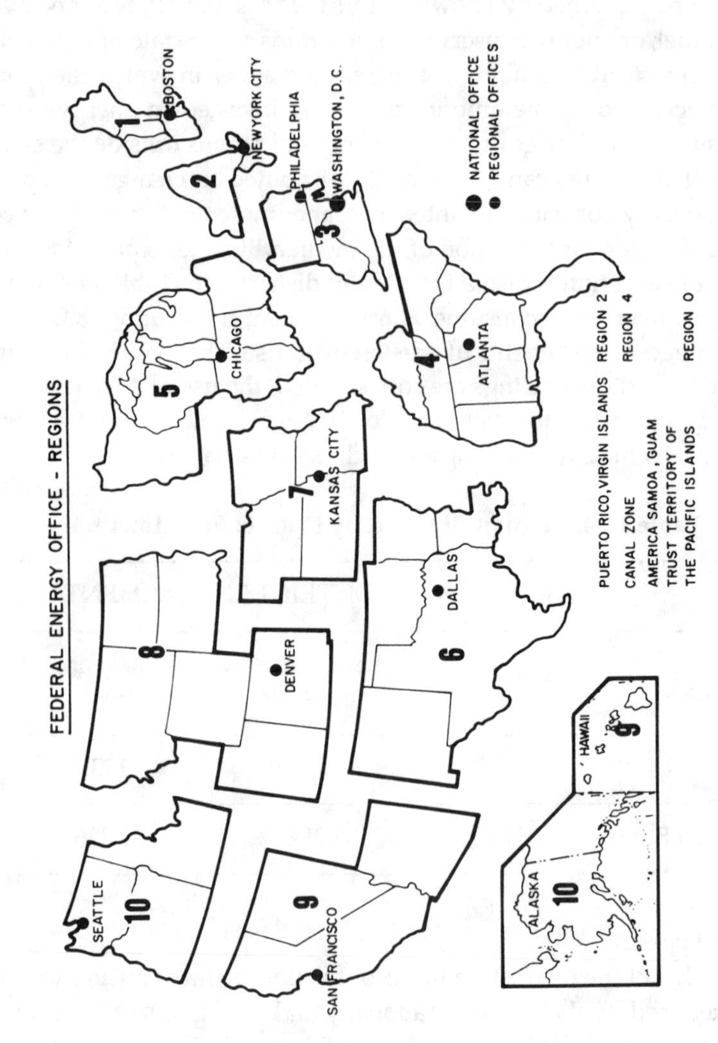

Figure 17-1. Federal Energy Office—Regions.

6's 3% in oil consumption is significant to those wishing to reduce oil imports. Any analysis of the economic impact of gas deregulation on the schools should consider that the impact will be much greater in Regions 4, 5, 6 and 9, and that relatively little effect would be seen in Regions 1, 2, and 10. Coal strikes would hit Regions 3, 4, and 5 hardest.

Table 17-5. Fuel Sources, All Buildings.

FUEL TYPE	NO. OF BLDGS. REPORTED AND PERCENTAGE *N*	%
Natural gas	5,999	54.89
Oil (all grades)	2,734	25.01
Electricity		
Htg/cooling	809	7.40
Htg only	962	8.80
Cooling only	852	
Propane	220	2.01
Coal	168	1.54
Butane 19	0.002	
Diesel	16	0.0015
Total value	10,927	99.65

Table 17-7 provides an analysis of fuel sources region by region. In Region 1, for example, oil supplies 79% of the schools' energy needs, while gas provides 11% and electricity 9%. Conversely, Region 6 relies on natural gas to serve 79% of its school needs, while oil is negligible and electricity contributes 18%.

The data also revealed that district size appears to influence some fuel usage. The larger districts use a disproportionate amount of natural gas, while small districts show greater reliance on electricity. Oil, on the other hand, seems to supply about 25% of the districts, regardless of size. Table 17-8 presents the percentage of certain fuels used relative to district size.

Table 17-6. Percentage Consumption by Region of Specified Fuels (by School Building Count).

		FUEL SOURCE						
FED. EN. REGION		NATURAL GAS	OIL	ELECTRICAL Htg/c	Htg/ONLY	Co/ONLY	COAL	PROPANE
1	*n*	34	236	13	16	9	-	4
	%	.6	8.6	2	2	1	0	2
2		81	293	9	37	13	-	1
		1	10.7	1	4	1	0	.4
3		255	854	97	58	125	53	1
		4	31	12	6	15	31	.4
4		1,072	248	182	220	205	36	99
		18	9	22	23	24	21	45
5		1,215	447	106	149	101	53	9
		20	16	13	15	12	31	4
6		1,370	9	206	104	168		25
		23	.3	25	11	20	0	11
7		456	164	51	45	33	6	42
		8	6	6	5	4	3	19
8		348	54	44	204	14	11	22
		6	2	5	21	2	6	10
9		1,044	45	54	73	165		13
		17	1.6	7	7	19	0	6
10		124	387	47	56	19	9	4
		2	14	6	6	2	5	2
Total		5,999	2,734	809	962	852	168	220

IV. ENERGY MANAGEMENT

It is impossible to limit the discussion of remedies for energy inefficiencies to the bricks and mortar of the physical facility. The school building and the people who use it are inexorably linked, for much of the energy that can be saved is saved through the vigilance and cooperation of individuals.

The most important elements in implement an effective energy management program are the commitment and teamwork of the people involved. School personnel are highly committed to provide quality education for students. They are equally committed to performing the other, supportive tasks necessary to ensure that quality education is provided. In recent years a new activity has been added to the long list of tasks for superintendents—energy management. School administrators have come to embrace energy management as a valuable tool to assist them in holding down operating expenditures and channeling money into quality

Table 17-7. Percentage of Fuel Sources, Federal Energy Region ($n < 10$ not treated, shown by X).

	FEDERAL ENERGY REGIONS									
	1	2	3	4	5	6	7	8	9	10
Fuel Type *n* Natural gas *n* = 5999%	34 11	81 19	255 19	1072 58	1215 61	1370 79	456 60	348 51	1044 84	124 19
Oil (all grades) *n* = 2734	236 79	290 69	854 65	248 13	447 23	9 X	164 21	54 8	45 4	387 61
Electricity htg/cooling *n* = 809	13 4	9 X	97 7	182 10	106 5	206 1	51 7	44 6	54 4	47 7
htg only *n* = 962	16 5	37 9	58 4	220 12	149 7	104 6	45 6	204 30	73 6	56 9
cooled only *n* = 852	*									
Propane *n* = 220	4 X	1 X	1 X	99 5	9 X	25 1	42 5	22 3	13 1	4 X
Coal *n* = 168	- X	- X	53 4	36 2	53 3	- X	6 X	11 1.6	- X	9 X
Butane *n* = 19	- X	- X	- X	- X	- X	16 .9	- X	- X	3 X	- X
Diesel *n* = 16	- X	- X	- X	- X	- X	- X	- X	- X	5 X	11 2
All Fuels Total *n* = 10929										

*No total figures are given for electrically cooled buildings because most of these buildings are also heated and thus would be counted twice.

education services.

The commitment begins at the top, with the school board and superintendent. They begin by making the school community aware of problems relative to energy and motivating them to assist in solving those problems. Everyone can have a role—from the school board to the principal, from the superintendent to parent-teacher groups, from the maintenance engineer to students and teachers in every classroom. Energy conservation goals should be set, and teamwork should be used because it is the best approach to goal accomplishment. Energy conservation teams appointed by the superintendent usually consist of an energy coordinator, the business official, the maintenance engineer, a principal, a PTA member, a teacher, and a student representative. In many school districts there are both district-wide teams and teams for each school.

The role of the team is to set out goals for energy conservation and to develop a plan to reach the established goals. Individual members of the team are then responsible for implementation of certain aspects of the plan. Typical roles might include the following:

Table 17-8. Fuel Source by District Size: Number of Buildings and Percentages.

	LESS THAN 600 $n = 419$		600 to 4,999 $n = 2,996$		5,000 OR MORE $n = 7,353$	
Fuel type	*n*	%	*n*	%	*n*	%
Natural gas $n = 5,999$	154	37	1,446	48	4,327	59
Oil (all grades) $n = 2,734$	103	25	732	24	1,853	25
Electricity htg/cooling $n = 809$	20	5	254	8	528	7
htg only $n = 962$	72	17	341	11	524	7
(cooling only) $n = 852$	(19)		(153)		(671)	
Propane $n = 220$	47	11	138	5	28	.4
Coal $n = 168$	11	3	78	3	77	1
Butane $n = 19$	9	2	—		10	.1
Diesel $n = 16$	3	.7	7	.2	6	.08

School board and superintendent	Establish policy
Energy coordinator	Coordinate the implementation of all energy conservation activities Report to superintendent and school board
Business official	Keep accurate records of energy expenditures and projected savings

	Monitor billings closely for an early detection of newly developing energy inefficiencies
Maintenance engineer	Keep all mechanical and electrical equipment in good operating condition
	Make necessary modification to equipment and operating procedures to ensure energy conservation
Principal	Set the tone of energy conservation in the school
	Post and enforce energy conservation guidelines and operating procedures for the school
Teacher	Implement and monitor energy conservation in the classroom
	Motivate students to learn about energy conservation
Students	Assist in implementing and monitoring energy conservation activities

The SEED Audit Process

In addition to identifying problems and opportunities, the energy audit and subsequent data analysis define the framework for an effective energy management program. The data analysis provides realistic quantifiable estimates of where energy and money may be saved. It points out the areas of greatest loss and therefore the areas of greatest energy savings opportunities. In this manner, this process assists in setting priorities for energy conservation activities.

The SEED energy audit process consists of three major steps: the pre-audit, the full audit survey, and a presentation of the results. The pre-audit is the data collection and preliminary analysis phase of the audit process. At this time, monthly fuel and power consumption records and billing schedules for at least the past year or two are assembled. Weather data for recent years are collected from the local weather station or the National Oceanic and Atmospheric Administration. Architectural drawings, including modifications and additions, are inspected to compute square footage, percentage of wall area, amount of window area, and the insulating properties of construction materials. Mechanical and electrical specifications are examined to determine power demands. A careful examination of the data collected will point out potential areas of energy inefficiencies. At this time, a boiler efficiency test is made to determine how much of the fuel consumed is actually used to heat the building.

Once the anomalies between the as-designed and actual energy consumption rates are defined, the full on-site audit inspection is conducted, with inspectors looking for opportunities to correct those anomalies and save energy. In addition, the audit team interviews the principal and maintenance personnel to determine building use and number of occupants. These interviews are extremely valuable not only for the information gathered but also for conveying the importance of energy conservation to those who operate the building.

Once the audit survey and data analysis are complete, a written report is prepared. The report clearly and precisely enumerates the problems and lists alternative recommendations. To assist the school administrator in justifying expenditures, cost estimates are provided for each proposed solution. The length of time required to recover the initial cost (payback period) is included as an additional tool to justify expenditures. Payback periods should be used with caution, as they are only a good yardstick for determining relative costs. The recommendations suggest priorities based upon the amount of energy saved, cost estimates, payback period, and the effect of each recommendation on other problems and/or proposed solutions.

Summary

In summary, there is a proper sequence, according to SEED, in which the remedies should be implemented:

1. Tighten the building envelope. This includes insulating windows, weatherstripping, caulking, and reducing ventilation and exhausts.

2. Lower the building temperature, especially the night and weekend settings.

3. Modify the mechanical systems.

The lighting modifications can be implemented at any point in the energy management program (refer to appendices A and B for energy audit checklist, and data collection forms.)

V. PROJECTED SAVINGS

The results of the 20 audits have been compiled in Table 17-9. The principal finding of the SEED 20-school study is that the average school can reduce its energy consumption by 48.6%, without the use of expensive capital equipment or exotic technology. To the extent that the selected schools are representative, there is a significant opportunity for school energy conservation.

In the audits, the engineers first identified sources of heat gain and loss and quantified each. This information, averaged for the 20 schools, is presented in the energy consumption profile, Table 17-10. The profile identifies the percent of heat lost through the most common circumstances, and shows the amount of heat gained from sources other than the furnace. It also divides electrical consumption into several categories. This analysis led to identification of low-cost opportunities for energy efficiency. With one or two exceptions, recommendations were limited to off-the-shelf materials and proven procedures.

Table 17-9. SEED Summary of Projected Savings.

	Percent of Fuel	Percent of Electricity	Percent of Total	Payback Period (months)
East Hartford. Connecticut	41.0	40.0	40.0	4
Newark, Delaware	52.3	23.6	36.3	4
Decatur, Georgia	75.0	16.3	34.4	17
Oak Lawn, Illinois	69.0	32.5	47.2	5
Baltimore, Maryland	58.2	30.5	41.2	19
Lexington, Massachusetts	48.5	34.2	42.3	18
Livonia, Michigan	56.4	24.4	44.0	10
Manchester, New Hampshire	65.4	32.3	48.0	8
Lancaster, New York	53.7	45.2	50.8	12
West Irondequoit, New York	51.5	30.8	45.8	9
Cincinnati, Ohio	63.6	49.1	55.2	13
Bethel Park, Pennsylvania	65.0	31.3	51.3	14
Havertown, Pennsylvania	75.8	64.9	67.6	18
Pawtucket, Rhode Island	75.0	21.0	63.0	9
Garland, Texas	59.5	57.6	58.0	7
Houston, Texas	51.0	44.6	46.3	5
Ft. Belvoir, Virginia	85.0	29.0	60.7	12
Newport News, Virginia	62.5	21.3	43.4	12
Salem, Virginia	58.2	59.8	55.2	21
Racine, Wisconsin	61.8	9.9	41.5	14
AVERAGE	61.4	34.9	48.6	11.6

Table 17-10. SEED Energy Consumption Profile.

HEATING FUEL LOSSES	PERCENT OF LOSS	PERCENT OF TOTAL
Transmission		
Roof	20.6	
Windows and doors	35.6	
Walls	20.4	
Floors	5.6	
		82.2
Ventilation		20.7
Infiltration		
Windows	9.8	
Doors	7.2	
		17.0
Potable hot water		7.6
System losses		2.2
Total heating losses		129.7
HEAT GAIN FROM SOURCES OTHER THAN HEATING FUEL	**PERCENT OF GAIN**	
Electric Power	15.3	
Occupants	7.2	
Daylight	7.2	
Total heating gains		29.7
Heating losses minus gains		100.0
ALLOCATION OF ELECTRIC POWER CONSUMPTION		
Fluorescent lighting		33.8
Incandescent lighting		19.8
Exterior lighting		3.2
Ventilation		11.6
Hydronic pumping		15.3
Refrigeration		2.2
Other kitchen		5.5
Miscellaneous		8.6
Total electric power consumption		100.0

Table 17-11 identifies projected savings by activity. This table is the best support for the overall conclusion that energy consumption in the selected schools can be cut nearly in half. Under heating fuel, five of the first six items, other than window insulation, account for more than half the projected savings. This is possible through such inexpensive and interrelated actions as night thermostat setback, tightening of the building envelope, and control of excessive air movements.

Table 17-11. SEED Projected Savings by Activity.

HEATING FUEL	PERCENT OF SAVINGS	PAYBACK PERIOD (MONTHS)
Caulk and seal windows, skylights and air conditioning window units	6.4	19
Weatherstrip and deactivate doors	4.9	2.5
Insulate windows and skylights	32.5	24
Reduce fresh air intake	7.3	14
Reduce ventilation rates	8.1	2.7
Tune thermostat system	30.7	3.1
Adjust and modify heating system	6.3	18
Install independent water heaters	2.1	65
Lower hot water temperature	1.7	7.6
ELECTRIC POWER		
Deactivate lights near windows	25.4	4
Use high efficiency fluorescent lamps and ballasts	15.0	4
Replace incandescent lamps	33.6	25
Reduce hydronic flow rate	12.7	20
Reduce ventilation rate	4.7	10
Implement energy saving kitchen procedures	2.0	12
Miscellaneous	6.6	8

VI. TRAINING ENERGY MANAGERS

The technical workshop, a vital part of the SEED program, was developed to accomplish three objectives: to explain the interrelationships of the energy-consuming aspects of the building; to illustrate where and how much energy is being lost and at what cost; and to demonstrate ways to conserve energy. The audience was comprised primarily of nonprofessional maintenance engineers and school business officials. These people are dedicated, conscientious, and experienced in sound, practical methods of school maintenance and operation. They are the "front line of defense" for any effective energy management program.

As the SEED program proceeded, the form and substance of the technical workshop were modified to meet the needs of the audience. One of the major purposes of the workshop was to share with the audience our findings and experiences in establishing energy management programs. We too learned from the SEED audit process, in particular from the audiences of the technical workshops. For example, in Livonia, Michigan, we learned that placing pegboard over the outside of fresh air inlets is an excellent way to reduce the amount of outside air entering the classroom. We encouraged the attendees to speak out, to describe their problems, and to identify the solutions they had developed. Problems most frequently discussed related to energy losses due to transmission, infiltration, ventilation, lighting, and heating and cooling systems. This real exchange of ideas and information was best accomplished in a small, informal getting wherein people were willing to become active participants. Also of great importance to us was the identification of barriers to conservation (i.e., why certain recommendations we make cannot be carried out in some schools). The dialogue in the workshops identified a number of barriers.

SEED, through its technical workshops, attempted to show that maintenance personnel are capable of implementing effective energy management programs. Nothing new or revolutionary is needed. Recommendations, presented in the technical workshops, included only simple, quick-fix, low-cost energy conservation measures. They included a quantification of the amount of energy and money saved. The recommendations were designed to assist schools in developing energy management programs by establishing priorities for implementation of energy-saving measures.

VII. DATA ANALYSIS

After fully defining the problems relative to energy inefficiencies and collecting all the data to describe a building energy use, the energy manager needs to know how to analyze the data. Three sets of calculations are given to help the energy manager in the analysis: (1) "rules of thumb," guidelines to determine where energy dollars are going; (2) a balance sheet that allocates energy loss and cost to specific problem areas; and (3) recommendations for problem solutions that are quantified.

"Rules of thumb" are general guidelines that put a price tag on things that use or misuse energy. These formulas can apply to any school in the district, and were derived to support the SEED audit process. They are easily understandable and provide emphasis for energy saving concepts.

For example, a 1° lowering of the night thermostat on a boiler has the same effect as a 3.5° reduction during the day. Why? Because the school building is normally unoccupied 3.5 times longer than it is occupied. Therefore, the energy manager can readily see that daytime temperature settings are not as important as night settings.

Other "rules of thumb" will provide answers to the following questions:

How much does an MMBtu (one million Btu) of heating cost?

$$\frac{(\text{Fuel costs} - \$\text{ per unit})}{(\text{MMBtu per unit}) \times (\text{boiler efficiency})} = \frac{\$}{\text{MMBtu}}$$

How much is the cost of heating by electricity?

$$\frac{\$/\text{kWh} \times 1\text{ million}}{3{,}412\text{ Btu/kWh}} = \frac{\$}{\text{MMBtu}}$$

How much money is lost by transmission through windows?

$$\frac{\#\text{ degree-days} \times 24\text{ hr} \times 1.4\text{ Btu/ft}^2/\text{hr/°F}}{1\text{ million}} \times \$/\text{MMBtu} = \$\text{ lost/ft}^2/\text{yr}$$

How much does heat contributed by occupants cost per year?

$$0.38 \text{ MMBtu/person/yr} \times \$/\text{MMBtu} = \$ \text{ gained/person/yr}$$

How much money does heat gained from sunlight cost per year?

$$0.1 \text{ MMBtu/ft}^2\text{/yr} \times \$/\text{MMBtu} = \$ \text{ gained/ft}^2\text{/yr}$$

How much do losses from infiltration cost?

$$\frac{1.08 \text{ Btu/°F hr cfm} \times \# \text{ degree-days} \times 24 \text{ hr}}{1 \text{ million}} \times \$/\text{MMBtu} = \$ \text{ lost/yr/cfm}$$

How much do losses from ventilation cost?

$$\frac{5.4 \times \# \text{ degree-days}}{1 \text{ million}} \times \$/\text{MMBtu} = \$ \text{ lost/kWh}$$

How much do gains from electric power contribute?

$$\frac{90° \text{ of annual electric power consumption} \times 3{,}412 \text{ Btu/kWh}}{1 \text{ million}} \times \$/\text{MMBtu} = \$ \text{ gained/kWh}$$

Balance Sheet

A balance sheet, much like an accountant's debit and credit sheet, illustrates where energy is lost or gained and how much it costs. The balance sheet is the basis for recommendations, which are in turn evaluated against cost of implementation to determine payback period. This identification of energy inefficiencies helps the energy manager determine where the greatest energy savings can occur. Anomalies (imbalance) in the balance sheet indicate that some things were not properly accounted for, or that there are some heretofore unknown consumers of energy, indicating a further search. It was these anomalies that prompted our "midnight raids." It is the allocation of dollars to energy loss that can assist the energy manager in calculating the time required to recover capital expenditures. This feature, the payback period, helps the energy manager convince those who control school budgets to make the expenditures necessary to implement energy conservation measures. When the energy manager combines the identification of large energy inefficiencies with the cost and payback period of conservation activities, an energy management plan can be defined.

Balance sheets can be prepared for energy expenditures for heating and electricity. An example of a condensed heating balance sheet is given below:

CONDENSED HEATING BALANCE SHEET

	LOSSES MMBtu	LOSSES $		GAINS MMBtu	GAINS $
Transmission			Electric power	2,000	6,000
Windows	8,000	24,000	Occupants	1,000	3,000
Others	10,000	30,000	Daylight	500	1,500
Ventilation	2,000	6,000		3,500	10,500
Infiltration	1,500	4,500	Net use, calc.	18,000	54,000
Total	21,500	64,500	Actual use	20,000	60,000
			Discrepancy	10%	

An energy manager can analyze this example and readily determine that the greatest opportunity for energy savings exists in the windows. Therefore, it will be necessary to figure out ways to eliminate heat transmission through windows.

Once general "rules of thumb" have been applied and a balance sheet prepared, the energy manager is in a position to define alternative means of correcting the various problems of energy loss, allocate a cost, and define an energy management program.

In the SEED technical workshops, recommendations were made that were specific to the particular school building that had been audited. In addition, general recommendations were made for most commonly found problems so that energy managers could apply them to other school situations.

The general approach to improving the energy efficiency of a building is to tighten the building envelope, that is, to insulate windows and seal cracks, holes, and other openings where energy can be wasted and heat lost. The result of "tightening up" the building is that the demand for energy drops. Then, one can make the necessary internal (temperature control) adjustments that will save even more energy.

In order to reduce heat loss due to transmission, infiltration, and ventilation—to tighten up the building envelope—the energy manager should consider the following actions:

1. Close and seal operable windows that are not needed.
2. Ensure that windows remaining operable seal well when closed.
3. Caulk and weatherstrip around windows and doors.
4. Insulate the upper two-thirds of windows with a translucent window covering.
5. Deactivate unused doors by issuing administrative procedures.
6. Keep doors to corridors closed.
7. Place pegboard over the outside of fresh air inlets.
8. Ensure that dampers on unit ventilators and roof-top exhaust fans are working properly.
9. Disconnect unnecessary roof-top fans.
10. Place remaining fans on timers.
11. On hydronic heating systems, reduce pump flow and power by shaving pump impellers.
12. Plant trees and shrubs as a windbreak.

Once the cracks in the building envelope have been reduced, the energy manager can make the necessary internal adjustments. The boiler provides the greatest opportunities for energy savings. Because the building now leaks less heat, the night setback temperature on the boiler can be turned back. The energy manager should experiment with temperature settings and determine the lowest night and weekend thermostat settings permissible for that particular building. In most cases, it will be from 45°F to 55°F, resulting in time-averaged night-weekend temperatures of around 60°F.

Room thermostats can probably be readjusted. The energy manager should make certain that all thermostats are calibrated properly. Keep in mind that the daytime temperatures are not as important as nighttime settings. This fact makes it easier for the energy manager to keep teachers and students happy and warm and still be able to implement energy savings.

CHAPTER 18

CASE STUDY OF A COMMERCIAL RETROFIT PROJECT: EAST TEXAS STATE UNIVERSITY, COMMERCE, TEXAS

Morris Backer. P.E.,
and
J. Phillip Upton, P.E.

I. INTRODUCTION

East Texas State University is a small, multipurpose, regional, state university of approximately 12,000 students, located in Commerce, Texas. The map of its physical plant identifies 100 buildings constructed since 1894. The recent rapid escalation of energy costs prompted a study to establish a conservation program. The buildings are served by almost every type of lighting and HVAC system marketed, and in many cases are old enough to be potential maintenance headaches.

II. PROPOSAL

The proposal suggested studying, initially, the 24 buildings using over 85% of the university's total energy expenditure. The scope of the work includes a survey of the mechanical and electrical systems to recommend modifications to energy usage and maintenance requirements.

III. WORK PLANNING

Larry Hardaway, the lead engineer assigned the work, planned his attack by creating a flow chart describing tasks, problems, and expected results. His chart, which follows, may be useful in other applications:

A. OBJECTIVE

1. Investigate and analyze energy usage.
2. Identify energy saving opportunities.
3. Recommend and evaluate energy conservation measures.
4. Develop appropriate funding.

B. PROBLEMS

1. Meter usage not available by buildings or systems.
2. Limited billing histories.
3. Limited equipment documentation.
4. Limited building plans.
5. Limited system flexibility.

C. INITIAL SOLUTION PLANNING

1. Initial meetings and survey to prepare general study approach.
2. Develop plans directed to meet objective and overcome problems.
   - (a) Gather past billing histories.
   - (b) Prepare survey sheets and collect data.
   - (c) Conduct on-site investigations.

D. REDEFINE SOLUTIONS

1. Review resource data and refine study direction.
2. Collect additional data necessary.

E. COMPUTE SOLUTIONS DATA

1. Using resource data, prepare energy and cost computations.
2. Prepare implementation cost estimates.
3. Evaluate payback periods.

F. DECISIONS

1. Recommend solutions and associated cost and savings.

G. PRESENT SOLUTIONS

1. Write report.
2. Print and prepare copies.

H. ADJUST AS NECESSARY

1. Incorporate owners' input and adjustments.

IV. BUILDING SURVEYS

The university's buildings are of all ages. Few have plans and specifications or even inventories of their mechanical and electrical systems. Information had to come from physical surveys and operators of the systems. There were limited utility billing histories to provide operating profiles. Metering devices were not conveniently arranged. It was decided a set of inventory forms to be completed by operations personnel could minimize the survey team's time at the site. (Example forms are included in Appendix C.)

The forms were given to the university's facilities section with the following instructions:

1. Fill out "Building Description form" for each building as listed in the Study Scope.

2. Select the appropriate equipment form for each type of equipment used in the building. Use as many equipment forms as there are pieces of equipment in use.

3. Add additional information in supplementing information sheets as necessary.

V. DISCUSSION

The university uses two forms of primary energy, electricity and natural gas, to operate the building systems. Electrical power, supplied by Texas Power and Light Company, enters the campus at a primary voltage of 12,470 volts through seven metering points, from which it is distributed to all buildings. Three of those points were the sources for the 24 buildings studied. Natural Gas, supplied by The Lone Star Gas Co., is also metered at seven points, two of which serve the buildings studied.

Since the metering points serving the study buildings also serve other buildings, billing history did not provide accurate fuel costs for each particular building. Therefore, the portion of energy used for each building was estimated from an individual building analysis based on equipment inventories and load profiles developed by operators. These data, collected on the survey forms, listed each component in the building mechanical systems along with types of electrical service and lighting systems. Field measurements were taken to determine the average rate of energy used by each system component. By utilizing this rate and known operating schedules, energy usage was computed for each building. This computed consumption was then compared with the composite billings to ensure realistic estimates.

Table 18-1. Use of Energy in a Building

Contractor	Source	To Decrease Energy
External	Ambient conditions solar loads	Modification to building envelope
Internal	Lighting People Equipment Operating Profiles	Change loads or Profiles
Inherent to System	Additional energy used for normal operation	Modify installed systems

The survey forms also documented specific system component details, including age, operating hours, previous operating history, present condition, estimated reliability, type of system, capacity, and other information. Each building was then examined by visual inspection, and interviews were arranged with the operating staff. All information was used to determine which buildings and systems could be retrofitted without producing secondary maintenance and/or environmental problems. Savings opportunities were listed and evaluated for cost-effectiveness. Table 18-1 describes how energy is used in a building. It also defines approaches to savings.

The survey identified many needs in each of the categories examined. There were leaky buildings, buildings that emptied but remained lighted, or buildings being furnished too much air. It became obvious that the two elements needing the most consideration were: (1) the system components, which needed better control in order to match their operation to the building needs; and (2) outdated equipment requiring excessive operating and maintenance costs.

A. Energy Management Control System

The effects of internal contributors as well as the inherent system contributors can be programmed to optimize energy efficiency. The extent of energy savings is dependent on how closely the system components can be controlled to match the needs. Computerized automation systems, often called energy management and control systems (EMCS), are manufactured for this specific application. They have varied capabilities and must be selected to match requirements. In this application it was necessary that the control system be capable of start-stop functions, temperature resetting, temperature monitoring, flow monitoring, and data acquisition, as well as recording. Additionally, the system had to be easily programmed and capable of controlling a variety of devices so that all of the campus buildings might be included.

Modular systems, which enable the user to arrange components to meet his present requirements and expand later, were examined. These systems may also be used to monitor equipment and print instructions as part of an equipment maintenance program, which may include logging of operating hours to produce printed notices of needed lubrication, filter maintenance, excess vibration, extreme temperature conditions, high water alarms, etc. Such maintenance can improve scheduling, to maintain more equipment with less manpower, while improving maintenance efficiency.

Technology has been developed that allows microprocessors in the field units, as shown in Figure 18-1, to handle most of the routine processing of data and control of equipment. Programs for the field units may be written in Fortran at the main computer and then downloaded to the field unit. This allows the user considerable freedom after the system is installed. The field units can be programmed to accomplished start-stop functions, automatic control, and alarm functions even if the central computer is not operating. This type of system is a reliable, flexible, and useful tool in managing energy usage in any installation.

To determine the optimum level of system sophistication, it was necessary to compute the expected energy savings. It was estimated that a savings of $171,700 could be realized by programming various system components to match building requirements. The expected cost for such

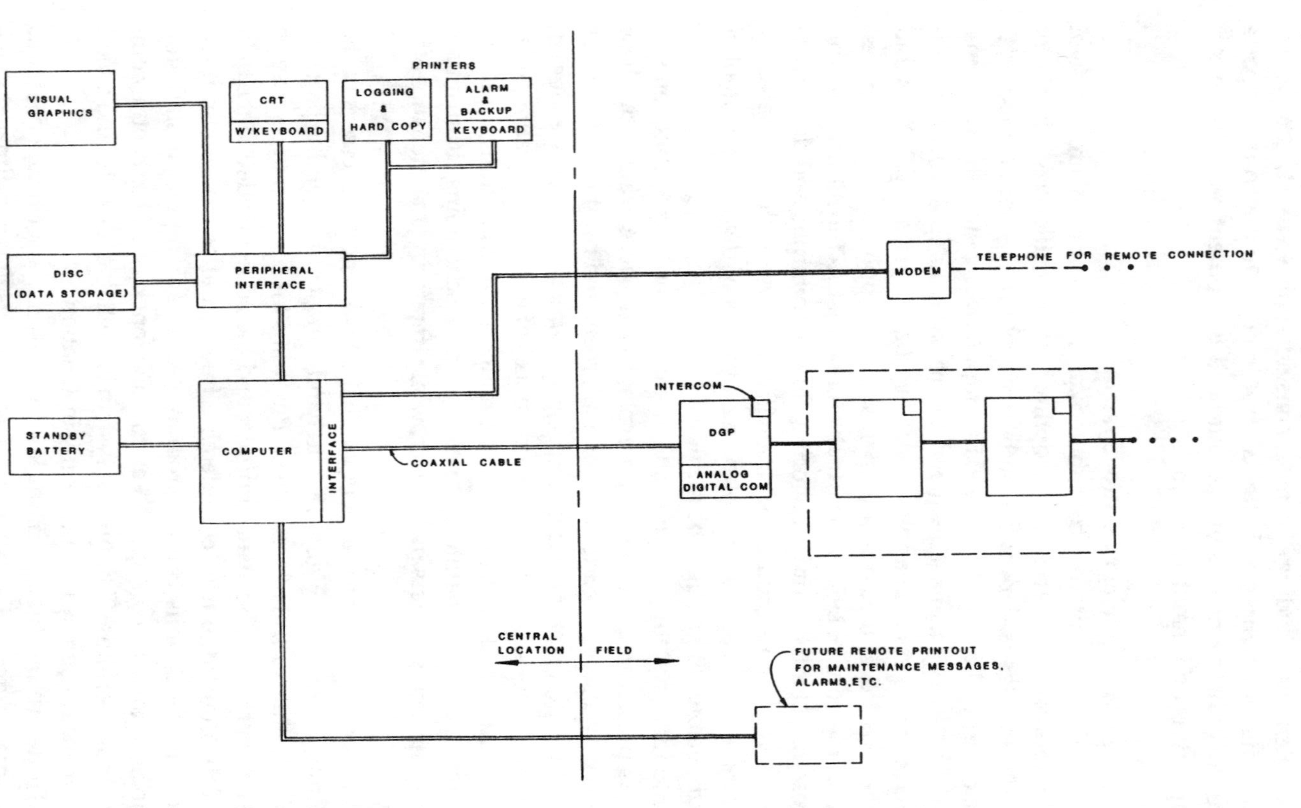

Figure 18-1. System of Main Computer and Field Units. (Courtesy of East Texas State University, Commerce, Texas

an automation system would be $1, 104,000, resulting in a 6 1/2 year simple payback.

B. Reduced Utility Cost Through Limiting Demand

An analysis of the billing history revealed that the university was paying more per kilowatt hour for electricity during the low usage months of the year because of the conditions of the Texas Power and Light Company Service Agreement. The billing is based upon actual energy used and maximum rate of demand, but the minimum billing demand quantity will be at least 65% of the maximum demand recorded in a 15-minute interval of the past June, July, August, and September. The demand during several of the past winter months fell below this cost, and the minimum demand provision was exercised.

As a large portion of the utility cost was demand charges, it was found effective to install demand-limiting systems. These systems monitor demand at all times and are capable of turning off equipment on a preselected priority schedule to establish a maximum acceptable level. The equipment is cycled on again when the demand level permits. Using this device, all equipment on the system is assigned a priority so that vital equipment always remains in operation. The estimated cost for this system at the time of our study was $80,000. It was estimated the system would save approximately $59,500 annually after a 1-1/2 year payback period.

C. Electrical Light Relamping

The building lighting and electrical systems were surveyed and light level measurements taken. Computations were then made to determine the savings that could be derived by revising the lighting in the building areas. The latest Illumination Engineers Society criteria were used for each task area. It was determined that a 13% reduction in the present lighting energy and demand was possible without any noticeable effect on illumination. The savings were to be produced by revising the lighting in stairways and corridors, and by reductions in lamp wattage used for aesthetic effect. Significant energy savings appeared possible in the field house and gymnasium by replacement of low-efficiency incandescent lighting with high-efficiency, high pressure sodium units. The estimated cost for these modifications was $288,400. Expected savings were $149,500 annually after a two-year payback.

D. Utility Distribution

The majority of the university's buildings are served by individual mechanical systems. However, two of the buildings' mechanical plants are capable of being expanded into central plants. The present individual systems vary from small window installations with electric heating coils, to large centrifugal liquid chillers and steam boilers. The total combined cooling capacity of these systems is more than 2,000 tons (7,040 kW) of refrigeration. Central plant installations with distribution mains and building connections offer advantages other than energy conservation, and higher inherent efficiencies of the system components used in central plants can save up to 50% of the energy required for individual systems. A lower initial investment required because less backup machinery is needed, the ability to take advantage of campus wide load diversities, reduced maintenance costs, decreased inventory through standardization of repair parts, improved equipment reliability, and added flexibility for alterations in buildings are all benefits of a central plant.

In analyzing the various buildings and the locations of the existing central plants, a scheme was developed by which most of the independent building systems could be connected. The plan included the addition of new chilled water and steam piping to the existing distribution systems, with connections to those buildings. The capacity of these plants was also to be increased. The cost for this conversion was estimated at $625,000. It was estimated that the improved efficiency in maintenance, utility conversion, operation, control, and life expectancy would save $50,200 per year.

E. Building Mechanical System Energy Reductions

Many different types of air-moving systems were being utilized on the campus. Depending on the type, the amount of fan and heating cooling energy varied significantly. Systems that vary the supply air quantity to maintain room conditions were comparatively low energy users. Those that supply constant volumes of air but vary the air temperature by controlling heating or cooling coil output were moderate energy users. Multizone and double duct systems, installed in about half of the buildings, were, however, larger users. Such systems maintain comfort conditions by supplying hot and cold air streams that are mixed to produce desirable air temperatures. Cold air must be used to cool hot air in order to produce the desired supply air temperature. This requires that both refrigeration and boiler energy be generated simultaneously and used at all times regardless of actual building loads.

In order to improve the performance of these systems, additional controls may be added to establish cold and hot air temperature settings that result in the least amount of wasted energy. Computations for the ten buildings using these types of systems indicated the cost for modifying the systems was approximately $100,000. The annual savings for these improvements were estimated at approximately $21,800, with a four-year payback period.

The moderate-energy-using systems could be converted to the least-energy-using systems by the addition of terminal control devices. These devices vary the air volume to satisfy room conditions and therefore save fan energy. The cost of retrofitting these systems was estimated at $78,000. The annual energy saved was estimated to reduce utility cost by approximately $13,400, with a six-year payback period.

F. Control Settings and Outside Air Balance

Many building systems were equipped with control cycles that utilized outside air for cooling when possible. Some buildings used fixed outside air quantities to maintain ventilation rates and odor control. Although the correct portions of outside air were set properly at the time of their installation, many dampers were no longer at their original settings. This increased energy consumption; recalibration was required. The improper adjustments increased the amount of cooling and heating needed annually to compensate for undesirable outside quantities. The recalibration was estimated to cost approximately $72,000. Reductions in cooling and heating energy were estimated to save $9,300 annually. An eight-year recovery period was envisioned.

G. Architectural Recommendations

The survey identified improvements that could be made to decrease loading for several air conditioning and heating systems. The improvements included the addition of solar screening devices, the addition of wall insulation, and modification to roofing systems in order to reduce mechanical system energy consumption. For a cost of $300,000, a reduction in utility consumption of $28,700 per year, with a 10-year payback period, was estimated.

H. Remodeling of Mechanical Systems

Several building systems called for complete remodeling in order to achieve a reliable operating condition. Most of these systems had served their expected useful lives, and their maintenance costs were very high.

VI. RECOMMENDATIONS

Table 18-2 is a summary of the recommended improvements that were described in Section V. The savings indicated are annual amounts based on the projected utility costs and include estimated construction.

Table 18-2. Recommendations.

Number	Item	Construction Costs	Annual Savings	Years Payback
(1)	Energy management computer systems	$1,104,000	$171,700	6-1/2
	Demand limiting	80,000	59,500	1-1/2
(2)	Relighting	288,400	149,500	2
(3)	Distribution	625,000	50,200	12-1/2
(4)	Building mechanical systems	100,000	21,800	4-1/2
	variable volume	78,000	13,400	6
(5)	Outdoor air adjustments	72,000	9,300	8
(6)	Architectural	300,000	28,700	10-1/2
Totals		$2,647,400	$501,100	5-1/2
Other				
(A)	Remodeling mechanical	500,000	—	—

VII. CONCLUSIONS

East Texas State University received the report of this study and proceeded to obtain funding. In the following year, part of the funding became available; the first phase of the design encompassed recommendation numbers (1) and (3). The first portion will include the use of the university's CDC computer and its in-house capability to provide the software for the energy management system. The design work on that portion includes the sensory and control elements needed to input the data required. The second portion of this first phase includes the addition of the distribution piping and central plant capabilities. Following phases will convert building systems and connect them to these distribution systems, as well as the energy management computer system.

CHAPTER 19

HEAT RECLAIM SYSTEM PROJECT FOR A CHEMICAL LABORATORY

Fritz A. Traugott

I. BASIC DATA

The biochemistry building, located at the University of Rochester River Campus, was provided with an air conditioning system for various occupancy and environmental usages (Figure 19-1). The center core of the building is occupied by laboratories and is five stories high. There are 17 building bays, approximately 40 ft × 19 ft (12.19 m × 5.79 m), on each side of a central pipe space. The air distribution for this building is provided by 17 air-handling units located in a penthouse, providing vertical air distribution through the pipe space into two reheat boxes per floor and bay. The original design made air available through laboratory hoods in each bay on the first four floors and through laboratory hoods in each bay on the top floor (Figure 19-2). Individual exhaust fans were installed in the penthouse for all the above-mentioned hoods, totaling 425 (Figure 19-3).

Environmental criteria dictated that the laboratory space be kept under negative pressure in relation to the corridor. The exhaust capacity for each fan was designed to be 1,390 cfm (656.08 L/s). Exhaust air, then, equaled 425 × 1,390 cfm (425 × 656.08 L/s) or 590,750 cfm (278,834 L/s). One hundred percent of the outdoor air for the entire center core of this building was required, encompassing some 500,000 cfm (236,000 L/s), or approximately 3 cfm per sq ft (15.24 L/s per m^2).

From the above information, it was obvious that a considerable amount of heating was expended to maintain proper indoor environment conditions when ambient conditions of 93°F (33.89°C) during the summer and –5°F (–20.56°C) during the winter needed to be considered. A supply temperature of 55°F (12.78°C) was selected for year-around use. This air was then reheated by the individual reheat boxes before delivery to the laboratories.

II. INITIATION OF ENERGY CONSERVATION PROGRAM

The utility invoices amounted to one million dollars for electric power and $350,000 for central steam supplied by the university's own plant. With these inflated bills, it was obvious that a study was needed to investigate the ways in which energy could be saved in this building. Several ideas were considered and put into practice:

- Rebalance supply air quantities in spaces, consistent with calculated loads, to reduce preheat requirements.
- Provide a means to manually shut off exhaust fans from hoods that are not being used and, at the same time, reduce the supply air quantity to match the requirements of that space.
- Reclaim energy from the exhaust air and transfer it to the supply system (Figure 19-4).

To prove the effectiveness of the energy conservation study, we instituted the changes on three of the existing systems and metered both the modified systems and the non-modified system to compare operating costs.

To permit the described procedure, the spaces conditioned by these three selected systems were surveyed to discover which could be effectively shut off. Provision was made to do this and to reduce the air flow to these spaces. The cooling load calculations for the spaces were reselected for the lower air quantities, and the reheat boxes were adjusted downward for the new capacities.

A coil run-around system (Figure 19-5) was designed and implemented with a coil installed in the outdoor air intake for each of three selected air-handling units. This coil was installed in front of the existing preheat coil and behind the filters. The existing exhaust fans for each hood of space were stacked three high in the penthouse, with individual outlet

Figure 19-1. Typical Laboratory Supply and Exhaust (All Figures in This Chapter Courtesy of the University of Rochester, Rochester, New York.)

stacks projecting through the roof in rows of three on each side of the central pipe space. Because of space limitations in the penthouse, the exhaust fan discharges needed to be discharged through the roof via an exhaust reclaim coil. A unique design combining the three exhaust fan outlets discharging through a single coil with appropriate baffles to separate the air flows was prepared. These coils were installed above the roof at a 45° angle, with the air flow making a 90° offset before being discharged vertically (Figures 19-6 and 19-7). A glycol solution, 40% by volume, was used in the piping distribution system to connect with each coil. Appropriate meters were installed to measure the heat removed by

Figure 19-2. Typical Fume Exhaust Fan System.

the chilled water coils, the heat added to the preheat coils, and heat transferred in the reclaim system (Figure 19-8). Electric meters were installed to measure power usage of the hot water circulating pumps, the supply air fans, and the reclaim heat circulation pump. The power usage of the exhaust fans was calculated from one-time meter readings. (For details on the coil system, refer to Figures 19-9, 19-10 and 19-11.)

III. CONSTRUCTION COSTS

The contractor making the system construction changes determined that the cost for energy conservation amounted to $52,000 per air-handling system. Of this, $1,477 was allotted for system upgrading maintenance items and air side balancing.

Figure 19-3. Typical Exhaust Ports on Roof Without Energy Recovery.

Figure 19-4. Typical Exhaust on Roof with Energy Recovery Units.

Figure 19-5. Typical Energy Recovery Unit Piping on Roof.

Figure 19-6. Singular Energy Recovery Unit Detail.

IV. ENERGY SAVINGS BY USE OF ACTUAL METERS

The energy usage was metered for 2-1/2 months. The net savings per system was calculated to be $7,787. This was based on electrical costs of $.025/kWh and steam costs of $3.92/1,000 lb ($8.63/1,000 kg) of steam. These costs were the actual costs paid by the university. The number of degree-days for this period was 3,371, which resulted in a savings of $2.31 per degree-day. The average heating degree-days for Rochester is 6,255, which equated to an annual savings of $14,450 per unit. The estimated cooling savings was $2,200 per year, resulting in a total annual heating/cooling operating cost savings of $16,650 per unit (refer to Table 19-1 for specific details on savings).

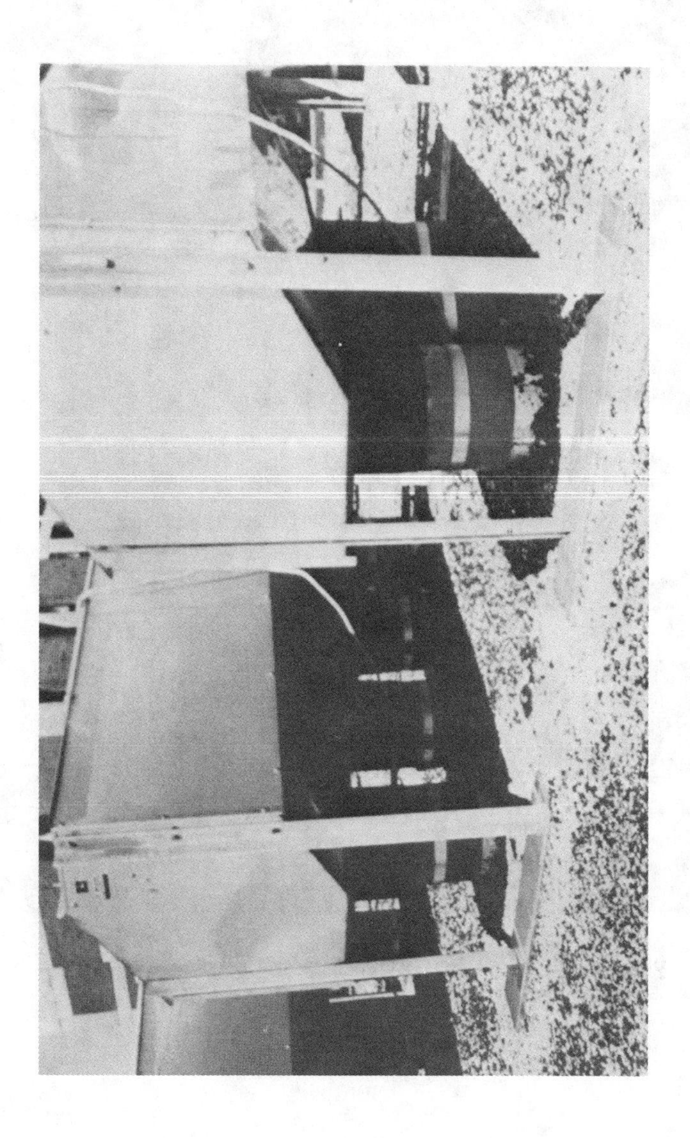

Figure 19-7. Exhaust Fume Connection to Energy Recovery Unit.

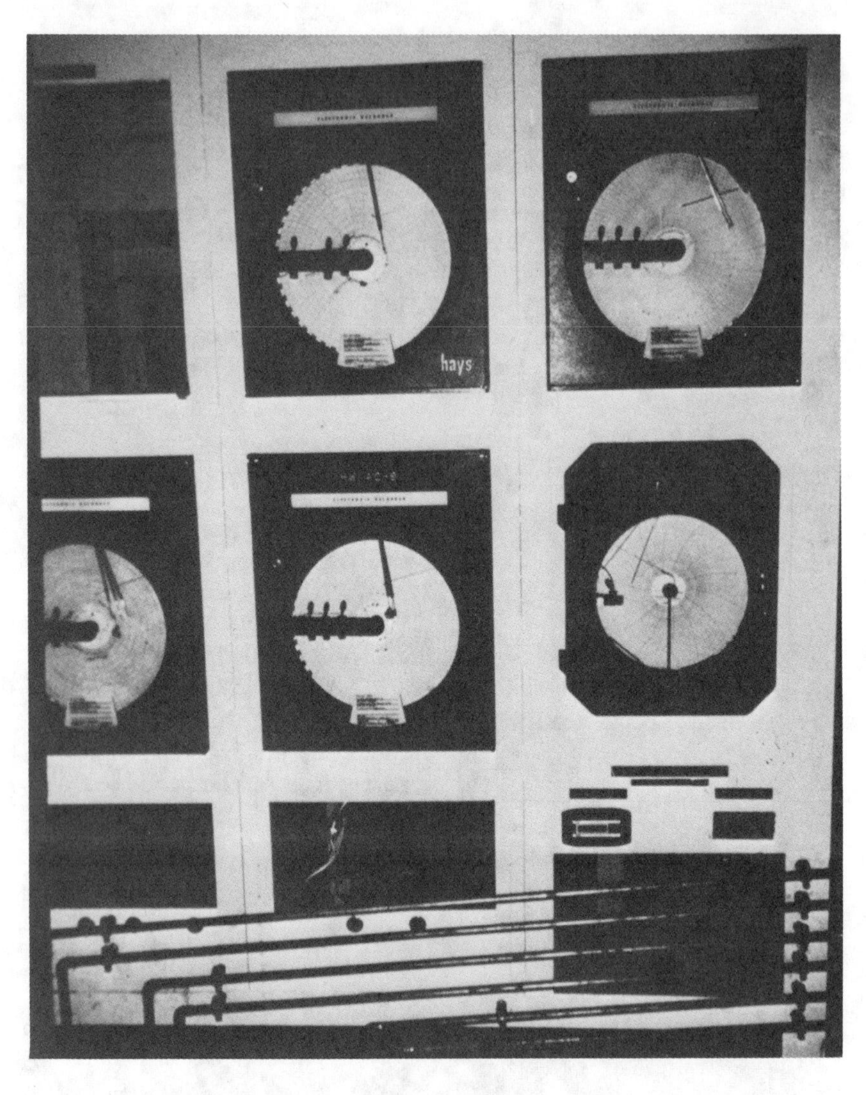

Figure 19-8. Test Instrumentation Control board.

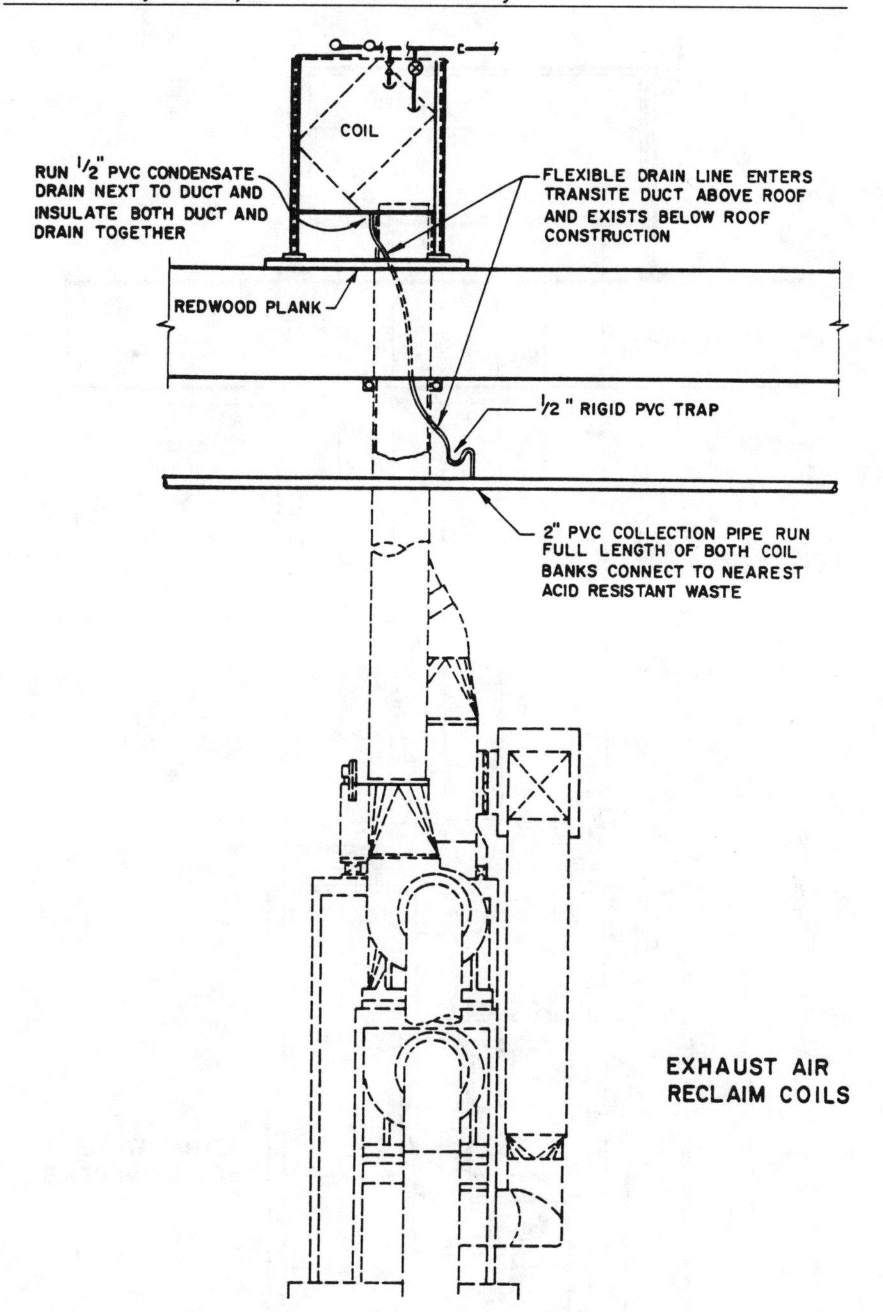

Figure 19-9. Exhaust Air Reclaim Coils, Side View.

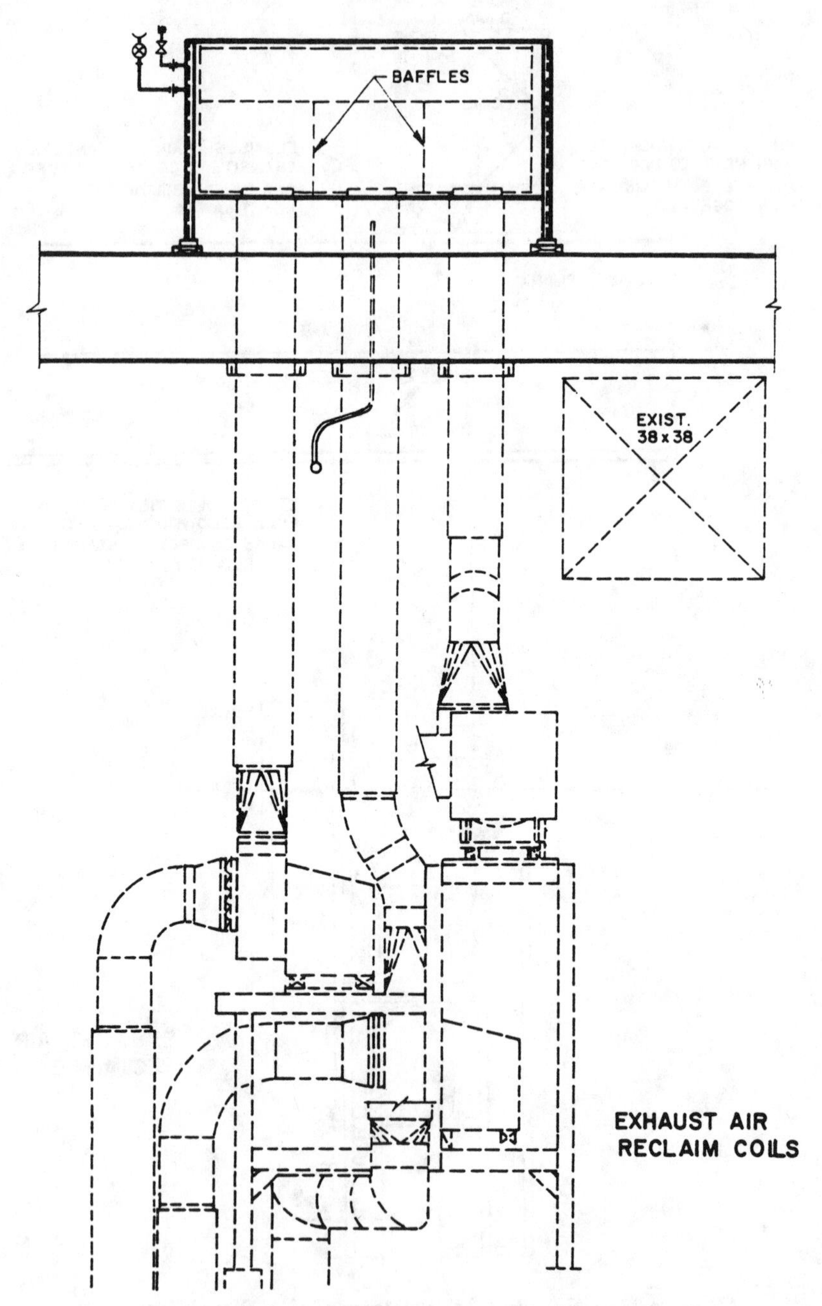

Figure 19-10. Exhaust Air Reclaim Coils, Front View.

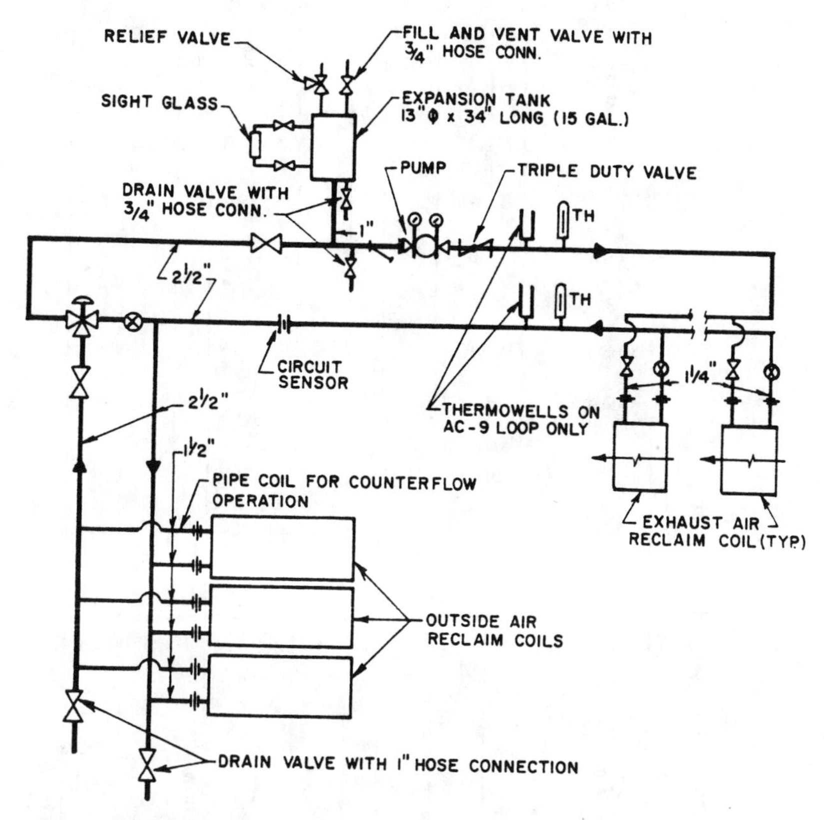

Figure 19-11. Heat Transfer System Piping Schematic, No Scale.

V. ECONOMIC REVIEW OF CONSTRUCTION COST VERSUS ENERGY SAVINGS

Using the projected annual savings of $16,650 per unit and the previously indicated construction cost of $52,000 per unit, the capital cost recovery can be obtained in less than four years, using an 8% interest rate. It is obvious that if we use today's higher energy costs, the capital recovery costs period will go down to probably between one and two years. Therefore, projects of this type become very lucrative in cost management savings. This type of retrofit permits capital recovery of first costs at a much greater rate than could be obtained by improvement of the building envelope and related techniques. This does not imply that the building envelope should not be improved; it simply points out that system retrofit modifications in general have a shorter payback than other related construction energy savings.

Table 19-1. Energy Calculation Sheet.

MONITORED ITEM	RUN #1 METER READINGS	RUN #2 METER READINGS	DIFFERENCE	METER MULTIPLIER	OTHER MULTIPLIER	QUANTITY SAVED	ENERGY RATE	COST SAVINGS LOSS	REMARKS
Date of reading	11-11-76	1-31-77							
Time of reading	8:30	10:30	1,946 hr						
Outdoor air temperature	32	13							
C.W. Btu's AC11 & AC12	—	—	—	15,000	—	—			
C.W. Btu's AC8 & AC9	—	—	—	15,000	—	—			
4 − 5 difference						—	—	—	Not applicable
H.W. Btu's AC12 (kJ)	039544	278797	239253	10,000	—	2,392 $\times 10^6$	2.52×10^9		
H.W. Btu's AC9 (kJ)	170471	190427	19956	6,000	—	120 $\times 10^6$			
7 − 8 difference						2,273 $\times 10^6$	3.46/10.6 Btu	$+7865	$8,910 @ $3.72/10^6 kJ $3.92/10^6 Btu
Reclaim Btu's AC9	NA	NA	—	3,000	—	—			
Supply fan kWh AC12	6163	9409	3246	10	—	32,460			
Supply fan kWh AC9	4886	9155	4269	10	—	42,690			
11 − 12 difference						− 10,230	.025/kWh	$ − 256	

H.W. pump kWh AC12	0488	1429	941	10	—	9,410			
H.W. pump kWh AC9	0268	0991	723	10	—	7,230			
14 − 15 difference						2,180			
Reclaim Pump kWh AC9	0299	0805	506	10	—	− 5,060			
16 − 17 difference						− 2,880	.025/kWh	$ − 72	
Supply air avg. CFM (l/s) AC12	398,297 843,850	915,667 1,939,974	517,370 1,096,125	2,400	1/2c/× 60	22,530	(10,634)		
Supply air avg. CFM (l/s) AC9	199,456 422,577	599,700 1,270,552	400,244 847,975	2,400	1/2c/× 60	17,430	(8,227.0)		
19 − 20 diff. (CFM)						5,100			
Avg. sup. air temp. AC9 & 12	50°F 10°C/ 58°F 14.4°C	50°F 10°C/ 58°F 14.4°C							
Reheat savings (Btu) (kJ)						164.4 $\times 10^6$	$3.46/$10^6$/ $3.28/$10^6$ kJ	−569	−$644 @ $3.72/$10^6$ kJ 3.92/10/6 Btu
Exhaust fan kWh AC9	Calculated		15.6 kW		1946	30,358			
Exhaust fan kWh AC12	Calculated		12.5 kW		1946	24,325			
24 − 25 difference						− 6,033	$.025/$20^6$	$−15i	131
Net savings for 1,946 hr								$+681T	One system only

Source: Courtesy of the University of Rochester, Rochester, New York.

Chapter 20

Industrial Energy Management

Paul W. O'Callaghan

I. ENERGY AUDITING

Before any worthwhile attempt can be made to identify cost-effective energy-conserving options within a manufacturing organization, a fully comprehensive audit of energy inputs, throughputs, and outputs must be constructed. Figure 20-1 demonstrates such an audit for the heating period of a typical large factory in the United Kingdom. The unit kW, adopted throughout, has been obtained by normalizing the data using the total number of heating hours per annum. The fuel energy delivered to the boilers, expressed in terms of gross calorific value, is modified by the aggregated annual mean conversion efficiency of the boiler plant, and discounted by the distribution losses through pipelines and heat transfer equipment. Sundry heat generators within the plant are identified, and the proportion of sundry heat gains (i.e., that fraction that is rejected into the internal environment) is added to the space heating heat delivered. All the electrical or process energy that is used to produce mechanical work or to provide lighting or high temperature process heating (including welding and flame cutting) within the factory ends up as a sundry heat gain, unless local cooling systems are provided. The net distribution losses shown in the audit have also been reduced by the amount of heat energy "lost" by high temperature pipework systems but accepted by the internal environment. The remaining sundry energy is accepted by the internal environment. The remaining sundry energy is rejected directly to the environment (e.g., from exhaust air extract fans, waste hot water, battery charging, the condensers of air conditioning and refrigerating systems, or the cooling banks of air compressors) and so does not add to the space heat supply.

It must be emphasized that, during the heating season, sundry gain energy performs two useful functions in tandem: it fulfills that purpose for which the energy is primarily released (e.g., process work or heating, machining, lighting, etc.) and also provides bonus heating. Where, however, the predominant space conditioning requirement is for cooling, the

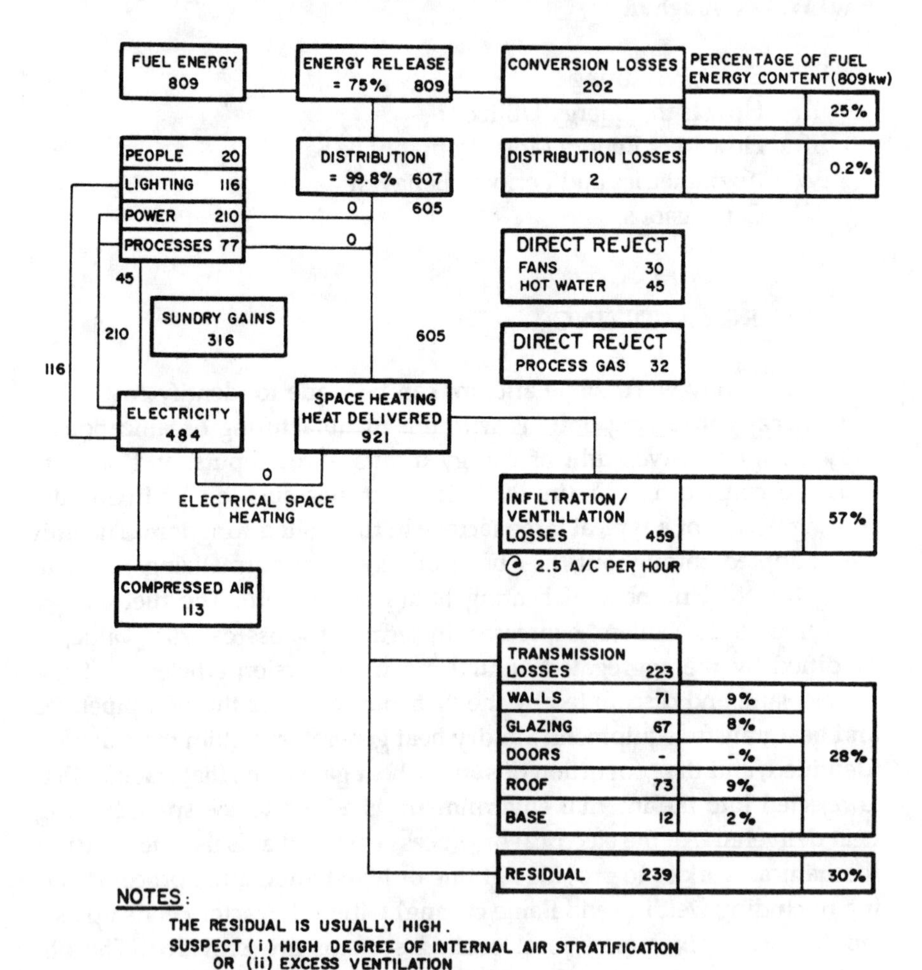

Figure 20-1. Summary of Heat Flows Through Building.

ing and air conditioning system, and are, therefore, a disbenefit. When this is the case, efforts should be made to remove excess heat at the source, using water or air at the prevailing outside environmental temperature. This course of action prevents the transfer of a sundry gain to the internal environment, which must then be cooled using higher energy grade chilled water or refrigerant.

After the total space heat delivered is estimated, infiltration/ventilation and fabric transmission losses are then computed from a survey of the building, its usage, and occupancy characteristics. In the ideal case, the total amount of heat delivered to the conditional space exactly matches the sum of the heat rejections via these two mechanisms.

II. HOW IS THE ENERGY UTILIZED?

Table 20-1 summarizes output data from five energy audits recently constructed for industrial manufacturing systems in the United Kingdom. It is seen that although the overall energy consumptions vary from 1,000 kW to 90,000 kW, the usage patterns are startlingly similar when normalized using the base area or, in particular, the enclosed volume (Table 20-1a).

This similarity is also evident in considering the G-value (the total energy rejection by fabric transmission and ventilation divided by the product obtained by multiplying the enclosed volume times the mean annual temperature difference between inside and outside environmental conditions), but is not indicated by the more commonly cited U-value. The ratio of transmission to ventilation losses indicates the relative advantages that may be gained by insulating or recovering "waste" heat from the rejected ventilating air.

By far the greatest proportion of energy is utilized in every case in offsetting ventilation losses. Typical losses are as follows:

Infiltration/ventilation losses	35-45%
Transmission losses	7-31%
Conversion and distribution losses	11-20%
Compressed air generation	1-12%
All other energy uses	0.4-22%

It is notable that all other uses of energy (i.e., those that do not add to the sundry heat gain) account for no more than 22% of the total energy consumed.

Table 20-1. Examples of Energy Audit Summary Data.

(a) *General Data*

Factory	Overall Energy Consumption	Mean Energy Consumption, $\overline{kW}$ per M² (Base Area)	Mean Energy Consumption, $\overline{kW}$ per Capita	Mean Energy Consumption, $\overline{kW}$ per Unit Volume
A	1,300	0.195	2.72	0.0268
B	11,000	0.209	4.21	0.0278
C	22,000	0.198	11.00	0.0274
D	90,000	0.377	6.6	0.0312
E	3,600	0.180	6.87	0.015

(b) *Energy Indices*

Factory	Overall Fabric U-Value, $Wm^{-2} K^{-1}$	Overall G-Value $Wm^{-2} K^{-1}$	Transmission/ Ventilation Loss
A	2.02	1.28	0.49
B	2.07	1.20	0.44
C	1.33	1.44	0.19
D	1.50	1.44	0.15
E	3.24	1.00	0.54

(c) *Input/Output Data*

Factory	Fuel Energy, $\overline{kW}$	Sundry Gains, $\overline{kW}$	% of Total Energy Usage					
			Conversion and Distribution Losses	Infiltration/ Ventilation Losses	Fabric Trans-mission Losses	com-pressed Air	Residual from Balance	All Other Uses
A	800	300	15	35	17	9	18	6
B	6,000	2,400	12	44	18	1	3	22
C	11,000	4,000	11	35	9	12	18	15
D	56,000	14,000	13	44	7	11	16	9
E	3,000	800	20	45	31	0	3.6	0.4

III. HOW MAY ENERGY MOST PROFITABLY BE SAVED?

A. Eliminating the Residual

A large discrepancy between the net space heat delivered and the total of the ventilation and fabric transmission losses indicates that the system operation does not correspond to the design specification. A negative residual implies that less than the design minimum air change rate is being achieved. A positive residual can result from:

(1) **The Maintenance of Excess Air Temperatures.** The temperatures are above those stipulated in the design. Temperature measurements should be made for substantiation. Closed-loop local thermostatic controls should be introduced. When heating and cooling services are provided, the "deadspace" bandwidth should be selected to be as wide as is compatible with thermal comfort considerations.

(2) **Overventing.** This problem is often difficult to rectify. Random infiltration must be minimized by weather stripping and attending to doors and windows. A balanced inlet/exhaust fan system should be provided and air flows continuously monitored and controlled to adhere to minimum ventilation requirements during the heating season. Excess capacity can be installed for summer use.

(3) **Internal Vertical Air Stratification.** This common occurrence can result in vertical temperature gradients of up to 20°C per 30 meters of height while the set design level is maintained at floor level. The existence of stratification causes far greater ventilation and transmission losses than the assumption of a uniform internal design temperature implies. The gradients may be destroyed by mixing the air, using fans, blowers, etc.

B. Insulation

The options for insulating should be identified and ranked in order of cost-effectiveness for sequential implementation.

C. Reclaiming Heat from Exhausts

The exhausts include all those items listed as direct reject in the audit, the heat from air conditioning or refrigerating condensers, the heat dissipated from air compression, and the ventilation exhaust. The last is by far the greatest single possibility for heat reclaim. Suitable heat recovery equipment (e.g., recuperators, run-around coils, regenerators, heat pipes, or heat pumps) is available to reclaim up to 70% of the heat in contaminated extract air. The major problem in installing a waste heat recovery system arises from the need to collect and redirect exhaust fluids to common ducts containing the fresh air heaters. Straight payback periods commonly quoted for waste heat recovery systems are often less than three years.

D. Improved Boiler Efficiency

Most boiler systems are correctly designed and well maintained. Thus it is often difficult to obtain substantially increased efficiency without introducing recuperative devices. Care must be taken that individual boilers do not overmodulate as a result of unbalanced system load factors. Close control of boiler groups is necessary, and short-term thermal accumulation might be beneficial. Distribution losses are commonly less than 2% of the total energy throughout. It should be appreciated that any energy savings accruing from the application of conservation techniques in the factory result in greater savings of fuel at the boiler house. The energy accountant should, however, also consider the effects upon boiler efficiency of the resulting reduced demand loads.

E. Compressed Air Services

Compressed air is the most expensive form of energy—3.5 times the cost of electricity, kilowatt for kilowatt at 100 psi (689 kPa)—and should thus be used with the utmost frugality. The operating pressures maintained should be questioned, and the possibility of heat recovery from coolers should be investigated.

F. All Other Uses

Only after the major single energy loss sectors have been dealt with should close attention be rendered to the various dispersed and different individual processes. Energy savings via improved monitoring and controls, insulating, and recovering waste heat are always possible. The optimal grade of energy (i.e., type, temperature, and pressure) should be adopted for each station in a process. Attempts should be made to cascade energy through the various levels of process systems, finally rejecting heat to become a sundry gain, if this is beneficial. Process savings that reduce bonus sundry gains are financially worthwhile only if the cost of primary process energy per useful kilowatt-hour exceeds that for space heating fuel. Descriptions and analyses of energy-conserving modifications to individual items of process equipment tend to be esoteric and thus defy generalization.

IV. ASYMMETRIES AND PERIODIC BEHAVIOR

The auditing technique described considers the system in an imaginary quasi-steady state where inputs and outgoings are balanced over a representative annual heating cycle. Further energy savings are possible by considering the periodicity of energy demand. The energy should be provided where and when it is required. An efficient distribution system and the elimination of maldistributed energy (i.e., such as results from stratification, heating unoccupied areas, employing blanket design parameters to include warehouse areas as well as workshops) is vital. Many control systems exist to regulate local air change rates, to modulate groups of boilers to maximize overall conversion efficiency, to peak lop the demand curve via load scheduling, and to optimize pull-down periods. Energy storage devices are being researched to further reduce energy waste as a result of load factor mismatches.

V. CONCLUSIONS

The importance of considering not merely the boiler plant or the load distribution system, or the building characteristics in isolation, but an entire energy-consuming system as an integrated whole has been stressed. A method of energy-auditing the overall system has been described that leads to the identification of conservation options in order of cost-effectiveness. Survey data from recent energy audits have been included and analyzed to highlight the major areas for further attention.

Chapter 21

Industrial Energy Observations and Opportunities

Robert C. LeMay

I. INTRODUCTION

The observations in this chapter are the result of hundreds of industrial plant energy surveys in widely diverse industries. Additional insights were gained from presenting 34 industrial management seminars. The findings herein have been augmented by the observations of ten other energy specialists, and it is hoped that the numerous recommendations we have included here will prove to be timely and useful in industrial applications.

II. STEAM BOILERS

Despite the fact that approximately half our industrial fuel is burned under steam boilers, and the fact that many well-operated steam plants exist, we are convinced that the majority of all industrial boiler operators neither know their boiler efficiencies nor have on hand the relatively simple instruments required to determine them. In other words, they are still "flying blind," hoping that those who service their boilers at infrequent intervals have properly adjusted boiler combustion, and that favorable adjustments have survived. Because fuel compositions and combustion conditions continually change, however, we suggest that boiler operators purchase and use these devices if they are not already employed.

A. Boiler Condition or Control Arrangement

Boiler condition or control arrangement is at times less than satisfactory. In a brass mill surveyed, our portable instruments showed much excess air at the stack, although the burner adjustment indicated no such condition. How could this be with a tight metal surrounding enclosure? A member of our survey team was resourceful enough to go down beneath the boiler, and found that the boiler floor was perforated and sucking in much air beyond the burner wall. This unwanted air chilled the tubes and severely reduced overall heating efficiency.

A brewery had boilers whose fuel-air ratio had always been manually adjusted. After new boilers were installed with automatic fuel-air ratio control, the chief engineer reported that he cut his fuel usage by one-half.

B. Boiler Location

Boiler location is often too far from the places where steam is used. In two large plants surveyed, we found that steam line and condensate return losses were so large that steam costs at the points of use were twice as high as at the boilers. In one old Connecticut plant, it was determined that $200,000 could be saved annually by abandoning the remote, overaged boilers and locating smaller, packaged boilers near the steam-using equipment.

C. Winter Comfort/Process Steam Boilers

Boilers providing steam for both winter comfort heating and smaller process steam uses are often too large for their comparatively small summer loads. With extensive line losses, they are consequently inefficient for summer use. In such cases we recommend either that smaller boilers be employed at points of steam use, or that the process be direct-heated. Either plan permits the large boiler(s) to be shut down during the inefficient summer use period.

D. Economizers and Air Preheaters

Economizers which transfer heat from departing flue gases to incoming boiler feedwater, and preheaters which transfer the heat to incoming combustion air, are all too frequently missing from many of the larger industrial water tube boilers where today's fuel prices dictate that they be employed. Serious thought should now be given to their addition, and an economic survey should be conducted.

E. Waste Heat Boilers

Waste heat boilers are not yet very numerous in the United States. In the past, this situation resulted partially from the relative unavailability of standard packaged units; most of those installed were custom-designed. Today, however, there are many standard catalog units available, and several organizations are prepared to size and sell them. These boilers come with or without auxiliary firing burners to take over when waste heat is not sufficiently available. Wherever high temperature gases are discharged and steam is needed, their use should be considered.

In one plant making heavy forgings, steam for hammers and presses came from a remote boiler plant and suffered serious line losses en route. An on-site study showed that flue gases from the forge furnaces could be run through waste heat boilers to generate all the steam needed by the forming equipment. Consequently, plant management concluded that its large central steam plant should be abandoned in favor of forge shop waste heat boilers. This change reduced manpower, maintenance, and fuel consumption.

III. STEAM DISTRIBUTION AND USE

All too frequently, steam systems fail to return condensate, or to have fully insulated steam lines, return lines, and condensate receivers. Generally, most plant steam lines have been insulated once. In one textile plant, we found 3,000 ft of high-pressure steam main that had never been covered. Hastily made extensions and repairs are often left uninsulated, and to find condensate return lines insulated at all is rather unusual. At one plant, slightly opened valves at the ends of steam lines substituted for both traps and return system!

The next most common shortcoming is the failure to check and repair steam traps with regularity. It is more common to wait until obvious trouble develops before traps are repaired or replaced. Unfortunately, many trap failures are not obvious. Recently, a large midwestern grain processing plant hired a new plant engineer who found that three-fourths of all steam traps were defective.

There are many uninsulated steam and hot solution vessels in industrial plants, most frequently in metal cleaning rooms, plating rooms, and dairies. Using the excuse that inspectors fear product contamination from loose insulation, most operators leave pasteurizers, evaporators, cheese vats, spray dryers, and heated storage vessels entirely uninsulated in dairies. This causes much heat loss, resulting in uncomfortable working conditions. We believe that all of these vessels can be safely insulated behind carefully fabricated stainless steel jackets, but we suspect that operators have been hesitant to make the necessary investment. Tire molds and other rubber molding presses are seldom insulated today, resulting both in high energy consumption and in less than comfortable working conditions.

An entirely unexpected situation was found in the plant of a well-known company that manufactures steam turbine generators at other locations. High pressure steam with superheat was, for some reason, reduced at the boiler room wall to 15 psig (103 kN/m^2) for distribution to other buildings, through a pressure regulating valve. This pressure reduction loses much energy that could be harnessed by pressure reduction through one of the company's own steam turbine generators.

IV. METAL MELTING

Much energy is wasted in metal melting, and this is increased by the fact that melting temperatures are generally high. Ferrous scrap preheating before melting can now be accomplished by drawing hot combustion gases from a burnered refractory cover down through a scrap-containing bucket into a heat fan. This is much preferred to firing burners against scrap piled beneath sheet metal covers. Yet, too many instances of the latter still exist.

Ladle preheating wastes much fuel in many foundries, because of the all too common practice of simply firing down into the ladles with open burners. In one Pennsylvania steel foundry, we observed this practice with ladles 8 ft (2.4 m) in diameter by 8 ft deep. Flame tails rose 10 ft (3 m) into the air above the ladles. Since our first visit, this foundry has correctly installed a burnered refractory cover on one of the ladles, and will presumably do this to confine the heat in the other two as well.

Most reverberatory melters are less thermally efficient than cupolas or several other improved furnace designs now available. A few years ago during the natural gas shortage, we were shown five reverberatory furnace-fed production lines in a large midwestern aluminum extrusion plant. The manager wanted to add a sixth line but could not then purchase more gas. After seeing that high temperature flue gases were being discharged from furnaces directly to the outside, we expressed the belief that metal flue recuperators would recover sufficient heat in combustion air to release enough fuel for that sixth production line. Since our visit, we have been informed that the recommended change has been carried out and the desired result achieved.

Too many melting furnaces have recently been observed with burners firing through open holes in their walls; this arrangement permits very little control of fuel-air ratios. These burners should be replaced with sealed-in units to bring fuel-air ratio control into an acceptable range.

V. HOT FORMING METALS

Steel mills customarily have full-time combustion and energy management engineers. Yet while visiting one of the larger Eastern mill's reheat furnace department, we recently observed large 2,000°F + (1,100°C +) furnaces with doors open 2 ft above their sills because operators could not be bothered to close them between one slab and the next. This, of course, reveals a labor situation that should be corrected.

High production forge shops such as those found in automotive plants are usually well set up; it is normally the job forging establishment whose furnaces and hot bending machines are large energy wasters. We were once called from Philadelphia to Ohio, where an automatic hot bending machine was failing to maintain rated production. Simply by climbing a ladder and removing a dirty air filter, we were quickly able to restore full production. Maintenance of heating equipment is always worthwhile.

Other thermal inefficiencies are more difficult to overcome. The job forging industry is largely energy-wasteful. Unnecessary furnace openings radiating at about 2,400°F (1,316°C), broken refractories, uninsulated linings, ill-fitting doors, crumbling walls, too few burners, maladjusted combustion, and poorly arranged operating schedules are some of the problems observed. We even saw a well known Pittsburgh-area plant with a forge furnace too short for its car hearth. Some enterprising person had knocked out the rear of the furnace and then piled loose bricks at the car's end in an attempt to reduce heat losses at that point. The resulting condition was both unsafe and a major source of heat/energy loss. We reached the conclusion that approximately half of the plant's $3 million annual natural gas purchases could be eliminated.

In another plant, long rolldown furnaces had been designed so that burners at the low ends would send hot gases up the sloping hearths, over the work in a counterflow manner, and out the flues at the high ends where the work was fed in. Along each side of the furnace, however, was a row of holes presumably closed with rolling cast iron covers. Holes were used by operators who inserted bars to nudge along the large work pieces. Careful inspection showed that the covers were sufficiently loose that upper covers discharged flue gases while lower covers admitted sufficient unwanted air to lower furnace temperature and to oxidize the ingots excessively. Our final report stated that one properly designed and insulated furnace with tight doors could do the work of both furnaces present, while using one half the fuel.

Not many forge shops yet employ heat recovery from those hot flue gases leaving at temperatures up to about 2,500°F (1,371°C). Where steam presses and hammers are used, waste heat boilers may perhaps be the best investment. In other cases, high temperature recuperators for combustion air preheating should be considered. And steel forge furnaces operate in a temperature range where oxygen additions to combustion air may possibly save both fuel and money: this is so because added oxygen reduces the necessity of raising nitrogen in the air to high temperatures needlessly.

VI. METAL HEAT TREATING

It is their use on metal heat treating furnaces for which flue recuperators have most widely been publicized—for the preheating of combustion air. Most frequently, they are attached to the ends of radiant tubes. They are helpful and should be encouraged, but most metal heat-treating furnaces today still continue to waste heat in flue products.

One useful and economical device too seldom employed today is thermal cascading, where hot flue products from high temperature processes are transferred to heat lower-temperature processes nearby, thus saving fuel at the lower-temperature equipment. In a Wisconsin plant with a long heat treating furnace aisle, steel hardening furnaces stood on the left and lower temperature "draw" furnaces on the right. Fortunately, management was preparing to cascade heat between one pair of furnaces

across the aisle. If this worked satisfactorily,
management would then duplicate the system all along the aisle.

Thermal cascading can convey heat to "beneficiaries" other than furnaces. In a Connecticut plant, we observed gas-fired comfort heaters being installed in a room adjacent to a heat treating room. In that instance, furnace flue heat should have been cascaded through the wall, either directly or via heat exchangers. Cascading requires only a heat fan, insulated ductwork, a bit of engineering, and sometimes a heat exchanger where flue atmosphere from the higher temperature unit is unacceptable in the downstream location. This has not been advertised widely, probably because expensive equipment is usually not required.

A. High Temperature Insulating Wool

Recently, a new family of high temperature insulating materials has been developed in the form of kaolin or alumina wool blankets or blocks, which can and often should replace hot face refractories in linings of furnaces where there is neither abrasion nor exposure to molten materials or liquids. Proper substitution of these wool materials has saved users significant amounts of energy and heat-up time. Such lining conversions should definitely be encouraged after proper installation has been assured. Make sure that any materials against the higher-temperature face of these wools will not overheat as a result of their addition and suffer resulting deterioration.

B. Flue Gas Analyses and Flame Adjustment

Gas analyses are taken on boiler flues with some regularity to assure proper combustion; so why not also at heat treating furnaces which usually operate at higher temperatures? Where multiple-burnered furnaces require individual burner ratio adjustments, this is usually more complex than adjusting single-burnered boiler furnaces. Yet this is no reason why the effort should not be made, since excess air or gas at high temperature units causes losses greater than in lower-temperature ones. Furthermore, excess air oxidizes away steel at higher temperatures. Once flue analyses are made, furnace fuel-air ratios should be adjusted accordingly. Excess flame should not be permitted to protrude from furnaces, because fuel that burns outside heating units does no good inside them.

VII. VITREOUS ENAMELING

Traditionally, most vitreous enameling furnaces have been indirectly fuel-fired or electrically heated, probably because early attempts with direct fuel firing did foul some of the enamels with incomplete combustion products. Yet we have in recent years conducted sufficient tests with good combustion equipment to be confident that a good line of fused colors can usually be produced in direct-fired furnaces. This has been proved now both in this country and in Europe. The only apparent limitation is that a few delicate colors cannot be exactly matched against those earlier fired out of contact with combustion products. Because much vitreous enameling can now be direct-gas-fired, radiant tube (indirect) furnaces for this purpose usually consume much more gas than is needed. This is so because radiant tubes are less efficient and more costly to purchase and maintain than direct firing.

One midwestern manufacturer relied upon radiant tube firing to fuse vitreous enamel on the interior walls of so-called glass-lined water heater tanks, but complained that he had to start heating his furnace at 5 P.M. the previous evening in order to reach the required 1,600 ° F (871°C) operating temperature the following morning. Yet a competitor not far away was performing the same operation in a direct-fired furnace that heated so quickly that operators shut it down during the lunch hour. To overcome this long heat-up problem, we referred the complaining operator to the fast-heating-equipment's manufacturer.

The most promising opportunity for thermal cascading we have seen recently was in a Wisconsin plant whose parallel production lines perform vitreous enameling and organic enameling plus hot solution washing and drying, with all units separately fired and flued through the roof. There appeared no good reason why exhaust gases from the higher-temperature radiant-tube-fired vitreous enameling furnace could not be employed to heat the lower-temperature organic enameling oven. Likewise, its exhaust gases, or those from the immersion-tube-fired solution tank, could be used to heat the drying oven. The only reason given for failure to cascade this heat was that no one had proposed it.

VIII. GLASSPLANTS

Large glass melting tanks have long employed checker work brick-type regenerators to preheat combustion air, as do those steel mill open hearth furnaces that remain in service. Usually, that is all that is done to recover heat from the melter, yet temperatures measured in flue gases departing glass plant regenerators often show readings of 1,000°F (538°C). This is too much heat to throw away, and we are delighted to learn that waste heat boilers have been added downstream from the regenerators in a few plants, with successful results. Batch dust carryover probably requires incorporation of "soot blowers" or some other means for cleaning boiler tubes, however.

Glass "day tanks," unit melters, pot melters, and other small melting units are all too seldom equipped with appropriate heat recovery devices. However, one West Virginia plant recently reported saving 21% of its fuel by adding recuperators and insulation to day tanks. Glass annealing and decorating lehrs are usually lined with insulating fire brick (IFB). We believe this brick should now be replaced with kaolin wool, with corresponding decreases in burner input ratings. Only inertia seems to have delayed this beneficial step. Glass fire finishing or "fire polishing" is one of the least scientific steps in glass making, and the oft employed "flame bath" is usually highly wasteful of energy. In the light of what is now known about proper fire finishing, we believe most of the "lazy" flame baths in use should give way to efficient modern equipment with better-controlled flames.

IX. ROTARY KILNS

Too many inclined rotary kilns neither preheat incoming feed nor recover heat for combustion air from the departing product. In a midwestern paper mill, the operators reported inability to calcine desired quantities of lime. Because they employed heat recovery at neither end of the kiln, the reason was quite apparent. The engineering and revisions to recover that heat would certainly cost far less than the contemplated purchase of a second kiln, and would require very little additional floor space.

In the United States, with labor costs high and past fuel costs low, most Portland cement plants have chosen the long, inclined rotary kilns with comparatively high shell heat losses. In Europe, where fuel costs have long been a prime consideration, the more energy-economic vertical kilns are often encountered. Near the firing ends of inclined rotary kilns, external shell temperatures around 500°F (260°C) are often measured. Because they usually operate in the open, this design loses much heat to passing winds and rain.

X. BAKERY OVENS

Whereas there are today many fine ovens in use, there are nevertheless some rather sad examples in industrial food plants. A row of older ovens in a Michigan plant was losing so much hot air and combustion product from both ends that a study of internal gas flow was indicated, or, alternatively, a study of heat recovery overhead. Most serious of all, however, was the fact that several of the oven's lateral pipe burners were discharging gas-air mixtures without ignition. That, of course, was both wasteful and hazardous. Some of the ovens employing central gas-air premixing were operating well, but those with burners separately ratio-controlled were poorly adjusted. Our recommendation was to call in the oven builder's service engineer, and possibly convert all ovens to complete gas-air premixing.

Success in several countries with an electrostatic device for increasing oven output and re-
ducing fuel usage appears to justify its increased use here. This so-called Acceletron device provides a simple and relatively inexpensive means for imposing an electrostatic field between metal baking pans and an overhead metal grid. This results in speed-up of moisture migration to the surface of the dough, where the usual oven heat transfer modes can take over. Bakers we visited were happy about the improvements, which they claimed also benefited product quality.

XI. DRYING PROCESSES GENERALLY

Production drying and curing processes require very large amounts of oil and natural gas. As a consequence, these processes challenge us to find ways to reduce fuel consumption. Applications can be divided between the group where water must be removed (such as for drying foods, ceramics, lumber, paper, textiles, and aqueous coatings) and the other group, where hydrocarbon solvents must be carried away (such as is required in baking enamels, foundry cores, and other materials that employ hydrocarbons).

Most older ovens curing organic coatings have built-in energy waste because they were designed to carry much excess air that would assure that oven atmospheres remained well below the lower explosive limit (LEL) for the solvent being used. There is now a way to operate such ovens much more economically, by employing instrumentation that will constantly determine the condition of the oven atmosphere and hold it closer to, yet safely below, the LEL. By resorting to this procedure, an East Coast textile plant was able to double product output with no increase in fuel. We are inclined to believe that this control system can be readily adapted for curing organic finishes on all types of materials.

The previously discussed thermal cascading can also be used to good advantage in many drying processes. Whereas ceramic plants have long employed kiln exhaust gases to predry and preheat incoming greenware, it is now time to look around every production dryer for sources of higher-temperature waste heat. Of course such hot waste gases must be adequately clean and well distributed into the dryer, but the dryer's automatic temperature controls can then be expected to cut back heating energy inputs and effect the desired savings.

Should the waste heat be too hot for the dryer to use directly, automatic dampers can admit enough room air to moderate its temperature. Most washer/dryer sets should employ thermal cascading; manufacturers of new systems are, for the most part, incorporating it, particularly where solution heating employs immersion tube firing.

In discussing bakery ovens, we described the Acceletron process, which speeds moisture migration to the surface of the dough. There is good reason to believe that comparable accelerated moisture migration could take place in drying green brick, other food products, wallboard, lumber, and other unfired ceramic items. The manufacturer of the process believes that the drying of water-based coatings would also benefit.

While conducting an energy survey in a Michigan food products plant, we discovered one drying room that was so hot as to be quite uncomfortable. While taking flue gas analyses from overhead air heaters, we observed that a pressurized hot duct had been left uncapped, causing the room and occupants to dry more effectively than the food product. We arranged for the duct to be capped, after which the hot air was redirected to the dryer. We wondered, of course, how long that heat had been misdirected.

In a poorly maintained Minnesota brick factory, hot kiln exhaust products were conveyed to drying ovens that seemed strangely ineffective. The hot gas transfer duct overhead was covered with fiber glass insulation. By climbing up alongside the overhead duct, we learned that the fiber glass insulation was water-soaked from a roof leak, and was consequently serving as an evaporative cooler for the gases being transported. This incident emphasizes the fact that the condition of insulation and refractories needs periodic checking.

XII. ELECTRICAL POWER FACTOR

Because uncorrected electrical power factors in an industrial plant often increases power bills, in-plant correction should be made by adding either capacitors or large synchronous motors as circumstances permit.

XIII. ELECTRICAL EOUIPMENT AND DEMAND

Many industrial plants employ a few pieces of equipment that create heavy electrical loads during relatively short operating cycles. If these high loads are not staggered, total electrical demands will be higher than otherwise. Load control equipment can be added that will substantially reduce demand charges, lower the power capacity needs, and yet, in most instances, allow the same productive capacity. Substantial power cost savings will result. Demand control equipment need not always be computerized; simple clock switches can do the job in many cases. Surveys also indicate that it is not uncommon to find electrical equipment turned on for extended periods when no work is being done. Unreliable manual operation can be eliminated by installing automatic switching equipment to control the situation.

Some plants continue to use obsolete equipment until it becomes impossible to purchase or fabricate replacement parts. While such a practice does postpone capital expenditure, in some cases it wastes as much as half of the electrical power. We surveyed one plant where obsolete equipment was generating electricity at costs approximating 15 cents per kilowatthour.

Transmitting power over a spread-out industrial complex can cause electrical energy waste. As the plant grows and electrical load increases, the voltage feeding the complex should be increased to offset distribution power losses.

XIV. ILLUMINATION

Many large commercial and industrial establishments allow their lights to operate much longer than necessary. While cleaning and maintenance are sometimes done outside of regular working hours, it is very wasteful to use the entire lighting system for these purposes when a fraction of it would provide sufficient illumination. Obsolete incandescent lighting systems, which emit relatively small amounts of illumination per watt, should be replaced by newer, more efficient fixtures that utilize halogens, mercury, or sodium vapors. At the least, incandescent bulbs should be replaced with fluorescent lamps.

XV. BUILDING STRUCTURES

This presentation would be incomplete without a discussion of the causes of excessive building heat losses. Factories have been surveyed, in northern states, with holes in their walls, broken windows, negative internal pressures, large amounts of single glass, unshielded loading docks, poorly fitting doors, and uninsulated sheet metal outer shells. Because of higher costs for energy, steps are now being taken to correct such defects or to erect better structures. Large single-glass roof monitors and some of the side-wall single glass type are being insulated, in the belief that more heat goes out than light comes in. Enclosures are being installed around loading docks, makeup air heaters are returning work areas to positive pressure, and sources of air leakage are being tightened.

A particularly helpful and inexpensive procedure used when factory heat collects near the tops of high bays is to install controllable-speed propeller fans to blow heat down to working levels. In new construction, convection heating is giving way to radiant overhead gas heaters that heat workers, machinery, and floors preferentially and with much lower fuel consumption. Improvement of comfort heating in industrial plants generally requires the application of common sense, and it is often accompanied by recovery and reuse of process heat now wasted.

CHAPTER 22

INDUSTRIAL COMBUSTION RETROFIT OPPORTUNITIES

John H. Hirt

I. INTRODUCTION

Everyone knows the price of fuel has skyrocketed, but that, in itself, is not a sufficiently dramatic statement to get the full attention of most plant management. In many plants, the energy bills exceed the net corporate profit, or the total plant labor costs. In some cases, energy costs are in excess of 50% of the total operating cost.

By far the most practical remedy for this is to perform a meticulous energy audit on each individual piece of energy-consuming machinery. This consists of analyzing every possible source of energy loss, and then deciding on the best way to solve the problems. In this way, it is possible to approach optimum operating conditions step by step.

Additionally, to increase the awareness of the individual departments, it is wise to charge each department with its own energy costs and allocate individual worker production bonuses on the basis of how well the department meets a predetermined energy goal.

The saving of energy is a hard concept to sell; arms often must be twisted to get the audit project under way, but there are immense savings (profits) to be made on almost any energy-consuming device. I have selected four practical cases to demonstrate the potential savings.

II. PROCESS CURING OVEN

Our first case is a typical process curing oven (Figure 22-1). It will generally have the following sources of energy loss, which must be justified if not remedied:

1. The conveyor belt itself conveys heat out of the oven.
2. The conveyor belt carries solvent or fume out of the oven where it must be collected and incinerated.
3. The belt entrance and exit area may be excessive, allowing excess air to enter the oven, adding to the heat load.
4. The burners may not be operating with the proper fuel-air ratio; they are burning too lean or too rich.
5. The recirculation may be inadequate or inefficient, requiring excessive air movement or operating temperature.
6. The oven shell insulation may not be adequate.
7. The oven shell seams/joints may be leaking, causing heat loss.
8. The exhaust rate may be excessive.

An energy audit of this oven would reveal each problem and its magnitude. When the problems are broken down into little problems, the solutions become more apparent.

III. OVEN WITH THERMAL OXIDIZER

Our second case is the same type of oven, to which has been added a thermal oxidizer with heat exchange provisions (Figure 22-2). It serves to preheat the incoming fume and to add to the heat of the recycled air from the cooling section to the curing section. Additionally, the thermal oxidizer is capable of burning natural gas, fuel oil, and liquid waste. The hydrocarbon/odor level in the exhaust has also been reduced to acceptable levels.

Instead of the thermal oxidizer's increasing the operating costs, it has actually reduced the total fuel consumption. On this particular oven, it

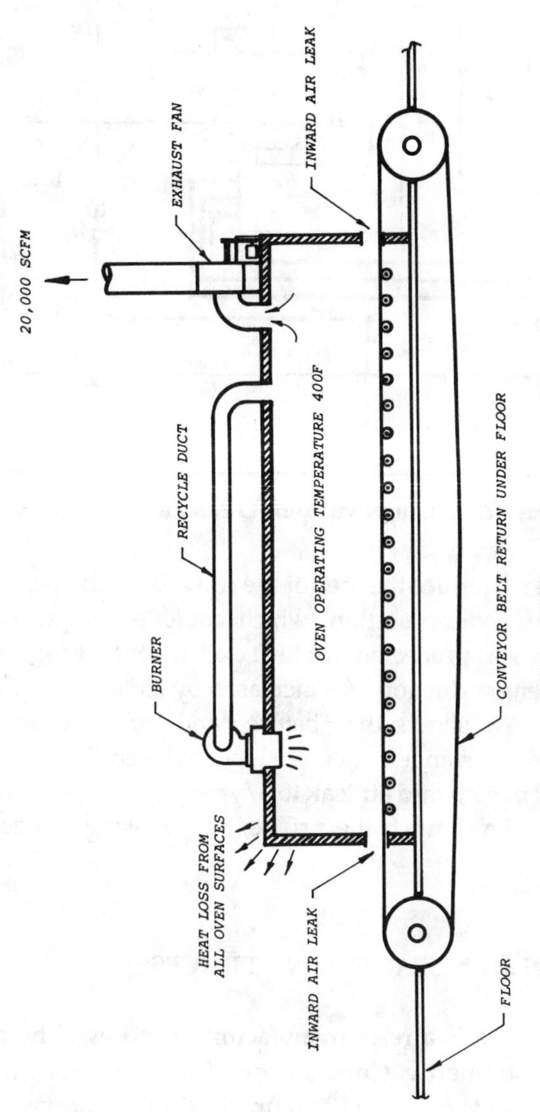

Figure 22-1. Typical Conveyor Belt Oven.

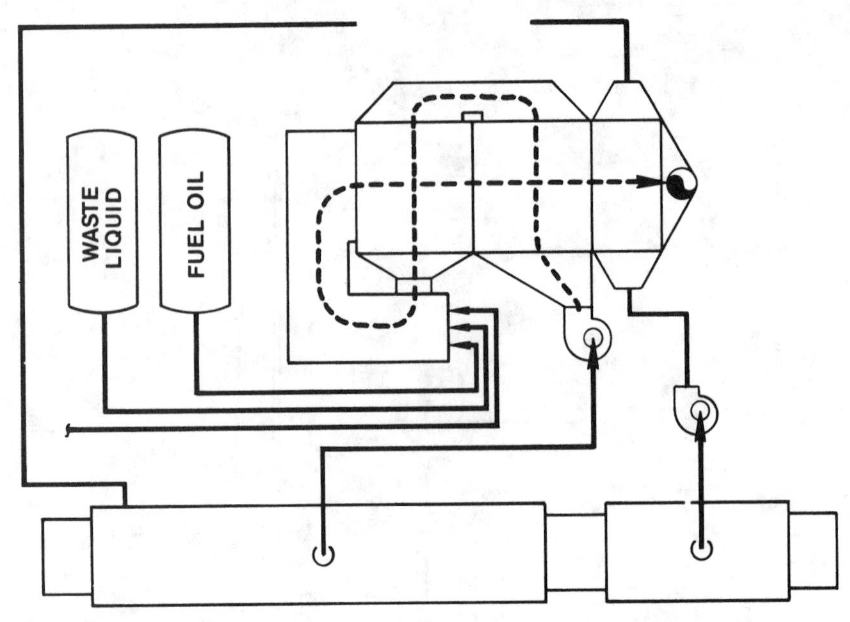

Figure 22-2. Conveyor Belt Oven and Thermal Oxidizer.

was possible to reduce the area of the conveyor belt openings by 50%. The air from the cooling section, which could no longer be exhausted to atmosphere, was preheated in a heat exchanger and recycled to the curing section. Oven production was increased by 25%.

Table 22-1 shows the "before" and "after" comparison between these two oven arrangements. We have reduced the cost of operating a 1 sq ft (0.0929 m^2) inward air leak to $/yr so that it is possible to point out to the operator how much the privilege of viewing the curing process is costing.

IV. RESIN MANUFACTURING PROCESS

Figure 22-3 is a resin manufacturing process. The cooking resin is blanketed with inert gas manufactured from natural gas. Inert gas produced this way is only 5¢/100 scfm (47.21/s), as opposed to 30¢ to 50¢/100 scfm (47.2 1/s) for nitrogen gas. Having this inert gas generator also allows the plant to blanket the hydrocarbon storage tanks, greatly reducing the fire hazard. Gases from the blend tank and storage tanks are diluted with air for safety reasons and then burned in the thermal oxidizer to eliminate noxious odors. The hydrocarbon content of these gases re-

Table 22-1. Conveyor Belt Oven.

	BEFORE	AFTER
Operating temperature, °F	400 (204.4°C)	400 (204.4°C)
Exhaust rate, scfm	20,000 (9,440 L/s)	10,000 (4,720 L/s)
Exhaust rate, MMBtu/hr	7.62 (2.23×10^6 W)	2.23 (0.67×10^6 W)
Lower explosive limit, %	3	6
Operating cost, $/yr	228,600	69,000
(Fuel $4.00/MMBtu—7,500 hr/yr		
Operating cost using liquid waste	—	-0-
1 sq ft inward air leak, $/yr	2,407	3,292
Same, without heat recovery		9,487

duces fuel consumption. In this system, the noncondensable vapors are also used as a fuel source, and the water of reaction is oxidized. Since they contain sufficient organic material to contribute to the net heat requirement, there is a cost benefit in burning them. Additional savings are realized by the elimination of disposal costs. Disposal costs are running as high as $50/barrel ($.31/L), with responsibility for transport and noncontamination of the dump site remaining with the disposee. Net results are that the customer has no air pollution, no waste requiring disposal, and a bonus of well over $100,000 worth of steam per year produced in the waste heat boiler.

Table 22-2 shows comparative costs before and after adding the thermal oxidizer/waste heat boiler.

Table 22-2. Resin Plant.

	BEFORE	AFTER
Misc. organic material disposal	$12,000	-0-
Water of reaction disposal	123,000	-0-
Air pollution odor complaints	Many	-0-
Steam production, lb/yr	-0-	3,450
Value at $6/M lb—7500 hr/yr	-0-	$155,000
Total savings/yr	—	$290,000

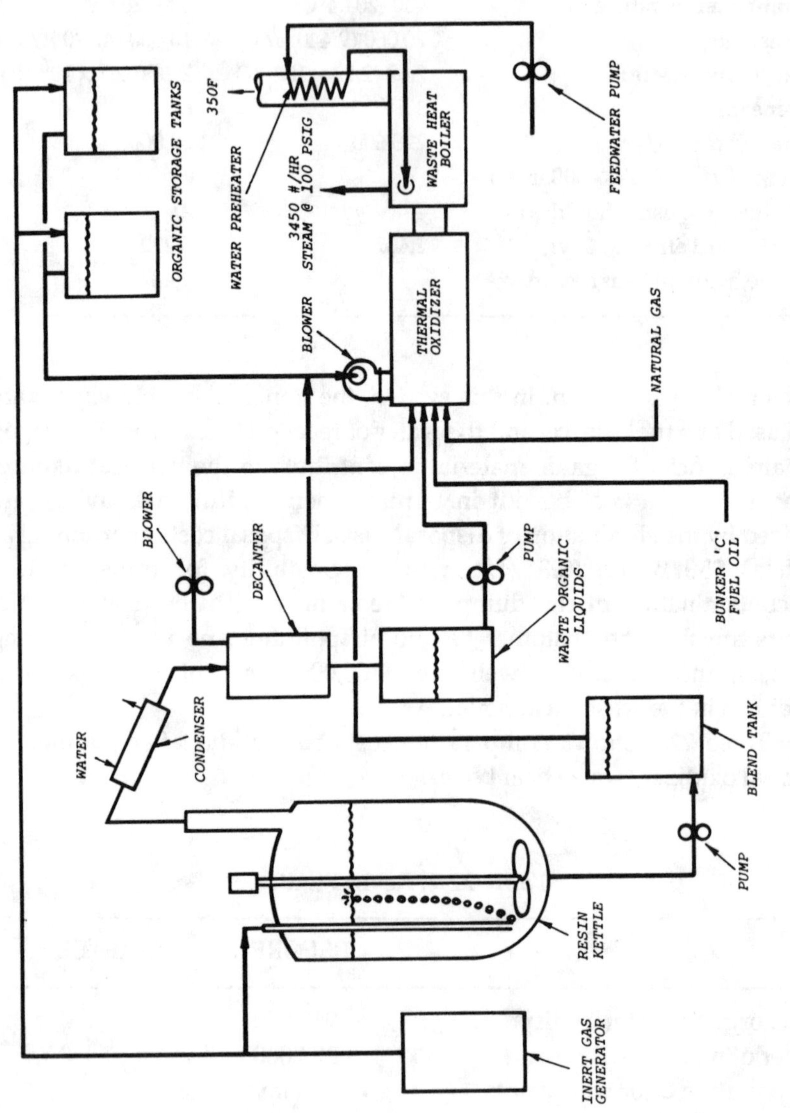

Figure 22-3. Resin Plant Process Schematic.

V. ALUMINUM SCRAP MELTING FURNACE

In Figure 22-4, we see a typical aluminum scrap melting furnace. The aluminum is melted in the well, often causing smoke problems. The furnace may or may not have proper hot gas circulation for proper heating of the metal. When firing natural gas, the burner should be capable of producing a 12% CO_2 Orsat with zero combustibles. As the gases approach this 12%, the excess air is reduced, increasing the thermal efficiency. Additionally, the gases are hotter, so they transfer their heat faster, causing a lower stack temperature. This further increases the thermal efficiency. A heat-exchanger can be added to the stack to preheat the combustion air, adding to the radiant effect of the flame and further increasing combustion efficiency.

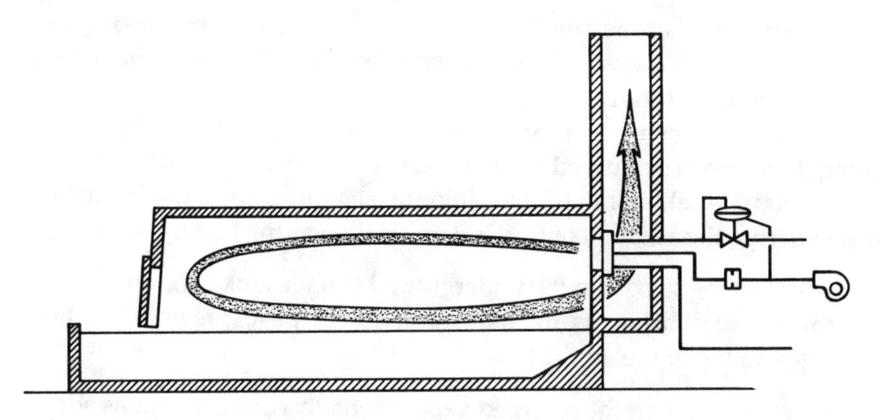

Figure 22-4. Aluminum Melting Furnace.

Table 22-3 shows the savings that can be achieved by proper burner and heat exchanger application. Obviously, the burner must be capable of accepting the preheated air. A large side benefit is the reduction of dross loss achievable by proper combustion and the elimination of inward air leaks.

VI. GASOLINE STATION SAVINGS

The application of air pollution control equipment often results in economic benefits. Figure 22-5 represents a typical gasoline service station equipped with vapor control devices. Environmental Protection Agency

Table 22-3. Aluminum Melting Furnace.

	BEFORE	AFTER
Operating hours per year	5,000	5,000
Firing rate, MMBtu/hr	22.5 (6.593×10^6)	16.9 (4.95×10^6)
CO_2, % of flue gas	10.5	12
O_2, same	2	0.2
Preheat, $/yr	450,000	337,000
Dross loss, %	3	1
Same, $/yr	1,688,000	563,000

has determined that the venting of fumes to the atmosphere during the filling of automobile gas tanks is unacceptable. By adding a vapor collection hose to the fill nozzle, the vapors are returned to the underground storage tanks. Similarly, as the tanker fills the storage tanks, the fumes being displaced are returned to the tanker.

A 200,000 gal/month (758 m^3/month) station achieves the following benefits from the addition of this system (known as the Hirt system):

1. By returning vapors to the underground storage tanks, the air-hydrocarbon equilibrium is maintained. This, in turn, prevents further evaporation of fuel in the tank.

2. In warm weather, the returned vapors actually condense out as additional gasoline.

3. The owner eliminates any future claims for disability from breathing the fumes.

Table 22-4 shows the savings possible with the system.

VII. CONCLUSION

It is not possible to cover in a brief chapter the entire spectrum of equipment to which the foregoing principles apply. I can only say that the energy audit approach is by far the best, and that the subsequent application of ingenuity and good engineering principles will reduce wasteful energy consumption to acceptable levels, helping all of us stay in business.

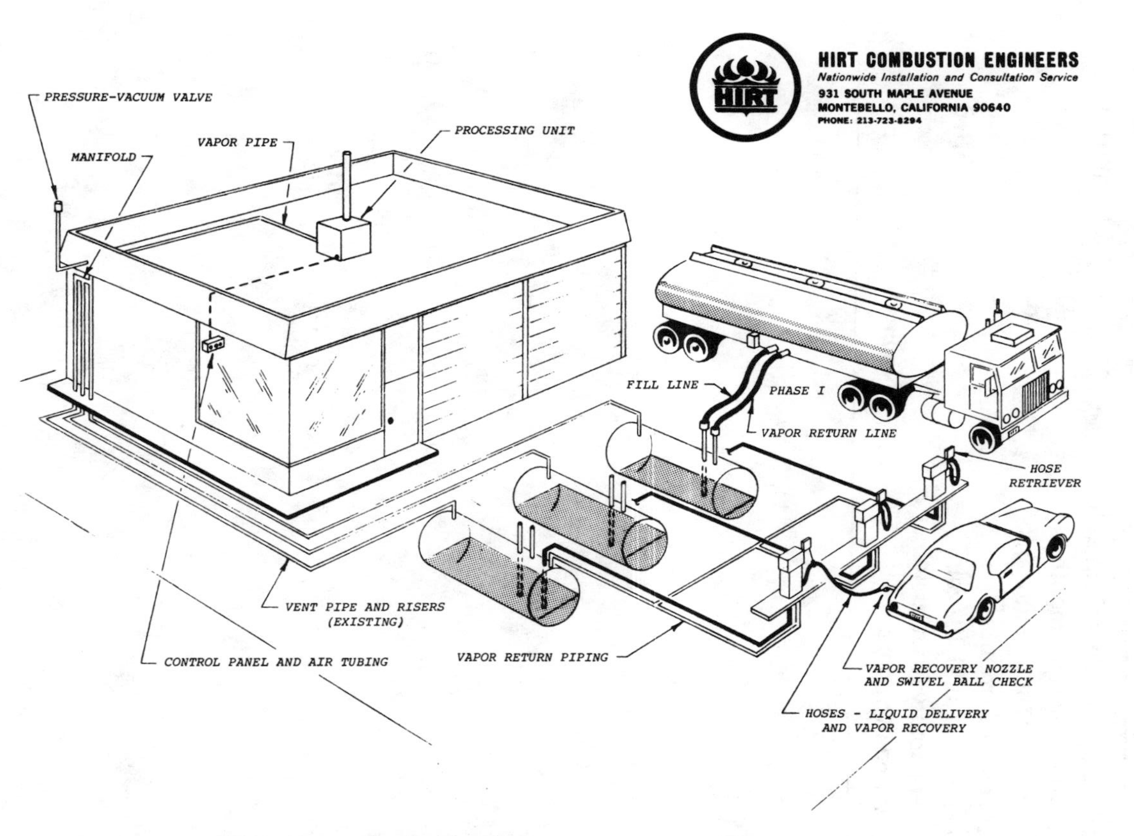

Figure 22-5. Gasoline Station Vapor Control System.

Table 22-4. Gasoline Service Station Vapor Destructor.

	BEFORE	AFTER
Station pumping rate, gal/month	200,000	200,000
Evaporation loss, gal/month	200	-0-
Recovery from filling auto tanks	-0-	100
Net gain, gal/month	—	300
Yearly net gain at $1.10/gal		$3,960
Potential liability from customer breathing of vapors	No limit	-0-

CHAPTER 23

INDUSTRIAL RETROFIT ECONOMICS

James M. Archibald

I. INTRODUCTION

With the advent of a shortage in fossil fuels and high costs for fuels and electrical power, it is incumbent upon plant operators to determine points of potential savings and to develop maximum fuel and power utilization for their facilities. The purpose of this chapter is to inform the reader about reductions in fuel and electrical energy usage that can be had by the full utilization of air-to-air, tube-type heat exchangers and the installation of equipment that can make use of heat now being wasted.

II. PLANT SURVEY

The first step in a fuel and power savings program is an obvious one: a plant survey. The person assigned to make or supervise the survey must be qualified and not too burdened with other responsibilities to devote a major portion of his or her time to doing a thorough job. If you do not have qualified personnel, there are many companies that can provide this service to you. Make certain, however, that the company you select has had experience in combustion, heat transfer, oven and furnace operation, air flow and duct design, fuels and fuel efficiencies, electrical energy savings potential, etc.

Any survey should start with a plant layout listing all of your energy- or fuel-consuming equipment. All data concerning each piece of equipment should be obtained and analyzed. When information is sketchy, meters should be installed. After you have examined combustion rates and correlated fuel use to production, you can determine the current equipment process efficiency and compare it to industry standards.

An integral part of the fuel usage evaluation of each process is a study of whether substitute methods or energy recovery will reduce energy use. Ask yourself any conceivable questions regarding a better way to use fuel. If you cannot answer them about specific subjects, do not hesitate to call in outside help.

After all the data are gathered and evaluated and you have answered all the questions, you are prepared to outline an energy use program. One method your program may call for is the use of a heat recovery system on one or more of your processes. A questionnaire for making an analysis of the system and what you want it to accomplish is given in Table 23-1.

III. TUBE-TYPE HEAT EXCHANGERS

If your evaluation leads to the conclusion that a tube-type heat exchanger can meet your needs, let us look at design considerations. The primary variables in the design of a tube-type heat exchanger are:

- The square foot area of the tubes in the exchanger
- The pressure drop in the exchanger of air over and through the tubes
- The heat transfer coefficient in Btu/sq ft/°F/hr (W/m^2°C)

Computer runs with the same air flow and percentage of heat recovery efficiency indicate the effect of surface area per tube on pressure drop and the size of the unit (refer to Figure 23-1). Final design is a compromise. Efficient design will use the optimum combination of pressure drop, spacing and arrangement of tubes, tube area, and size of unit to fit the

Figure 23-1. Information Needed for Design of Heat Exchanger for Given Application.

Name ______________________ Title ______________________
Company ______________________ Address ______________________
City ______________ State _____ Zip _______ Phone __________

1. Amount of air from heat source ______________________ SCFM.
2. A. Temperature of air ______________ MAX. ______________ MIN.
   B. Desired exhaust temperature from exchange __________.
3. Room available for installation.
4. Room available from heat source, " of water column _____.
5. Amount of air required by using system ______________ SCFM.
6. Rise in temperature of air required by using system __________ °F.
   Temperature in ____________ Temperature out ____________.
7. Typical chemical analysis of fume source. Especially interested in any form of chlorides, fluorides, or sulfides. Any alkalies that may tend to become sticky or gummy at high temperature or any other fume contaminant that could create problems in the exchanger.
8. Any abrasive particles in the fume source and the grain loading and size in micron.
9. Heat recovery desired in % recovery ______________________.
10. Other possible uses of heat ______________________
    ____.
11. General description of proposed system ______________________
    ____.
12. Other comments relative to the particular instllation ______________
    ____.

user's installation requirements.

Pressure drop is partially based on velocity and, therefore, in addition to area or footage of tube, can be adjusted by changing the diameter of tubes (pressure drop through the tubes) and the spacing of tubes (pressure drop over the tubes).

A good design will have a film coefficient of approximately 4 to 5 Btu/hr/sq ft/°F (22.72 to 28.4 W/m^2°C). Adjusting the variable factor to arrive at this film coefficient provides the most economical design. Heat exchangers are designed for percentage of recovery. The higher the per-

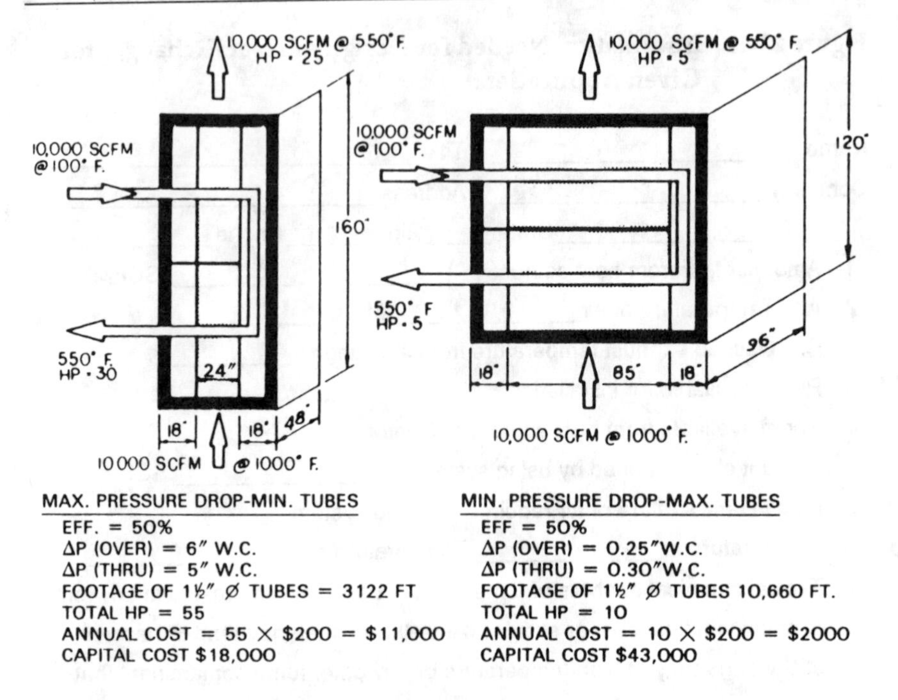

Figure 23-1.
Effect of Surface Area Per Tube on Pressure Drop and Size of Unit.

centage of recovery, the higher the Btu savings. If we compare a 10,000 scfm (472 L/s), 1,000°F (537.78°C) rise, 80% efficient heat exchanger with the 50% efficient heat exchanger, the Btu savings are increased from 5,800,000 to 9,280,000 Btu (6,119,000 to 9,790,400 kJ). The tube will increase four times, so the cost of the tube in the unit will increase by four times, and the cost of the unit will increase three times from, say, $25,000 to $75,000.

A typical curve for tube usage versus heat exchanger efficiency is shown in Figure 23-2. In Figure 23-3, a curve for fuel savings versus percent efficiency is shown. The effect of electrical power cost versus pressure drop is also indicated.

In evaluating heat exchanger units, check pressure-drop or horsepower when comparing cost. In other words, if your final design saves 50% of your fuel but increases horsepower 50%, you have made no real savings. The advantage of the typical tube-type heat exchanger is that the design lends itself to variable efficiency of heat recovery. By using various alloy steels, temperatures as high as 1,800°F (982.22°C) can be satisfactorily recovered. As a user, you are interested in long, maintenance-free life

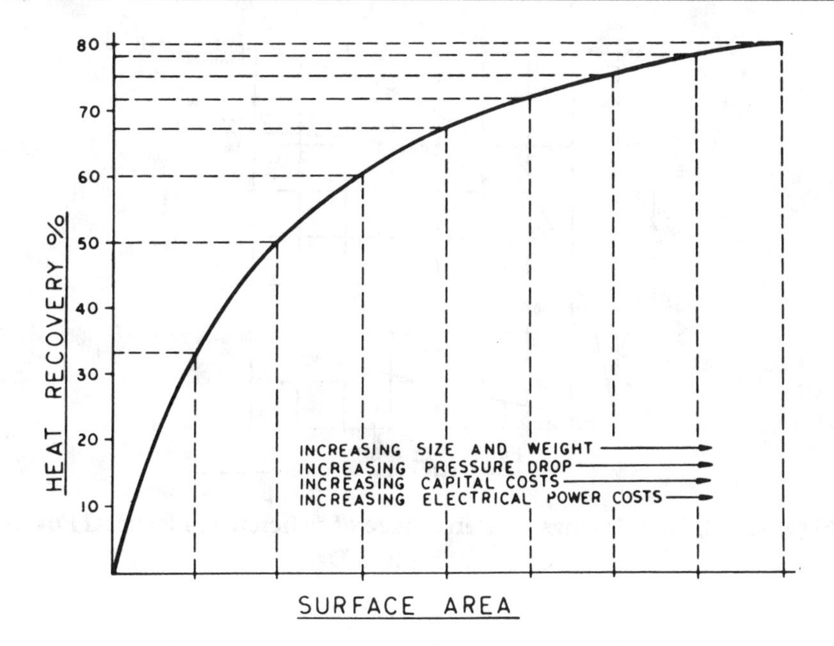

Figure 23-2. Tube Usage vs. Heat Exchanger Efficiency.

for your heat-exchanger installation. It must be designed correctly to withstand your process problems. Potential problems that will affect exchanger life, along with suggested solutions, are shown in Table 23-2. Units can be custom-designed to meet your requirements. Next, let us examine some typical applications of tube-type heat exchangers. The examples shown will have theoretical air quantities, air temperatures, etc., to make the computations easier to follow.

IV. DIRECT FURNACE GASES

Examine the use of direct furnaces or oven gases to heat your ware. An example of this would be a paint spray operation using barrels, buckets, or parts prior to painting cans. The hot products of combustion from your oven or furnace dry the cans from the washer prior to painting. In most cases, the products of combustion will not contaminate your finished ware. Such a system can be an economical application of heat; all that is needed is ductwork and some form of minor temperature control.

Other applications of direct furnace gases might be the preheating of process materials such as steel scraps prior to dumping into an electric

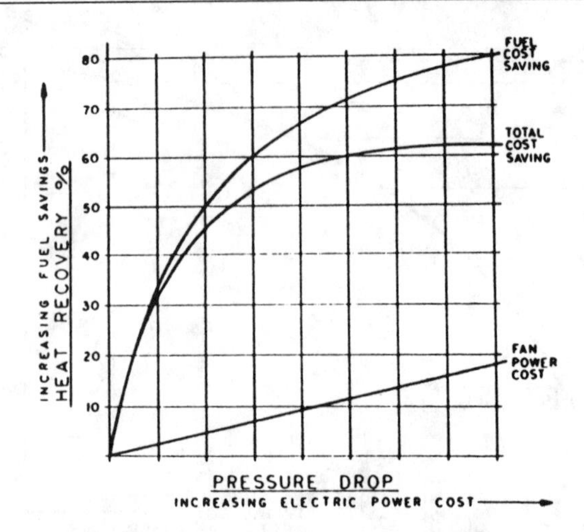

Figure 23-3. Fuel Savings vs. Percentage of Efficiency, Electrical Power Cost vs. Pressure Drop.

furnace, preheating ceramic ware prior to entering a firing kiln, or placing raw materials for cement manufacturing in a preheater design system.

If a heat exchanger is placed on a furnace that discharges to the atmosphere, it is desirable to keep pressure-drop through the tubes to a minimum so that no fan is necessary on the furnace discharge. This saves horsepower (kW) and keeps the cost of installation to a minimum. Extra footage of tube in the exchanger will be warranted to accomplish these savings. Figure 23-4 illustrates these savings. Your supplier can accomplish this design by reducing pressure-drop through tubes to less than 0.25 inch (62.25 Pa) water column.

V. SPACE HEAT

The heat from your furnaces or ovens or other heat devices can be recovered by installing a heat exchanger on the exhaust gas stack. The exhaust gases will discharge through the tubes of the exchanger, and fresh outside air will be preheated by passing over the tubes prior to entering the plant. Refer to Figure 23-5 for details.

In the winter, it is important that plant atmospheres be kept under positive rather than negative pressure so that the building is expelling air rather than bleeding in cold air, which requires makeup heaters to maintain temperature.

Table 23-2. Process and Design Problems.

Problem	Solutions
1. Expansion	1. Bellows-allowance for movement in all directions. 2. Individual tube expanders. 3. Fix one end. Let other end float.
2. Abrasion material	1. Run dirty gases through the tubes. 2. Use straight wall tubing. 3. Use Smith patented replaceable wear inserts. 4. Reduce velocity through the tubes. 5. Increase wall thickness of tube.
3. High temperature	1. Alloy steel to match temperature requirements. 2. Bleed in air to cool to meet temperature of tube material. 3. Bypass in event of furnace upset condition.
4. Carbide precipitation	1. Run unit continuous. Reduce cycling of furnace. 2. Use material to withstand carbide precipitation. 3. Install higher temperature Alloy steel.
5. Tube plugging due to sticky materials or condensation	1. Keep temperature to tube wall above dew point. 2. Increase velocity through tubes. 3. Run dirty gas through plain wall tubes. 4. Preheat first pass over tubes to prevent condensation. 5. Cool all gases below sticky temperature. 6. Provide positive mechanical cleaning.
6. Deterioration of tubes caused by contamination of corrosive material	1. Determine composition of gas stream. 2. Use corrosive resistant materials. 3. Use pilot test model. 4. Bypass to non-corrosive operating temperature. 5. Build replaceable sections to reduce maintenance. 6. Use thicker wall tubes. 7. Run dirty gases through tubes. 8. Neutralize contaminate with special additions.

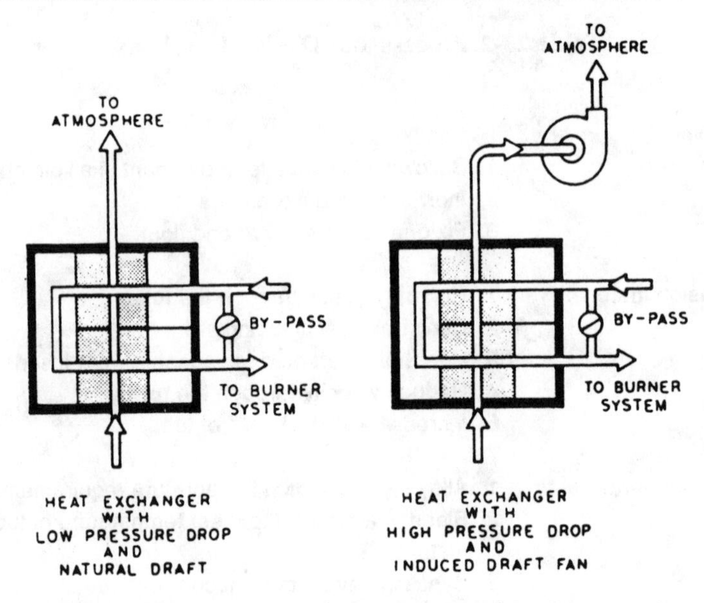

Figure 23-4. Heat Exchangers on Furnaces that Discharge to Atmosphere.

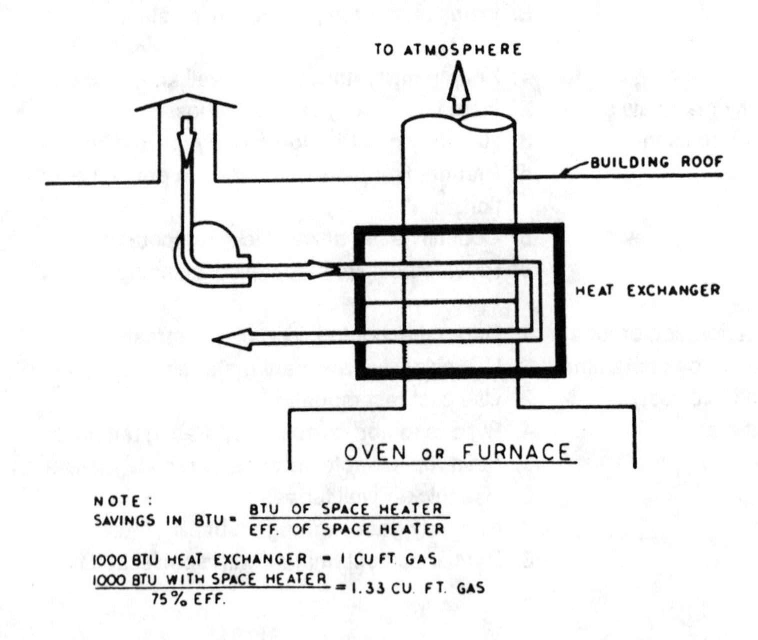

Figure 23-5. Recovering Heat by Use of a Heat Exchanger on the Exhaust Gas Stack.

Many plants have potential for fuel savings in the heating of combustion air for furnaces, ovens, etc. Furnaces now in operation usually have burners that can utilize preheated air up to 600°F (315.56°C). New furnace manufacturers can make available systems using 1,000°F (537.78°C) to 1,200°F (648.89°C) combustion air.

Potential savings are shown in Figure 23-6. A typical plant layout showing small furnaces is indicated in Figure 23-7. Special design of controls is necessary. This must be included with the heat exchanger at the time of installation.

VI. RECOVERY BACK TO PROCESS

Incineration of hydrocarbon fumes from paint baking ovens is required to meet air pollution requirements. In a typical system, the oven and coater discharge solvent-laden air to a thermal oxidizer (refer to

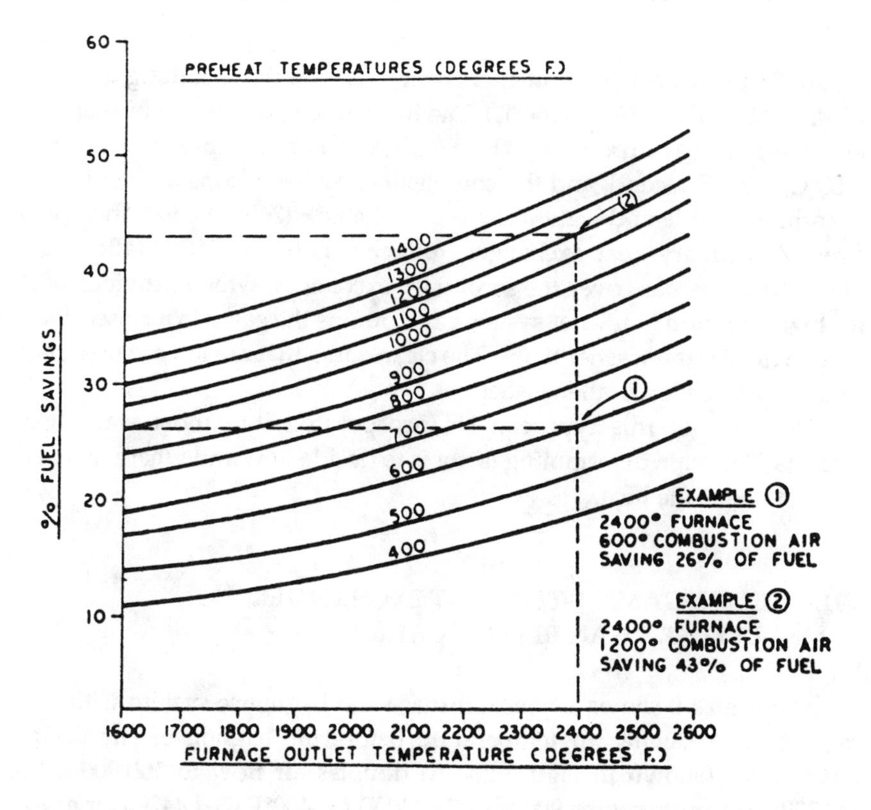

Figure 23-6. Potential Savings from Heating of Combustion Air.

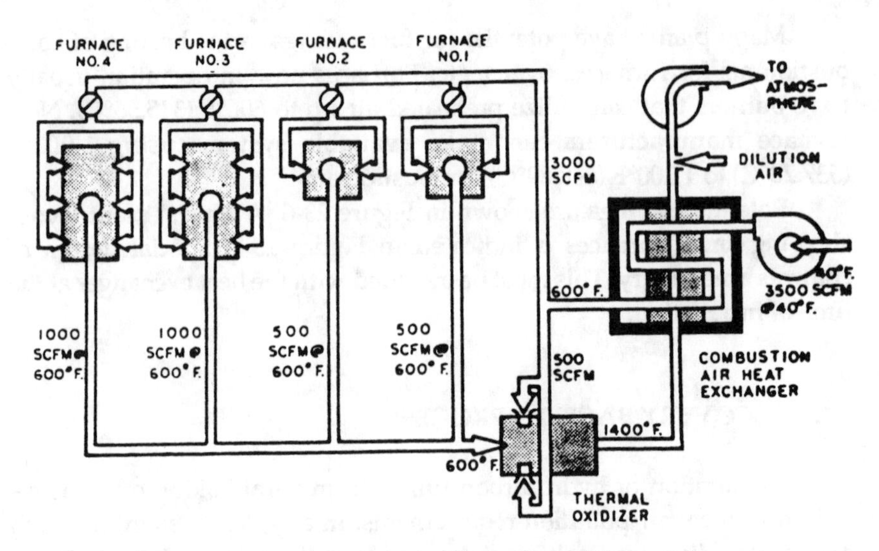

Figure 23-7. Wax and Plastic Burnout Furnaces.

Figure 23-8). A primary heat exchanger preheats the incoming air from 270°F (132.22°C) to 860°F (460°C). The fumes then pass over a burner that raises the temperature to 1,400°F (760°C). It is retained at this 1,400°F (760°C) for 0.5 second, and the combustible hydrocarbons are converted to carbon dioxide and water. The heated 1,400°F (760°C) gases then pass over the primary heat exchanger, reducing gases to 760°F (404.44°C). Next, the gases pass over the secondary exchanger, which provides heat for high-pressure hot water systems and pumps the gases in a closed-loop system to heat the washer tanks. The clean gases that are left are then sent to the dryer section of the washer.

Systems of this type can be incorporated into most waste gas streams. The main determining factor is to decide how and where to make the best use of the heat.

VII. COOL GASES WITH HEAT EXCHANGER RATHER THAN BLEED-IN AIR

Figure 23-9 shows a typical furnace and baghouse system with two methods of reducing air temperature before the baghouse. The use of bleed-in air (shown in Figure 23-9A) doubles air flow to 100,000 scfm (47,200 L/s) and requires 900 hp (671.4 kW) for 400°F (204.44°C) air at 18" SP (4484 Pa). Installing a heat-dispersing exchanger or exhaust-gas-cool-

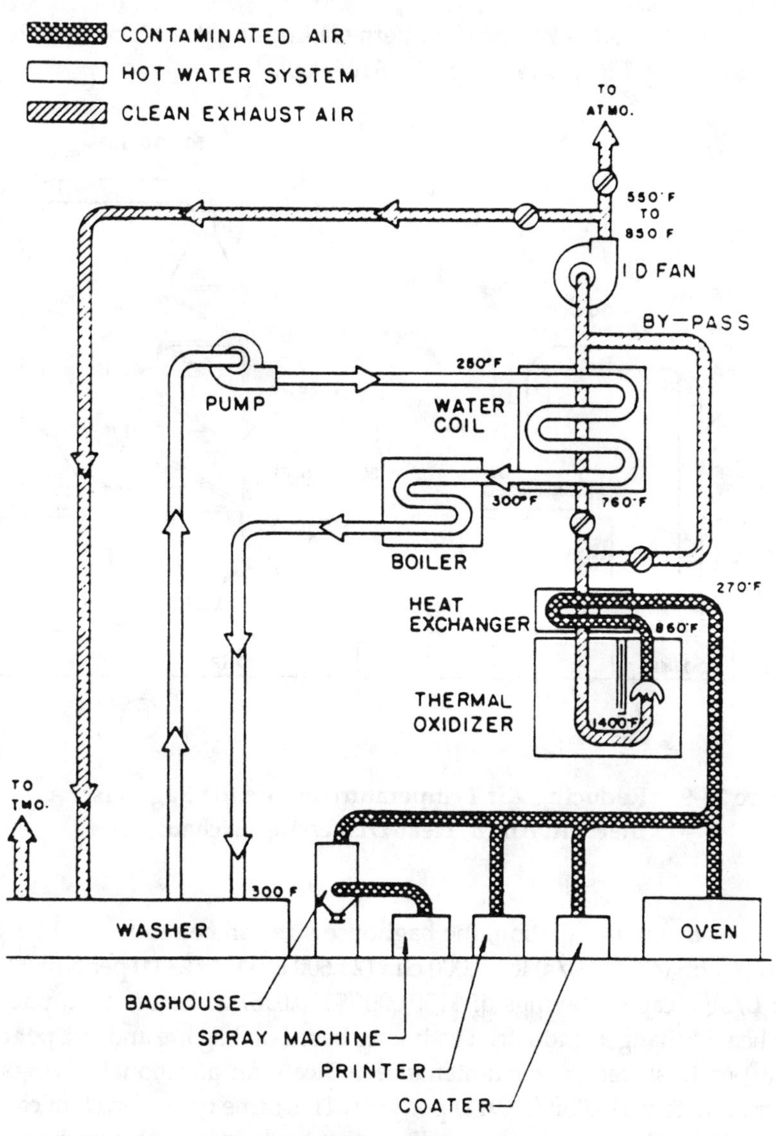

Figure 23-8.
Systems for Incineration of Fumes from Paint-Baking Oven.

ing exchanger (Figure 23-9B) adds only 2" SP (498 Pa) yet maintains original air flow at 50,000 scfm (23,600 L/s), reducing horsepower to 300 (223.8 kW). By adding 15 hp (11.19 kW) each for the four cooling fans, the total horsepower becomes 360 (268.56 kW), a savings of 540 hp (402.84 kW) over the traditional bleed-in air system. Annual savings, estimated at

\$200 per hp (\$268.1/kW) (per Southern California Edison), become 200 × 530 = \$106,000 (\$268.1 kW × 395.38 = \$106,000).

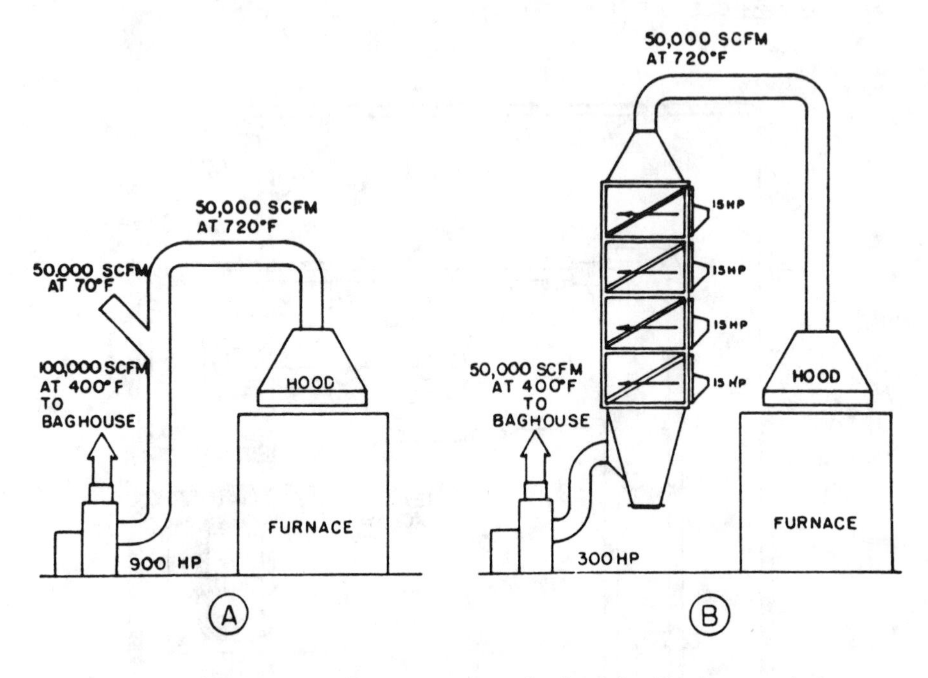

Figure 23-9. Reducing Air Temperature before the Baghouse. A, Bleed-In Air. B, Heat-Dispersing-Exchanger.

For a new installation, the baghouse size can be reduced 50% from 100,000 cfm (47,200 L/s) to 50,000 cfm (23,600 L/s). At \$3.00 per cfm (\$6.36 per L/s), a capital savings of \$150,000 (\$150,096) will more than pay for the heat exchanger. Industrial estimates based on bag life and 20¢ per cfm (42¢ per L/s) per year maintenance indicate an additional savings of approximately \$10,000 (\$9,912) per year. This same type of system can be applied to melting furnaces as well as other heating processes such as lime and cement manufacturing.

VIII. IRON FOUNDRY CUPOLA

An iron foundry cupola is shown in Figure 23-10. Its function is to melt scrap iron, and it is an excellent candidate for heat recovery. In recent years, foundries have had serious problems with air pollution, hydrocar-

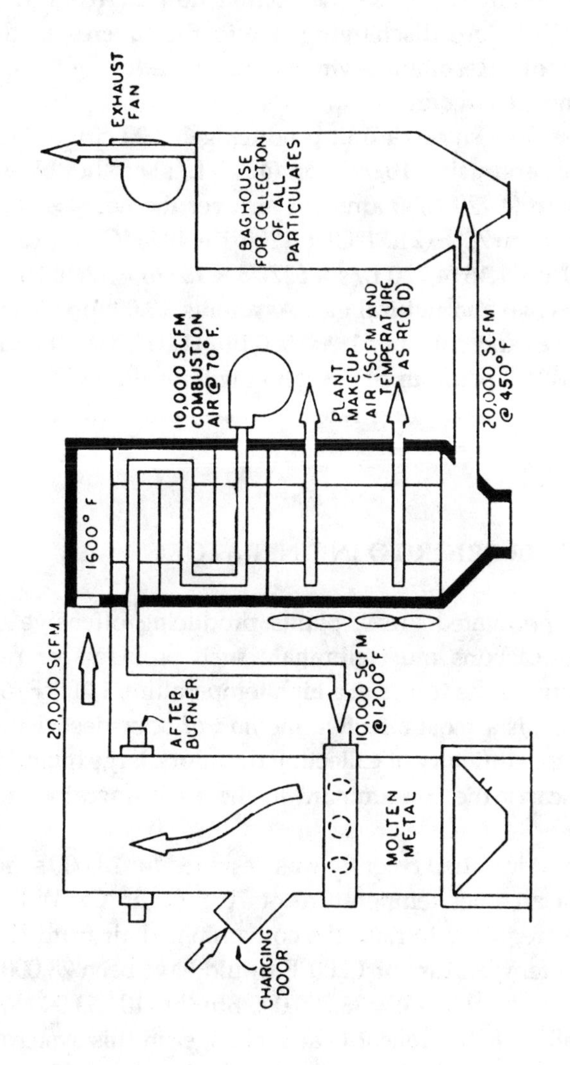

Figure 23-10. Iron Foundry Cupola.

bons, carbon monoxide, and particulates. Correcting these problems necessitated the use of additional Btus of fuel. A method used to recover this heat and also to cool the exhaust air from the cupola before entering the baghouse is the installation of an air-to-air tube-type exchanger in the system. Proper design can raise the combustion air from ambient to 1,200°F (648.89°C) before discharging it into the tuyeres in the lower section of the cupola. Resultant savings occur by reducing the quantities of natural gas and coke required.

The process flow shows a unit producing 20,000 cfm (9,440 L/s) of hot exhaust gases and using 10,000 cfm (4,720 L/s) of "hot blast" air. By passing 10,000 cfm (4,720 L/s) ambient air over the heat exchanger and thereby raising it from 70°F (21.11°C) to 1200°F (648.89°C), the savings will be 10,000 × 1130°F × 1.16 (4,720 L/s × 627.78 × 1.296) = 13,108,000 Btu/hr (3,840,206 W) in coke and natural gas. Assuming $3.00 for the combined price of natural gas and coke per 1,000,000 Btu (1,055,000 kJ) a savings of approximately $40.00 per hour of operation will result. A side benefit can also be realized; as the melting rate increases, productivity can be increased as well.

IX. PREHEAT AIR PRIOR TO INCINERATOR

In heavily populated areas, plants producing offensive odors or exhausting hydrocarbons must eliminate such problems or risk being closed down. Burning the fumes at a high temperature, 1,200°F (648.89°C) to 1,400°F (760°C), is a most effective method of odor destruction; however, because of the difficulty of collecting the odors, large quantities of air are required to carry the contaminant to the air incinerator or thermal oxidizer.

A typical unit installed recently was designed for 28,000 scfm (13,216 L/s) of air with an inlet temperature of 70°F (21.11°C). Without heat recovery, the gas required to raise the contaminated air from 70°F to the odor-destruction temperature of 1,200°F would have been 28,000 × 1,130° × 1.16 (13,216 × 627.78° × I.296) = 36,700,000 Btu/hr (10,753,000 W). It was decided to Install an 80% preheat heat exchanger in this system with an hourly savings of 29,300,000 Btu (8,584,900 W). Assuming $3.00 per 1,000 cu ft (28,300L) of gas, the savings were 29.4 × 3.00 = $88.20 per hour. Operating 6,000 hours per year, the annual savings averaged $529,200.00. For details of this system, refer to Figure 23-11.

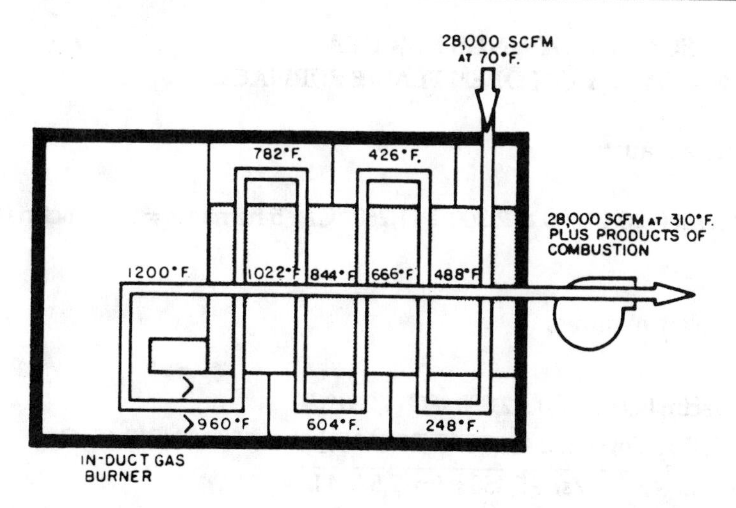

Figure 23-11. Thermal Oxidizer with 80% Heat Exchanger.

X. WASTE HEAT BOILERS

Another method of using available waste heat is in firing a waste heat boiler. Here, we use the exhaust from several furnaces and dilute the temperature to a range more acceptable for standard boiler construction. Each furnace has damper controls so as not to increase the exhaust flow with the induced draft fan at the boiler exhaust. The fan is equipped with a vortex damper that can maintain static pressure control throughout the system.

The process flow shown in Figure 23-12 indicates that five furnaces have an exhaust flow of 9,750 scfm (4,602L/s) at 2,300°F (1,260°C). With dilution with ambient air, there is 15,100 scfm (7,127.2 L/s) of 1,500°F (815.56°C) air available to the boiler.

When we transfer these Btu's to the boiler, assuming approximately an 80% effective transfer into steam, we have a boiler capable of producing 583 hp (434.92 kW).

Relating this cost back to the cost of firing the boiler with natural gas at a cost of $3.00 per 1,000 cu ft, (28,300L) this is equivalent to $73.46 per hour or, at 2,000 hours per year, $146,920.00 per year.

XI. FUEL COST SAVINGS FOR HEAT RECOVERY ON OPEN FLAME FURNACE

Furnace exhaust:

1,950 scfm (920.4 L/s) at 2,300°F (1,260 °C) × 5 furnaces = 9,750 scfm (4,602 L/s).

Dilution air required:

$$\frac{[9{,}750 \text{ scfm } (4{,}602 \text{ L/s}) \times 2{,}300 \text{ °FG } (1{,}260 \text{ °C})] + [5{,}450 \text{ scfm } (2{,}572.4) \times 70 \text{ °F } (21.11 \text{ °C})]}{9{,}750 \text{ scfm } (4{,}602 \text{ L/s}) + 5{,}450 \text{ scfm } (2{,}572.4 \text{ L/s}} = 1500 \text{ °F } (815.45 \text{ °C}$$

Exhaust to boiler:

15,200 scfm (7,174.4 L/s) at 1,500 °F (815.56 °C)

Exhaust from boiler:

15,200 scfm (7.174.4L/s at 420 °F (215.56 °C)

Heat content in 1,500 °F (815.56 °C) air:

15,200 scfm (7,174.4 L/s) × 1.1 (4.52)
× 1,500 °F (815.56 °C)
= 25,080,000 Btu (26,459,400 kJ)

Heat content in 420 °F (215.56 °C) air:

15,200 scfm (7,174.4 L/s × 1.1 (452) × 420 °F (215.56 °C)
= 5,544,000 Btu (5,848,920 kJ)

Assuming natural gas cost is $3.00 MCF:

$$\frac{420 \text{ °F/hr } (215.56 \text{ °C/hr}) \times 583 \text{ hp } (434.92 \text{ kW}) \times \$3.00}{1{,}000 \ (28{,}300)} = \$73.46 \text{ per hour}$$

Assuming only 40 hr/wk × 50 wk/yr = 2,000 hr/yr, yearly savings:

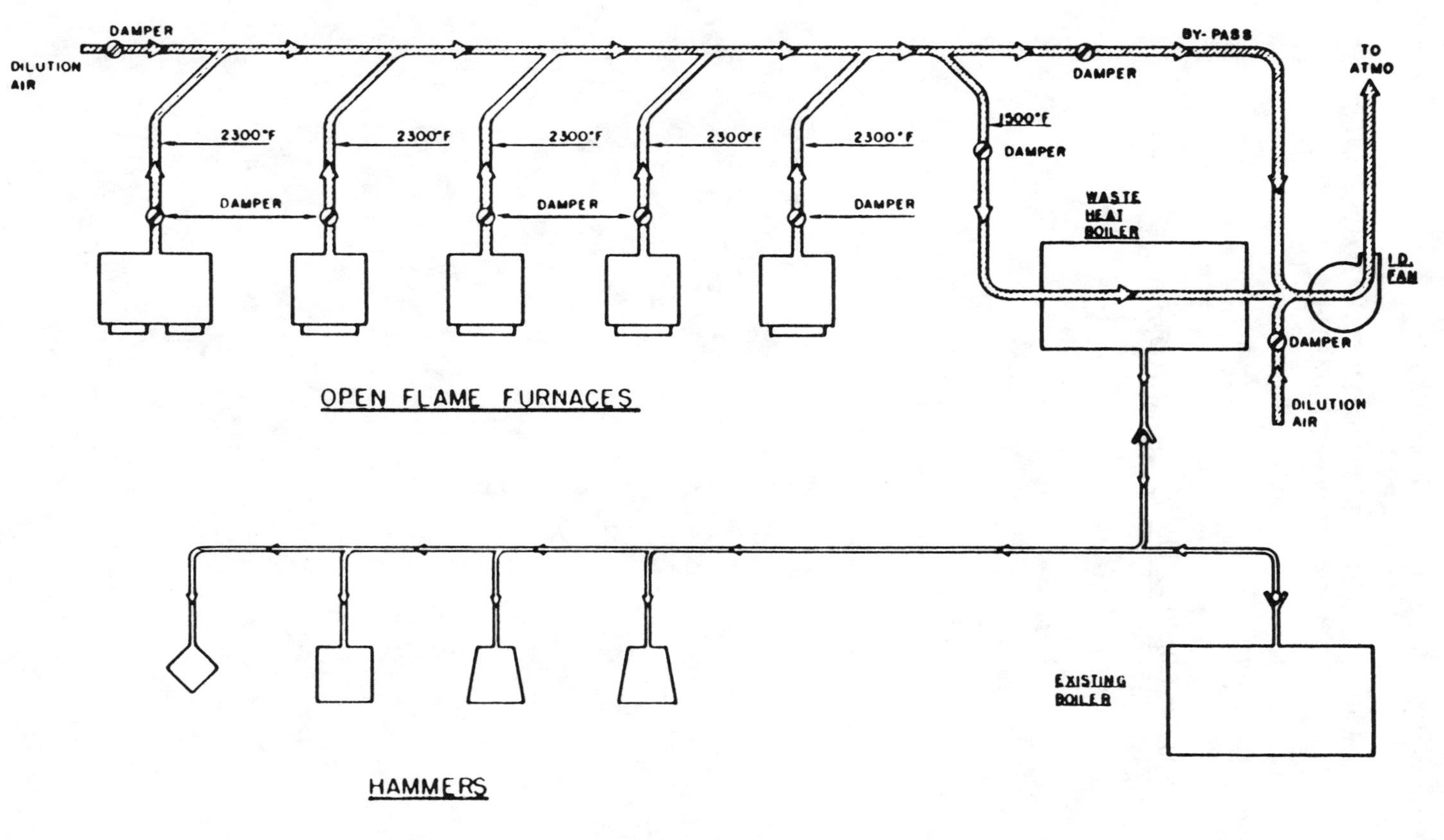

Figure 23-12. Heat Recovery from Open Flame Furnace.

2,000 hr/yr × \$73.46 per hour = \$146,920.00/yr

Btu transferred: 25,080,000 (26,459,400 kJ) – 5,544,000 (5,848,920)
= 19.536,000 Btu (20.610,480 kJ)

Boiler horsepower:

$$\frac{19{,}536{,}000\,\text{Btu}\ (20{,}610{,}480\,\text{kJ})}{33{,}475\,\text{Btu/hp}\ (47{,}389.13\,\text{kJ/kW})} = 583\,\text{bhp}\ (434.92\,\text{kW})$$

Appendix A

SEED School Energy Audit Checklist

This section is designed as a guide to provide an energy management team with a convenient, organized listing of areas of energy waste and inefficiency most commonly found in school buildings.

While conducting a tour through the building, use each checklist INDICATOR in the order it appears here to help quickly identify conditions of energy misuse. If TRUE is checked, the INDICATOR implies good energy management; no remedial action is necessary. If, however, the INDICATOR is FALSE, then remedial action is required.

The first remedial action taken should be the LOWER COST CORRECTIONS, operational and maintenance (O&M) opportunities suggested in the center. Where applicable, these suggestions should be assigned to specific personnel and implemented immediately. By themselves they can reduce energy consumption by 30% to 50% in buildings where no improvements in energy efficiency have yet been made.

Finally, consider each applicable suggestion under HIGHER COST CORRECTIONS. These are redesign, retrofit, or energy conservation measures (ECMs), and will usually require capital expenditures for implementation. Professional architects and/or engineers should be consulted prior to their implementation. Payback and life-cycle cost analysis should also be conducted before any of these decisions are made in order to determine which will save the most energy dollars.

A. HUMAN SYSTEMS

PATTERNS OF USE INDICATOR	LOWER COST CORRECTIONS (O & M's)	HIGHER COST CORRECTIONS (ECM's)
#1. The change of seasons has been reflected in thermostat settings. TRUE ☐ FALSE ☐ NOTES: ____________	☐ Adjust thermostat settings so that room temperatures of occupied areas are 68°F in heating season and 78°F during cooling season.	☐ Replace existing thermostat with a thermostat which has a separate setting for cooling and a separate setting for heating, or use one thermostat to control heating and one thermostat to control cooling.

PATTERNS OF USE INDICATOR	LOWER COST CORRECTIONS (O & M's)	HIGHER COST CORRECTIONS (ECM's)
#2. Location of all thermostats provides for moderate temperature fluctuations. TRUE ☐ FALSE ☐ NOTES: ____________	☐ Move thermostats from areas next to windows, doors, heating or cooling units to areas which reflect temperatures of conditioned spaces more accurately. ☐ If thermostat cannot be moved, protect it from the source creating fluctuation.	☐ Relocate thermostat in HVAC return system, where possible.

A. HUMAN SYSTEMS

PATTERNS OF USE INDICATOR	LOWER COST CORRECTIONS (O & M's)	HIGHER COST CORRECTIONS (ECM's)
#3. Areas which are unoccupied or minimally used have modified heating/cooling temperature settings. TRUE ☐ FALSE ☐ NOTES: ______	☐ Turn off heating system or close heat register if nothing in space can freeze. ☐ If area has its own thermostat, reduce heat setting to 55°F. ☐ Turn off cooling system or close registers. ☐ Use infra-red spot heaters in large spaces with low occupancy.	☐ Reduce heating/cooling demands of unoccupied areas with appropriate control systems.

PATTERNS OF USE INDICATOR	LOWER COST CORRECTIONS (O & M's)	HIGHER COST CORRECTIONS (ECM's)
#4. Activities, including custodial services, not normally considered part of the school day, are scheduled to reflect wise energy management. TRUE ☐ FALSE ☐ NOTES: ______	☐ Reschedule activities to accommodate partial reduction of energy demand. ☐ Consolidate activities to reduce building usage. ☐ Reschedule cleaning and maintenance activities during daylight working hours where possible. ☐ Where rescheduling is not possible, use lights and equipment only in those areas where individuals are working.	☐ Consider installing an automated energy control system that will predetermine the environment of occupied spaces.

A. HUMAN SYSTEMS

PATTERNS OF USE INDICATOR	LOWER COST CORRECTIONS (O & M's)	HIGHER COST CORRECTIONS (ECM's)
#5. Temperatures have been adjusted to reflect temporary usage of lobbies, corridors, vestibules and other public places. TRUE ☐ FALSE ☐ NOTES: ______	☐ Reduce heat supply to these areas. ☐ Disconnect or turn off electric heaters in these areas unless it creates a freeze-up problem. ☐ Discontinue air conditioning of these areas.	None practical

PATTERNS OF USE INDICATOR	LOWER COST CORRECTIONS (O & M's)	HIGHER COST CORRECTIONS (ECM's)
# 6. Building temperatures have been reduced to reflect unoccupied periods. TRUE ☐ FALSE ☐ NOTES: ______	☐ Reduce thermostat settings to 50°F when building is unoccupied during heating season. Begin reduction of heating during last hour of occupancy. ☐ Turn off all air conditioning units when building is unoccupied. ☐ Experiment with heating "turn-on" times to determine length of pre-heat period required for satisfactory comfort levels of occupants. ☐ Install appropriate automatic controls; e.g., time clocks.	None practical

A. HUMAN SYSTEMS

OCCUPANT IMPACT INDICATOR	LOWER COST CORRECTIONS (O & M's)	HIGHER COST CORRECTIONS (ECM's)
#1. Thermostats are locked to eliminate unauthorized adjustment. TRUE ☐ FALSE ☐ NOTES: ________	☐ Install locking screws. ☐ Post signs displaying proper thermostat setting. ☐ Replace defective thermostat with non-adjustable pre-set thermostats. ☐ Install tamper-proof covers.	☐ Relocate thermostat in HVAC return system, where possible.

OCCUPANT IMPACT INDICATOR	LOWER COST CORRECTIONS (O & M's)	HIGHER COST CORRECTIONS (ECM's)
#2. Staff utilizes blinds, curtains and other window-covering devices to reduce energy usage. TRUE ☐ FALSE ☐ NOTES: ________	☐ Instruct staff on techniques of reducing energy consumption with blinds, curtains and other window-covering devices. ☐ Inform staff to use natural lighting and solar heat gain when appropriate. ☐ Repair damaged shading devices. ☐ Install devices where needed.	☐ Install reflective or heat absorbing films in areas with excessive heat gain in summer.

A. HUMAN SYSTEMS

OCCUPANT IMPACT INDICATOR	LOWER COST CORRECTIONS (O & M's)	HIGHER COST CORRECTIONS (ECM's)
#3. The school provides a suitable inservice training program for building staff in the importance and techniques of wise energy use. TRUE ☐ FALSE ☐ NOTES: ______	☐ Appoint interested individuals to develop and implement an energy management program for staff. ☐ Extend the training program to include students through appropriate instruction. ☐ Remind staff and students of good energy conservation techniques through the utilization of Energy Conservation Plaques.	None practical.

OCCUPANT IMPACT INDICATOR	LOWER COST CORRECTIONS (O & M's)	HIGHER COST CORRECTIONS (ECM's)
#4. Occupants turn off lights in unoccupied areas. TRUE ☐ FALSE ☐ NOTES: ______	☐ Instruct staff on importance of conservation of electrical energy. ☐ Utilize only necessary lighting in large areas such as libraries, gymnasiums, auditoriums and cafeterias. ☐ Color-code switches to indicate minimum levels of light needed for normal activities.	☐ Install automatic timing switches in areas used for short periods, such as storerooms and janitors' closets. ☐ Rewire switches so that one switch does not control all fixtures in multiple task areas.

A. HUMAN SYSTEMS

OCCUPANT IMPACT INDICATOR	LOWER COST CORRECTIONS (O & M's)	HIGHER COST CORRECTIONS (ECM's)
#5. Doors and windows remain closed while building is being heated or cooled. TRUE ☐ FALSE ☐ NOTES: ______	☐ Instruct staff to keep doors and windows closed while building air is being conditioned. ☐ Adjust door mechanism for faster closing. ☐ Remove device which allows doors to be locked open. ☐ Permanently seal unnecessary operable windows.	None practical.

B. STRUCTURAL SYSTEMS

DOORS INDICATORS	LOWER COST CORRECTIONS (O & M's)	HIGHER COST CORRECTIONS (ECM's)
#1. All exterior doors are aligned properly, fit tightly and operate efficiently. TRUE ☐ FALSE ☐ NOTES: ______	☐ Realign doors that do not close properly. ☐ Replace or readjust automatic door closing. ☐ Repair or replace threshholds and/or gaskets. ☐ Repair or replace weatherstripping.	☐ Consider installing vestibule doors at heavily used entrances. ☐ Install mechanisms which automatically close doors to unconditioned spaces. ☐ Consider installing removable center posts with proper gaskets for double doors with large gaps. ☐ Consider installing wind screens to protect exterior doors from direct blasts of prevailing winds. ☐ Insulate large exterior doors.

B. STRUCTURAL SYSTEMS

WINDOWS INDICATORS	LOWER COST CORRECTIONS (O & M's)	HIGHER COST CORRECTIONS (ECM's)
#1. All windows on exterior walls are aligned properly, fit tightly and operate effectively. TRUE ☐ FALSE ☐ NOTES: ______	☐ Realign windows that do not close properly. ☐ Permanently seal those which cannot be properly aligned. ☐ Caulk and weatherstrip windows. ☐ Replace any broken or cracked windows.	☐ Consider replacing faulty or non-essential windows with walls or insulated panels. ☐ Consider replacing deteriorated or badly fitting frames with new energy efficient models. ☐ Install wall or insulated panel if window is not essential.

WINDOWS INDICATOR	LOWER COST CORRECTIONS (O & M's)	HIGHER COST CORRECTIONS (ECM's)
#2. Exterior walls have little excessive glassed areas. TRUE ☐ FALSE ☐ NOTES: ______	☐ Insulate the upper portion of each window.	☐ Consider replacing windows with walls or insulated panels. ☐ Install double pane windows if cost-effective. ☐ Investigate other products such as reflective or heat absorbing film. (Be sure to anticipate maintenance problems.) ☐ Use thermopane windows (utilizing the same casings) when replacing windows.

B. STRUCTURAL SYSTEMS

WALLS INDICATOR	LOWER COST CORRECTIONS (O&M's)	HIGHER COST CORRECTIONS (ECM's)
#1. Penetrations in exterior walls and the joints where different wall materials join have been properly caulked. TRUE ☐ FALSE ☐ NOTES: ______	☐ Ensure that all areas of potential air infiltration are sealed, using quality caulking materials. ☐ Cover all window cooling units when not in use to prevent air leakage through the units.	None practical

WALLS INDICATOR	LOWER COST CORRECTIONS (O·& M's)	HIGHER COST CORRECTIONS (ECM's)
#2. Wall insulation is adequate for local weather conditions. TRUE ☐ FALSE ☐ NOTES: ______	None practical	☐ Consider adding insulation to increase the R-factor, when remodeling or when replacing segments of the exterior wall.

B. STRUCTURAL SYSTEMS

ROOFS INDICATOR	LOWER COST CORRECTIONS (O & M's)	HIGHER COST CORRECTIONS (ECM's)
1. Roof is adequately insulated for local weather conditions. TRUE ☐ FALSE ☐ NOTES: ______	☐ Ensure that the vapor barrier faces the conditioned space.	☐ Bring the current insulation level up to standard recommended for your climate. ☐ Replace any water-damaged insulation.

ROOFS INDICATOR	LOWER COST CORRECTIONS (O & M's)	HIGHER COST CORRECTIONS (ECM's)
#2. Skylights have been modified to reflect wise energy management. TRUE ☐ FALSE ☐ NOTES: ______	☐ Caulk and seal around entire skylight. ☐ Be certain glass or plastic is in good condition to prevent air infiltration and heat loss. ☐ Install an additional glass or plastic barrier to reduce heat loss. ☐ Insulate skylight with a translucent insulating material.	☐ Replace skylight with insulating materials.

C. LIGHTING SYSTEMS

INTERIOR INDICATOR	LOWER COST CORRECTIONS (O & M's)	HIGHER COST CORRECTIONS (ECM's)
#1. Incandescent lamps are used rarely except for decorative purposes. TRUE ☐ FALSE☐ NOTES:	☐ Replace burned-out incandescent lamps with lower wattage types, where possible. ☐ Consider replacing incandescent lamps with lower-wattage, self-ballasting mercury vapor bulbs in large areas such as gymnasiums, cafeterias and auditoriums. (Note: Be sure at least one incandescent bulb is left in to provide a safety light should the mercury vapor lamps be accidentally turned off.) ☐ Discontinue use of extended service lamps in locations where they can be replaced easily.	☐ Substitute energy conserving lamps for non-decorative incandescent lamps. The former provides more light for less wattage at lower cost.

INTERIOR INDICATOR	LOWER COST CORRECTIONS (O & M's)	HIGHER COST CORRECTIONS (ECM's)
#2. Wherever fluorescent lamps have been removed, the ballasts have been properly disconnected. TRUE ☐ FALSE ☐ NOTES:	☐ Remove or disconnect ballasts. (Note: Although they appear inactive, ballasts consume significant amounts of electricity even though the lamps have been removed.)	☐ Investigate replacing unnecessary tubes with models which draw a small current, yet provide a uniform lighting effect.

C. LIGHTING SYSTEMS

INTERIOR INDICATOR	LOWER COST CORRECTIONS (O & M's)	HIGHER COST CORRECTIONS (ECM's)
#3. The entire school has been relamped with energy efficient fluorescent tubes, and ballasts have been replaced as needed with energy conserving types. TRUE ☐ FALSE ☐ NOTES: ____	☐ Replace defective fluorescent tubes with more efficient and lower wattage types. ☐ Replace burned-out ballasts with more efficient, lower wattage models.	☐ Install more efficient, lower wattage fluorescent tubes in all fixtures.

INTERIOR INDICATOR	LOWER COST CORRECTIONS (O & M's)	HIGHER COST CORRECTIONS (ECM's)
#4. Proper delamping, the removal of unnecessary fluorescent tubes and incandescent bulbs, has been completed. TRUE ☐ FALSE ☐ NOTES: ____	☐ Do not replace defective fluorescent tubes in areas where delamping is feasible. In four-lamp fixtures allow two lamps to remain, disconnecting appropriate ballasts. ☐ Do not replace burned-out incandescent bulbs in areas where delamping is feasible.	☐ Consider lowering fixtures in order to increase illumination levels on task areas. Higher illumination levels will allow a delamping program. ☐ Consider removing entire fixtures in areas where illumination is excessive. ☐ Install three-tube fixtures which allow three levels of illumination for varying tasks, when remodeling.

C. LIGHTING SYSTEMS

INTERIOR INDICATOR	LOWER COST CORRECTIONS (O & M's)	HIGHER COST CORRECTIONS (ECM's)
#5. Routine cleaning of lamps, tubes and fixtures is part of the energy management program. TRUE ☐ FALSE ☐ NOTES: ______	☐ Clean lamps, tubes and fixtures regularly. ☐ Replace yellowed, cracked or defective light diffusers as needed.	None practical

INTERIOR INDICATOR	LOWER COST CORRECTIONS (O & M's)	HIGHER COST CORRECTIONS (ECM's)
#6. Natural lighting is optimized. TRUE ☐ FALSE ☐ NOTES: ______	☐ Clean walls regularly. ☐ Repaint with light reflective, non-glossy colors to enhance illumination. ☐ Clean windows on a routine basis. ☐ Use drapes, blinds and curtains to increase natural light gain.	None practical

C. LIGHTING SYSTEMS

EXTERIOR INDICATOR	**LOWER COST CORRECTIONS** (O & M's)	**HIGHER COST CORRECTIONS** (ECM's)
#1. Defective exterior lamps and/or fixtures have been replaced with more energy efficient types. TRUE ☐ FALSE ☐ NOTES: ________	☐ Replace 150-watt security flood lights with 75-watt lamps. ☐ Consider blacking-out building and grounds during unoccupied night time periods. (Note: Some school districts have had success with this program.) ☐ Consider using remotely monitored intrusion alarms.	☐ Replace defective exterior lamps with mercury vapor, metal halide or high-pressure sodium lamps.

EXTERIOR INDICATOR	**LOWER COST CORRECTIONS** (O & M's)	**HIGHER COST CORRECTIONS** (ECM's)
#2. Exterior lighting is automatically controlled. TRUE ☐ FALSE ☐ NOTES: ________	☐ Reduce hours of operation of exterior lights. ☐ Ensure proper functioning of photo-cell controls.	☐ Install automatic lighting controls if needed. ☐ Modify the control system to include photocell "turn-on" and time clock "turn-off".

C. LIGHTING SYSTEMS

EXTERIOR INDICATOR	**LOWER COST CORRECTIONS** (O & M's)	**HIGHER COST CORRECTIONS** (ECM's)
#3. Playing field lighting system has been upgraded to save energy dollars. TRUE ☐ FALSE ☐ NOTES: ______	☐ Clean lighting fixtures regularly. ☐ Maintain and refinish lighting reflectors, as required, to obtain maximum illumination. ☐ Use only lights required during night practice periods.	☐ Replace incandescent lamps with more efficient and effective models such as high-pressure sodium or metal halides. ☐ Install separate meter for playing field lights to reduce total electrical demand, if applicable. (Note: This procedure will not save energy but can save dollars.)

D. MECHANICAL SYSTEMS

HEATING INDICATOR	LOWER COST CORRECTIONS (O & M's)	HIGHER COST CORRECTIONS (ECM's)
#1. Multiple boilers or heaters have been modified to prevent simultaneous firing. TRUE ☐ FALSE ☐ NOTES: ______	☐ Adjust boiler or furnace controls so that unit #2 will not fire until unit #1 can no longer satisfy demand.	☐ Install automatic staging controls, if needed.

HEATING INDICATOR	LOWER COST CORRECTIONS (O & M's)	HIGHER COST CORRECTIONS (ECM's)
#2. Stack temperature is in normal range as verified by routine flue gas analysis. TRUE ☐ FALSE ☐ NOTES: ______	☐ Perform regular flue gas testing to ensure proper air to fuel ratio and make necessary corrections. (Example: Clean air intake filters; check that spuds and nozzles are properly sized and not clogged; ensure that fuel pressures are not excessive, and there is an adequate supply of combustion air.) ☐ Reduce the boiler's firing rate.	☐ Purchase kit for flue gas analysis, or digital flue gas analyzer.

D. MECHANICAL SYSTEMS

HEATING INDICATOR	LOWER COST CORRECTIONS (O & M's)	HIGHER COST CORRECTIONS (ECM's)
#3. Thermostat settings accurately reflect room temperatures. TRUE ☐ FALSE ☐ NOTES: ______	☐ Recalibrate thermostats and controllers. ☐ Bleed and clean pneumatic lines, if applicable. ☐ Clean contacts of electrical control system, if applicable. ☐ Ensure that control valves and dampers are modulated properly and that heat distribution to space is unobstructed.	☐ Investigate new types of thermostatic control devices for application in specific situations. Install suitable type.

D. MECHANICAL SYSTEMS

HEATING INDICATOR	LOWER COST CORRECTIONS (O & M's)	HIGHER COST CORRECTIONS (ECM's)
#4. Automatic controls schedule heating water temperature in accordance with outdoor temperature. TRUE ☐ FALSE ☐ NOTES: ______	☐ Check temperature of hot water in heating system. If it appears excessive during periods of mild weather: a. Ensure that reset controls are functioning properly. b. Experiment with hot water temperature reduction when an acceptable comfort level is attained. ☐ Turn off boiler, pumps or heat source during summer. ☐ If the building has been properly caulked, sealed, weatherstripped and insulated, turn the boilers off at night. Exercise caution if the outside temperature is expected to go below freezing. In addition, make certain that the boiler will tolerate being turned off.	☐ Install temperature controls which automatically program water temperature according to outdoor temperature and will turn off heating unit when outside temperature reaches 60°F.

D. MECHANICAL SYSTEMS

HEATING INDICATOR	LOWER COST CORRECTIONS (O & M's)	HIGHER COST CORRECTIONS (ECM's)
#5. Heating pilot lights are scheduled to be turned off during summer. TRUE ☐ FALSE ☐ NOTES ________	☐ Turn pilot(s) off on prescheduled date. ☐ Post a reminder in boiler/furnace room of pilot reactivation date.	☐ Replace pilot lights with new electronic ignition models whenever feasible.

HEATING INDICATOR	LOWER COST CORRECTIONS (O & M's)	HIGHER COST CORRECTIONS (ECM's)
#6. Steam radiators and other steam equipment are maintained regularly for proper functioning. TRUE ☐ FALSE ☐ NOTES: ________	☐ Replace defective air vent valves. ☐ Clean thermostatic control valves on radiators. Replace if necessary. ☐ Clean or replace bellows element in defective thermostatic traps. ☐ Check the temperature on the down stream side of steam traps. Unless the pipe is moderately hot (as hot as a water pipe), it is malfunctioning. Take appropriate action.	☐ Reclaim steam trap heat losses by diverting the steam-vacuum system exhaust through a potable hot water pre-heat tank.

D. MECHANICAL SYSTEMS

HEATING INDICATOR	LOWER COST CORRECTIONS (O & M's)	HIGHER COST CORRECTIONS (ECM's)
#7. Insulation on hot water pipes has been inspected, is adequate and is in good condition. TRUE ☐ FALSE ☐ NOTES: ____________	☐ Replace damaged or missing insulation.	☐ Install additional pipe insulation in accordance with good energy conservation practices.

HEATING INDICATOR	LOWER COST CORRECTIONS (O & M's)	HIGHER COST CORRECTIONS (ECM's)
#8. Oil burner is operating efficiently without excessive smoke or sooting. TRUE ☐ FALSE ☐ NOTES: ____________	☐ Check for proper oil pressure. Adjust as required. ☐ Confirm that oil is at proper temperature and free-flowing. ☐ Inspect burner nozzles for cleanliness and correct spray angles. ☐ Verify proper air to fuel ratio by routine flue gas analysis. ☐ Reduce firing rate.	☐ Purchase vent for flue gas analysis or a digital flue gas analyzer.

D. MECHANICAL SYSTEMS

HEATING INDICATOR	LOWER COST CORRECTIONS (O & M's)	HIGHER COST CORRECTIONS (ECM's)
#9. Routine maintenance is employed to insure efficient operating of heating units. TRUE ☐ FALSE ☐ NOTES: ______	Boilers: ☐ Remove soot from tubes during routine maintenance. ☐ Remove scale deposits, sediments and precipitates on water-side surfaces. (Note: Rear portion of boiler is most susceptible to scale formation.) ☐ Repair all damaged or worn boiler insulation, refractory, brickwork and boiler casings. Furnaces: ☐ Repair or replace solenoid valve if fire does not cut off immediately when unit shuts down. Both: ☐ Reset hot water temperature. Limit switch to a higher setting if burner short cycles. ☐ Reduce firing rate.	☐ Replace dangerous or ineffective units with more efficient models.

D. MECHANICAL SYSTEMS

HEATING INDICATOR	**LOWER COST CORRECTIONS** (O & M's)	**HIGHER COST CORRECTIONS** (ECM's)
#10. Hot water/steam radiation units are operating efficiently. TRUE ☐ FALSE ☐ NOTES: ____________	☐ Open air vents and bleed off air until water appears. ☐ Repair or replace faulty thermostats. ☐ Check pneumatic lines for proper functioning. ☐ Replace faulty valves. ☐ Check water pumps for proper operation. Make necessary adjustments. ☐ Check boiler for correct operating temperature. Correct as required. ☐ Remove objects which may be obstructing heating units.	☐ Clean heating elements.

D. MECHANICAL SYSTEMS

VENTILATION INDICATOR	LOWER COST CORRECTIONS (O & M's)	HIGHER COST CORRECTIONS (ECM's)
#1. A quantity of outdoor air which does not exceed code requirements is used to ventilate the building. TRUE ☐ FALSE ☐ NOTES:	☐ Reduce outdoor air quantity to the minimum allowed by codes by adjusting appropriate dampers. ☐ Be sure that outdoor air dampers are closed when building is unoccupied. ☐ Check all outdoor dampers for defective seals and for proper closure. Repair, if needed. ☐ Place pegboard over the outside of the fresh air inlet.	☐ Replace defective or damaged dampers with new opposed-blade models. ☐ Install automatic controls to close dampers when building is unoccupied.

VENTILATION INDICATOR	LOWER COST CORRECTIONS (O & M's)	HIGHER COST CORRECTIONS (ECM's)
#2. The ventilation systems are programmed to use natural cooling whenever possible. TRUE ☐ FALSE ☐ NOTES:	☐ Utilize outside air for cooling rather than refrigeration units whenever possible. ☐ Be sure the economizer cycle is operating properly.	☐ Install an economizer cycle with enthalpy control.

D. MECHANICAL SYSTEMS

VENTILATION INDICATOR	LOWER COST CORRECTIONS (O & M's)	HIGHER COST CORRECTIONS (ECM's)
#3. The building's exhaust system functions in accordance with occupancy patterns. TRUE ☐ FALSE ☐ NOTES: ______	☐ Disconnect unnecessary exhaust fans and cover grill to prevent conditioned air loss. ☐ Re-wire special exhaust fans to operate only when room is occupied. ☐ Schedule all other exhaust fans to operate only when needed. ☐ Program exhaust fans with time clocks or other controls.	☐ Install variable speed motors (when replacing exhaust fans) to modulate fan speed, allowing only required ventilation. ☐ Modify all exterior exhaust ducts with controlled or gravity dampers.

VENTILATION INDICATOR	LOWER COST CORRECTIONS (O & M's)	HIGHER COST CORRECTIONS (ECM's)
#4. Return, outdoor air and exhaust dampers are sequencing properly. TRUE ☐ FALSE ☐ NOTES: ______	☐ Check damper linkage. Adjust for proper closure. ☐ Readjust indicators to indicate damper positions.	☐ Replace defective or damaged damper with new opposed-blade models.

D. MECHANICAL SYSTEMS

VENTILATION INDICATOR	LOWER COST CORRECTIONS (O & M's)	HIGHER COST CORRECTIONS (ECM's)
#5. Temperature of the air entering through the ducts during the heating season feels comfortably warm. TRUE ☐ FALSE ☐ NOTES: ____________	☐ Reduce air volume to eliminate "draft effect". ☐ Raise supply temperature to 65°F in perimeter zones (60°F in interior zones) during heating season only.	None practical

VENTILATION INDICATOR	LOWER COST CORRECTIONS (O & M's)	HIGHER COST CORRECTIONS (ECM's)
#6. The air flow is well-balanced and consistent throughout the building. TRUE ☐ FALSE ☐ NOTES: ____________	☐ Check air filters. Clean or replace regularly. ☐ Remove objects obstructing air flow; clean diffusers, registers and grilles. ☐ Balance air flow system, if required.	None practical

D. MECHANICAL SYSTEMS

AIR CONDITIONING INDICATOR	LOWER COST CORRECTIONS (O & M's)	HIGHER COST CORRECTIONS (ECM's)
#1. Thermostat settings accurately reflect room temperatures. TRUE ☐ FALSE ☐ NOTES: ______	☐ Restrict outdoor air intake when not using economizer cycle. ☐ Insure that control dampers and valves (especially the economizer cycle) are working properly. ☐ Calibrate thermostats and controllers. ☐ Clean thermostatic controls.	☐ Investigate new types of thermostatic control devices for application in specific situations. Install suitable type.

AIR CONDITIONING INDICATOR	LOWER COST CORRECTIONS (O & M's)	HIGHER COST CORRECTIONS (ECM's)
#2. Zone temperatures are maintained without the use of reheat coils. TRUE ☐ FALSE ☐ NOTES: ______	☐ Determine if temperatures remain in comfort zone when boilers are shut down during cooling.	☐ Convert to variable air volume system, if practical.

D. MECHANICAL SYSTEMS

AIR CONDITIONING INDICATOR	LOWER COST CORRECTIONS (O & M's)	HIGHER COST CORRECTIONS (ECM's)
#3. Multiple air conditioning compressors have been modified to prevent simultaneous start-up.	☐ Adjust compressor controls so that unit #2 will not start-up until unit #1 can no longer satisfy cooling demand.	☐ Install automatic staging controls, if needed.
TRUE ☐ FALSE ☐		
NOTES: ______		

AIR CONDITIONING INDICATOR	LOWER COST CORRECTIONS (O & M's)	HIGHER COST CORRECTIONS (ECM's)
#4. Insulation on cooling lines, pipes and ducts has been inspected, is adequate and is in good condition.	☐ Replace damaged or missing insulation.	☐ Install additional insulation on all delivery lines and ducts in accordance with good energy conservation practices.
TRUE ☐ FALSE ☐		
NOTES: ______		

D. MECHANICAL SYSTEMS

AIR CONDITIONING INDICATOR	LOWER COST CORRECTIONS (O & M's)	HIGHER COST CORRECTIONS (ECM's)
#5. Cool air in adequate volume is discharged into space. TRUE # FALSE ☐ NOTES: ____	☐ Clean condensor and evaporator coils, fins and tubes. Remove if necessary. ☐ Clean or replace air filters. ☐ Ensure that fire and balancing dampers are open and in correct positions. ☐ Verify that fan is rotating in proper direction.	None practical

AIR CONDITIONING INDICATOR	LOWER COST CORRECTIONS (O & M's)	HIGHER COST CORRECTIONS (ECM's)
#6. Refrigerated air is at recommended temperature levels. TRUE ☐ FALSE ☐ NOTES: ____	☐ Remove and clean the strainer if frost or sweat is visible at the strainer outlet. ☐ Inspect and clean all coils (including dehumidification coils) regularly. ☐ Clean dirty condensers. Dirty condensors increase system pressure, causing a decrease in system efficiency. ☐ Repair or adjust defective compressor valves. These can be identified by high discharge temperatures.	None practical.

D. MECHANICAL SYSTEMS

AIR CONDITIONING INDICATOR	LOWER COST CORRECTIONS (O & M's)	HIGHER COST CORRECTIONS (ECM's)
#7. Evidence indicates that chilled water piping, valves and fittings are intact.	☐ Repair all leaks. ☐ Check valves and fittings. Replace as required.	None practical.

TRUE ☐ FALSE ☐

NOTES: ______________________________

D. MECHANICAL SYSTEMS

AIR CONDITIONING INDICATOR	LOWER COST CORRECTIONS (O & M's)	HIGHER COST CORRECTIONS (ECM's)
#8. Refrigeration compressor is operating efficiently. TRUE ☐ FALSE ☐ NOTES: ______	☐ Check refrigerant charge level. Adjust to equipment specifications. Repair leaks. ☐ Check for faulty or fused electrical control circuits. Repair or replace as required. ☐ Ensure that liquid line solenoid valve is not leaking or stuck open. ☐ Inspect all compressor valves. Take corrective action if needed. ☐ Clean evaporation and condensor coils. Clean liquid line strainer if clogged. ☐ Reset high/low pressure control differential settings if needed.	None practical.

E. SPECIAL SYSTEMS

WATER INDICATOR	**LOWER COST CORRECTIONS** (O & M's)	**HIGHER COST CORRECTIONS** (ECM's)
#1. Hot water storage tanks, piping and heaters are operating efficiently. TRUE ☐ FALSE ☐ NOTES: ______	☐ Reset heater thermostat to 105°F-115°F or to local code requirement. This may require installing a "booster" for the kitchen. ☐ Turn off recirculating pump(s) when building is unoccupied. ☐ Repair all leaks. ☐ Replace damaged or missing insulation.	☐ Install adequate insulation on all lines and tanks containing hot water. ☐ Consider installing smaller domestic water heaters to maintain desired temperature in storage tanks. ☐ Install separate gas fired water heater for use in cafeteria and kitchen.

WATER INDICATOR	**LOWER COST CORRECTIONS** (O & M's)	**HIGHER COST CORRECTIONS** (ECM's)
#2. Heating cycle on electric water heater is restricted to low electrical demand periods. TRUE ☐ FALSE ☐ NOTES: ______	☐ Utilize heater's "vacation cycle" during extended vacation periods. ☐ Use a time clock or automatic controls to restrict the duty cycle.	None practical

E. SPECIAL SYSTEMS

WATER INDICATOR	LOWER COST CORRECTIONS (O & M's)	HIGHER COST CORRECTIONS (ECM's)
#3. Devices which restrict hot water usage have been utilized. TRUE ☐ FALSE ☐ NOTES:	☐ Install inexpensive flow restrictors in lines or faucets.	☐ Install mixing valves that predetermine water temperature. ☐ Re-plumb locker room showers to include a master valve to control shower time. ☐ Replace standard faucets with spring-loaded shut-off types. ☐ Consider installing, if practical, a solar water heater to assist in meeting hot water needs.

WATER INDICATOR	LOWER COST CORRECTIONS (O & M's)	HIGHER COST CORRECTIONS (ECM's)
#4. Swimming pool is energy efficient. TRUE ☐ FALSE ☐ NOTES:	☐ Maintain water temperature in maximum range of 80°F-84°F. ☐ Clean filters as required. ☐ Ensure that pool heating equipment is maintained regularly. ☐ Use a flue gas analysis to maintain proper air-to-fuel ratio. ☐ Experiment with ventilation/exhaust system to attain minimum air displacement and still meet code. ☐ Use a buoyant plastic sheet to cover water when not in use.	☐ Consider installing enclosure on outdoor pools. ☐ Investigate solar water heaters to maintain pool temperature.

E. SPECIAL SYSTEMS

KITCHEN CAFETERIA INDICATOR	**LOWER COST CORRECTIONS** (O & M's)	**HIGHER COST CORRECTIONS** (ECM's)
#1. Exhaust hoods and fans operate efficiently and only when needed. TRUE ☐ FALSE ☐ NOTES: ______	☐ Program operating of exhaust fan. ☐ Ensure that fans and hoods are properly sized to prevent excessive ventilation of other areas of building. Reduce fan speed if necessary.	☐ Investigate heat recovery options of exhausted air.

KITCHEN CAFETERIA INDICATOR	**LOWER COST CORRECTIONS** (O & M's)	**HIGHER COST CORRECTIONS** (ECM's)
#2. Food preparation equipment is used wisely. TRUE ☐ FALSE ☐ NOTES: ______	☐ Turn on electrical equipment only when needed; keep it off when not required. ☐ Clean refrigeration coils regularly. Be sure coils have sufficient air circulation space. ☐ Move refrigerator and/or freezer away from any heat source. ☐ Check (and replace, as required) gaskets on refrigerator, freezer and oven doors for proper seal. ☐ Keep refrigerators full.	☐ Install newer energy efficient models when replacing damaged or defective equipment.

E. SPECIAL SYSTEMS

CAFETERIA INDICATOR	LOWER COST CORRECTIONS (O & M's)	HIGHER COST CORRECTIONS (ECM's)
#3. Food service staff is cognizant of good energy conservation techniques. TRUE ☐ FALSE ☐ NOTES: ____	☐ Operate dishwasher only with full-loads. ☐ Train employees to conserve hot water. ☐ Cook with lids in place on pots and kettles. ☐ Thaw frozen foods in refrigerated compartments. ☐ Use fans to cool people not heat sources. ☐ Do not allow refrigerator/freezer doors to remain open. ☐ Avoid preheating ovens. (Studies indicate preheating is unnecessary even for baked goods, if food is allowed to remain in oven several minutes after oven is turned off.)	None practical.

E. SPECIAL SYSTEMS

LAUNDRY INDICATOR	**LOWER COST CORRECTIONS** (O & M's)	**HIGHER COST CORRECTIONS** (ECM's)
#1. Laundry functions are carried out in an energy efficient manner. TRUE ☐ FALSE ☐ NOTES: ______	☐ Train staff to keep lint filters in dryers and the exhaust hoods clean. ☐ Reschedule laundry operation to avoid electrical peak demand hours. ☐ Wash and dry full loads only. ☐ Develop concise operating instruction for each piece of equipment. Post Energy Conservation Plaques. ☐ Reduce washing and drying cycles. Studies indicate that accepted washing/drying times are excessive for most items.	☐ Investigate heat recovery options.

E. SPECIAL SYSTEMS

OFFICE MACHINES ELECTRICAL EQUIPMENT INDICATOR	LOWER COST CORRECTIONS (O & M's)	HIGHER COST CORRECTIONS (ECM's)
#1. Operation of electrical devices is carefully monitored. TRUE ☐ FALSE ☐ NOTES: ____	☐ Turn on office machines only when needed, e.g., copiers, coffee urns, typewriters. ☐ Use special high-demand equipment such as kilns and electric welders in low demand periods.	☐ Install a demand limiter to prevent excessive electrical load.

APPENDIX B

ENERGY DATA COLLECTION FORMS

This section organizes important school data according to the following categories:

A. General Administrative Information
B. Human Systems
C. Structural Systems
D. Energy Systems
   A. Lighting
   B. Mechanical
   C. Special Systems
E. Energy Consumption Summary
F. Solar and Renewable Resource Potential

A. GENERAL ADMINISTRATIVE INFORMATION

1. Name of Building or Complex: ____________________
   Owner: ____________________
   Public ____________________ Private ____________________
   Non-Profit ____________________ Indian Tribe ____________________
2. Building Category:
   ☐ Elementary ☐ Vocational
   ☐ Secondary ☐ LEA Admin.
   ☐ Junior College ☐ Other, Specify
   ☐ College or Univ. ____________________
3. Building Address: ____________________ City ____________________
   State ____________________ Zip ____________ Telephone Number ____________________
4. Year constructed: ____________________ Year of last major addition or modification: ____________________
5. Principal/Manager: ____________________ Telephone Number: ____________________
   Head Custodian/Operator: ____________________ Telephone Number: ____________________
6. Energy Management Coordinator designated: ☐ Yes ☐ No
7. Anticipated building modifications: ____________________
   ____
   ____
8. Previous energy audit completed: ☐ Yes ☐ No Specify. ____________________
   ____
9. Conservation Measures (retrofit) already implemented or under consideration:
   Yes No Specify project, cost and expected energy savings: ____________________
   ____
   ____
10. Previous architectural/engineering studies: ☐ Yes ☐ No Specify: ____________________
   ____

B. HUMAN SYSTEMS

1. Complete occupancy schedule. If the school operates on a seasonal schedule, or has other periods of at least a week's duration when the building is only partially occupied, the number of weeks partial use by calendar quarter should be entered, along with the approximate percentage of total gross square feet in use during such periods.
2. Note U.S. heating and cooling zone in which building is located (see maps below) and climate information unique to your location. Call your utility or local weather bureau.

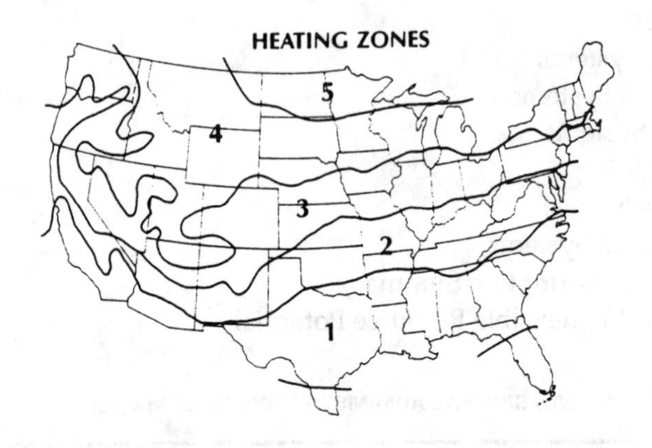

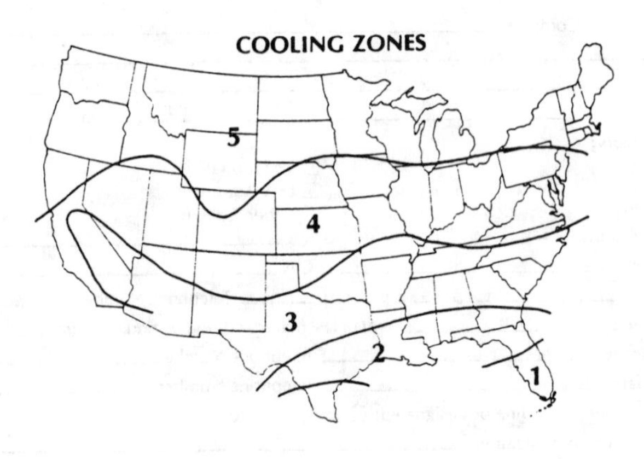

B. HUMAN SYSTEMS

Day(s)	Time Period	Average Occupancy	or % gsf Occupied	No. of Hours	Weeks/Year
Mon-Fri	day				
	evening				
	night				
Saturday	day				
	evening				
	night				
Sunday	day				
	evening				
	night				

Quarterly Partial Usage:

Quarter	Weeks	% Gross Square Feet (gsf)
1st		
2nd		
3rd		
4th		

2. Building Location:

U.S. Heating Zone # ______________ U.S. Cooling Zone # ______________

Annual Heating Degree Days ______________ Annual Cooling Degree Days ______________
(Base 65°F)

C. STRUCTURAL SYSTEMS

1. To calculate gross square feet (gsf) multiply the outside dimensions or measure from the centerline of common walls and multiply by the number of floors. If the building has wings, or the number of floors varies in one part from another, divide the building into sections, calculate the area of each section, and total. (Deduct from the total area any parking garages or other areas which are neither heated nor cooled.)

2-3. Note type and thickness of insulation material in roof, walls and floors. Note if there is none.

4. Sketch position of facility on site.
5. Briefly describe general building conditions. Indicate any structural flaws or deterioration, if applicable. If building is in good condition, so state.

1. Gross Floor Area: ______________________________ Gross Sq. Ft.
2. Insulation Type: Roof __________ Wall __________ Floor __________
3. Insulation Thickness: Roof __________ Wall __________ Floor __________
4. What is orientation of building on site? (Draw sketch.)

N

W — E

S

5. Description of general building conditions:

____

____

____

D. ENERGY SYSTEMS

A. Lighting

1-2. For each lighting type, fluorescent and incandescent, note the percentage of gross square feet of the building illuminated. Include an estimate of average usage in hours per week and hours per year.

3. Determine the total wattage presently used to illuminate the building's interior and divide by the gross floor area to compute an average lighting level in watts per square foot.

4. Indicate potential for lower lighting levels: classrooms, corridors, offices, gymnasium(s), auditorium(s), library, cafeteria, shops, others.

1. Fluorescent:
   Percentage of Gross Sq.Ft: ____________% Usage: ____________ Hr/Wk ____________ Hr/Yr
2. Incandescent:
   Percentage of Gross Sq.Ft: ____________ % Usage: ____________ Hr/Wk ____________ Hr/Yr
3. Average Interior Lighting Level: ____________________________ Watts/SqFt
4. Potential for lower lighting levels (check box if yes):

☐ Classrooms
☐ Corridors
☐ Offices
☐ Gymnasiums

☐ Auditoriums
☐ Library
☐ Cafeteria
☐ Shop(s)

☐ Other (Specify)
____
____

D. ENERGY SYSTEMS

B. Mechanical

1-10. Check the type and capacity of HVAC systems found in your school. If knowledge of the system is not available, obtain the information from the mechanical engineer, blueprints, specifications or nameplates. Total the cubic feet per minute (CFM) of air that the air systems supply to the building. Note what percentage of outside air is used. Note the heating and cooling capacities and fan horsepower.

11. Note the principal fuels used by the heating and cooling systems.

12-13. Larger buildings tend to have boilers or purchase hot water or steam. Small buildings tend to have unitary direct fired equipment. Determine your system type. Estimate the number of hours per day that the heating plant operates during winter. Examine the cooling equipment to determine system(s) type. Estimate the number of hours per day that the cooling system(s) operates during summer.

System Type	Total CFM	Minimum % Outside Air	Capacity BTU/hr	Fan Horsepower
1. Terminal Reheat				
2. Multizone				
3. Dual duct				
4. Variable Air Volume				
5. Induction				
6. Fan Coil	N/A			
7. Heat Pump				
8. Air Exhaust		N/A	N/A	
9. Radiation	N/A	N/A		N/A
10. Other (Specify) ____________				

11. Heating System

____

Fuel Type

12. Systems Types: (Check ✓)

____ Boilers
____ Purchased water or steam
____ Unitary Direct Fired
____ Furnaces
____ Package Equipment

13. Operation Profile:

____ hrs/weekday
____ hrs/Saturday
____ hrs/Sunday

Cooling System

____

Fuel Type

System Types: (Check ✓)

____ Absorption
____ Electric Drive
____ Steam Turbine Drive
____ Water Cooled Packaged Unit
____ Air Cooled Packaged Unit

Operation Profile:

____ hrs/weekday
____ hrs/Saturday
____ hrs/Sunday

D. ENERGY SYSTEMS

C. Special Systems

1. Indicate water heating source.
2. Complete data for swimming pool, if applicable.
3. Are laundry services provided? Enter fuel type.
4. Are food services provided?
5. List type and number of office equipment.
6. List kind and wattage of special high electrical demand equipment; e.g., kilns, electric welders.

1. Domestic Water Heated By: Electricity ______ Natural Gas ________
   Other ______________________
   (Specify)
2. Swimming Pool:
   ☐ Indoor ☐ Outdoor
   Heating Fuel: ____________________
   Normal Water Temperature: ____________ °F
3. Laundry Services ☐ Yes ☐ No
   Fuel Type: Washers ________ Dryers ________
4. Food Services ☐ Yes ☐ No
   Major cooking fuel: ________________
5. Office Machines (kind and number of each)

   ____ #____________

   ____ #____________

   ____ #____________

   ____ #____________

6. High Demand Electrical Equipment (kind and wattage)

   ____ ____________ watts

   ____ ____________ watts

   ____ ____________ watts

   ____ ____________ watts

E. ENERGY CONSUMPTION SUMMARY

1. Complete fuel use summary for base year (or last 12 months if no base year has been established), using utility records and Energy Usage Forms in Section 2. (If no past records have been kept, call your utility.) Multiply by the conversion factors (as required by the *Federal Register*, Section 450.42(11), April 2, 1979) and enter the results in Column D as annual BTU consumption.

   Transfer annual cost for each fuel from the appropriate Energy Usage Form to Column E.

   Compute consumption in BTU's per gross square foot per year by dividing the total of the entries in Column D by (C-1) gross square feet.

   Compute energy dollars per gross square foot per year by dividing the total of the entries in Column E by gsf. (Obtain gross floor area from previous chart (C-1) and energy costs from Energy Usage Forms.)

   Notes:
   - No. 2 oil should include other distillate fuel oils.
   - No. 6 oil should include other residual fuel oils.
   - Use a standard engineering manual or conversion factors provided by the State for other fuels (Row 8, Column C).

2. Based on your past year's utility bills, complete peak electrical demand data. For buildings or complexes over 200,000 gsf or if the electric rate contains a demand charge, determine if demand is recorded. Note times at which typical peaks occur during daily operation. Also note whether demand fluctuates on a seasonal basis, indicating month in which the highest demand occurs.

3. *If data is available*, indicate the fuel used by each of the major energy using systems listed and the annual consumption of each. For #5 "special", indicate special purpose facilities (e.g., food service, laundry) which use significant amounts of energy, fuel type used and annual usage.

E. ENERGY CONSUMPTION SUMMARY

A	B	C	D	E	F	G
Fuel	Previous 12 Month Totals	Conversion Factor	BTU's Consumed	Annual Cost	BTU's/Gross SqFt/Year	$/Gross SqFt/Yr
1. Electricity	KWH x	11,600 =				
2. Natural Gas	CCF x	103,000 =				
3. #2 Oil	gallons x	138,690 =				
4. #6 Oil	gallons x	149,690 =				
5. Steam	pounds x	1,390 =				
6. Coal	tons x	24,500,000 =				
7. Propane	gallons x	95,475 =				
8. Other, Specify	x	=				
9. TOTALS	N/A	N/A				

2. Peak Electrical demand:

Daily: __________KW   Annual: __________ KW

Time: __________   Month: __________

3. Fuel Use by Major Energy-Using Systems:

SYSTEM	FUEL TYPE	ANNUAL USE
Heating	__________	__________
Cooling	__________	__________
Hot Water	__________	__________
Lighting	__________	__________
Special, specify	__________	__________

F. SOLAR AND RENEWABLE RESOURCE POTENTIAL

1. Check characteristics of adjacent property.
2. Indicate nature of location.
3. Indicate building shape as square, rectangular, E-shaped, H-shaped, L-shaped or attach a rough sketch of the configuration. Note whether roof and southern wall are shaded or unshaded.
4. Note roof design. For the orientation of a pitched roof, indicate the *compass direction* of a line perpendicular to the ridgeline in the direction of the down slope. Note presence of roof obstructions such as chimneys, space conditioning equipment, water towers, mechanical rooms and stairwells. Identify the principal structural material of the roof; e.g., steel, concrete, or wood structural components. Also identify the type of roofing such as shingle, slate or built-up.
5. Indicate structure composition of southern facing wall. Determine percentage of wall area covered by glass.
6. Check type and location of space and water heating equipment.
7. Using information from the National Weather Service or your State's energy office enter monthly average solar insolation and wind speeds.
8. Note any special conditions or characteristics related to potential for solar or other renewable resource application.

APPENDIX C

INVENTORY FORMS

APPENDIX A. Inventory Forms

SYSTEM NO.	SYSTEM DESCRIPTION	IP/KW	PE		AMPS		
			EFF		VOLTS		
DAY TYPE __ OF __							

	AM												PM											
ORIG.	1	2	3	4	5	6	7	8	9	10	11	12	1	2	3	4	5	6	7	8	9	10	11	12
OPER.																								
% LOAD																								
KWH																								
REV.																								
OPER.																								
% LOAD																								
KWH																								
ENERGY TOTALS																								
OPER.																								
% LOAD																								
KWH																								

DAILY ENERGY USAGE	DAILY ENERGY SAVINGS	ANNUAL ENERGY USAGE	ANNUAL ENERGY SAVINGS

	DATE:	METER READINGS						
	HR.	MONDAY	TUESDAY	WEDNESDAY	THURSDAY	FRIDAY	SATURDAY	SUNDAY
AM	1							
	2							
	3							
	4							
	5							
	6							
	7							
	8							
	9							
	10							
	11							
	12							
PM	1							
	2							
	3							
	4							
	5							
	6							
	7							
	8							
	9							
	10							
	11							
	12							

INDUSTRIAL EQUIPMENT DATA

SHOP__________ ROOM__________ EQUIP. NO.__________

I. Electrical Data:

H.P.__________ ETF__________

Voltage__________ Phase__________ Hz__________

KVA Input__________ P.F.__________ KVA Output__________

II. HVAC Data:

1) Total KW__________ x 3.415 = __________ MBTUH
EFT x __________

2) Space KW__________ x 3.415 = __________ MBTUH

3) Spec. Air Cooling__________ cfm(1.08)_____ *TD- __________ MBTUH
1,000

4) Ind. Water__________ gpm(.5)_____ TD- __________ MBTUH

5) Net Heat to Space = (2-3-4) __________ MBTUH

6) Heat transfered to space No.__________ (1-2)__________ MBTUH

7) Estimated Hours Operated Daily (Diversity)__________ HRS

III. Industrial Equipment Data:

1) Ind. Cooling__________ gpm, __________ EWT, __________ LWT

2) Ind. Gases - Type__________, __________ SCFM __________ PSIG

3) Compressed Air - __________ Peak SCFM, __________ Steady SCFM
__________ PSIG

4) Steam - __________ Peak lbs/hr, __________ Steady lbs/hr
__________ PSIG

5) Misc. Req.__________

6) Special Treatment:
   a) Ventilation - Scrubbing __________
   b) Industrial Waste Treatment __________

IV. Plumbing - Industrial Waste:

1) Potable Water __________ gpm

2) Sanitary Waste
   a) Hub Drains __________ size
   b) Floor Drains __________ size

3) Industrial Waste
   a) Hub Drains __________ size
   b) Floor Drains __________ size
   c) Flow __________ gph

*Indicate Entering and Leaving Air Temperature

CAMPUS TOTAL INFORMATION

(All information from previous fiscal years) Annual electrical utilities cost previous 5 years______________

Monthly electrical cost immediate past year electrical service contract

Annual Natural gas utilities cost previous 5 years__________________

Monthly gas cost immediate past year gas service contract.

Total maintenance budget previous 5 years.

Total cost of sub-contracted work

Present energy conservation program

Answers

_____ Yes

_____ No

_____ Does not apply

_____ Information presently not available'

_____ Approximate value

GENERAL INSTRUCTIONS

1. Fill out "Building description Form" for each building as listed in study scope.
2. Select the appropriate equipment form for each type of equipment used in the building. Use as many equipment forms as there are pieces of equipment in use.
3. Add additional information on supplementary information sheets as necessary.

LIGHTING SYSTEMS

Building_______________________

Fluorescent__________, Wattage_________, No. of Units_______

Lighting task____

Operation - ____AM to_____PM, Continuous__________

Incandescent________Wattage__________, No. of Units________

Lighting task____

Operation______AM to _______PM, Continuous

Fluorescent________, Wattage________, No. of Units__________

Lighting task____

Operation _______AM to _______PM, Continuous________

Incandescent______________Wattage__________, No. of Units_______

Lighting task____

Operation ______AM to _______PM, Continuous__________

Fluorescent______, Wattage_______, No. of Units_________

Lighting Task____________________________________

Operation _____AM to _____PM, Continuous______

Fluorescent______, Wattage_______, No. of Units______

Lighting Task____________________________________

Operation ____AM to ____PM, Continuous

Period for Lamp replacement______________

Period for Cleaning lenses and bulbs_________

MULTIZONE AHU SYSTEM

Building Served_______________, AHU NO. _______________

MFR. Name_____________________ Model No._________________, Age_____________

Design Total Supply CFM_______________________, SP______________________

Design Total Outside Air CFM_____________________

Motor Nameplate HP_____________ Volt______________ Phase_____________

Motor Nameplates Amps__________________

Actual Amps L_1_________________, L_2_________________ L_3________________

Actual Voltage L_1_______________, L_2_________________ L_3_________________

No. of Zones____________________________

Type Damper controls - Pneumatic______________, Electric______________

Hot Deck Reset___________, Cold Deck Reset_______________

Type Cooling Coil - Dx_______________, CHW______________

Type Heating - Steam______________, Hot water_____________, Electric__________

Preheat coil____________

Economizer cycle_________________

Humidity control_________________

Previous operation problems:

Control failures__________, Zones_____________, Coils______________

Zones too cold in winter___________________

Zones too hot in summer______________________

Zones hot and cold alternately_________________

Present operation:

Manual start_________, Time Clock start______________

Operating hours__________AM to __________PM, Continuous_______________

Week-end operation required____________________

SINGLE ZONE AHU SYSTEM

Building Served ____________________, AHU No. ____________

MFR. Name ________________ Model No. ________________ Age______________

Design Total Supply CFM ______________ S.P.______________

Design Total Outside Air CFM ____________________

Motor Nameplate Amps________________

Actual Amps, L_1________________ L_2________________ L_3________________

Actual Voltage L_1________________ L_2________________ L_3________________

Single Zone__________, Terminal ReHeat____________

Type Cooling Coil - Dx__________, CHW ____________

Type Heating - Steam _________________, Hot Water____________, Electric__________

Preheat Coil______________

Economizer Cycle_________________

Humidity Control___________________

Previous operating problems:______________________

Control Failures_____________, Coils_______________

Space too Cold in winter_________________

Space Too Hot in Summer__________________

Space Hot and Cold Alternately________________

Present Operation:

Manual Start____________, Time Clock Start_____________

Operating Hours____________AM to __________PM, Continuous_____________

Week-end Operation Required__________________

DOUBLE DUCT AHU SYSTEM

Building Served_______________ AHU No._______________

MFR. Name_______________ Model No._______________ Age_______

Design Total Supply CFM_______________, SP_______________

Design Total Outside Air CFM_______________

Motor nameplate HP_______________ Volt_______________ Phase_______________

Motor nameplate amps_______________

Actual amps L_1_______________, L_2_______________ L_3_______________

Actual voltage L_1_______________ L2_______________ L_3_______________

No. of Mixing Boxes_______________

Type cooling coil - Dx_______________, CHW_______________

Type heating - Steam_______________, Hot water_______________, electirc_______________

Hot Deck Temp. Setting_______________ ^oF

Cold Deck Temp. Setting_______________ ^oF

Hot Deck Reset_______________

Economizer Cycle_______________

Preheat Coil_______________

Humidity Control_______________

Previous Operating Problems:

Control Failures_______________, Coils_______________, Air_______________

Spaces Too Hot in Summer_______________

Spaces Too Cold in Winter_______________

Spaces Hot and Cold Alternately_______________

Present Operation:

Manual Start_______________, Time Clock Start_______________

Operating Hours_______________ AM to _______________ PM, Continuous_______________

Week-end Operation Required_______________

FAN COIL UNITS

Building Served________________________ No. of Units________________

MFR. Name____________________ Model No.____________ Age_______

Design total supply CFM____________________, S.P.________________

Design Total outside air CFM____________________

Motor HP or watts________________, Volts______________

Type Control:

Fan speed_________, Coil Modulating Valves______________

Electric___________, Pneumatic____________

Previous operating problems:

Control Failures_______________. Coil Valves_____________

Control Signal___________, Fan Speed Switch______________

Space Too Hot in summer____________

Space Too Cold in Winter_____________

Space Hot and Cold Alternately_______________

Present Operation:

Operating hours________AM to ________PM, Continuous__________

Week-end Operation Required______________

HEATING-VENTILATION UNIT

Building Served________________, AHU No.____________

MFR. Name_____________, Model No._______________, Age___________

Design Total Supply CFM________________, S.P.____________

Design Outside Air CFM___________

Motor Nameplate HP___________, Volt__________, Phase______________

Motor Nameplate Amps_____________

Actual Amps L_1___________, L_2______________, L_3____________

Actual Voltage L_1___________, L_2____________, L_3_____________

Type Heating - Steam___________, Hot Water________, Electric_________

Min-max outside air dampers__________

Previous Operating Problems:

Control failures__________, Coils_______, Air________

Space to cold in winter__________

Space to hot in summer___________

Present Operation:

Manual Start_______, Time Clock Start__________

Operating Hours___________AM to _____PM, Continuous_________

Week-end Operation required____________

VENTILATION FAN UNIT

Building Served________________ Fan NO.___________

Space Served_______________

MFR Name____________________ Model No.___________ Age_____

Motor nameplate HP_________, Volt_________, Phase_______

Actual Amps L_1________, L_2_________ L_3______________

Type Drive - Belt__________, Direct___________

Type Blades - Propellar_________, Centrifugal_________

Present Operation:

Manual Start__________, Time Clock Start______________

Operating Hours__________AM to ______PM, Continuous__________

Week-end operation required________________

EXHAUST FAN SYSTEMS

Building Served____________________, Exhaust Fan No.__________

System Served________________

MFR. Name__________________, Model NO.______________ Age________

Design Total CFM______________, Static Pressure_____________

Motor Nameplate HP___________, Volts__________, Phase__________

Actual motor Amps L_1__________, L_2__________, L_3___________

Type Drive - Belt____________, Direct______________

Present Operation:

Manual Start__________, Time Clock______________

Operating Hours________AM to _________PM, Continuous___________

Week-end Operation Required__________________

COOLING TOWERS

Building served______________ Unit NO.________

System served_______________ Age___________

MFR.____________, Model No. _____________

Fan HP_________, Volts__________, Phase________

Motor nameplate amps____________

Actual amps L_1_________, L_2_________, L_3__________

Control temperature setting _______°F

Control method:

Fan sequenced__________, Water by-pass_______

Present condition:

Poor______, Fair_____, Good_______

CENTRIFUGAL WATER CHILLERS

Building name______________ Unit No.__________

Tons____________ Age____________

MFR.name_____________________________ Model No.______________

Compressor KW_____________, Volt__________, Phase____________

Hot gas by-pass option_______________

Load limiter option________________

Chilled water control setting_______________

Present Operation:

Operating hours________AM to _______PM Continuous____________

Seasonal Operation - months per year__________

WATER COOLED RECIPROCATING CHILLERS

Building Served________________________ Unit No.__________

Mechanical system served__________________

MFR. Name____________________________ Model No.________

Tons__________ Age__________

No. of compressors________

Compressor KW________. Voltage________, Phase__________

Condenser water pump HP________, Volt________, Phase________

Condenser water pump amps L_1______, L_2________, L_3________

Operating hours______AM to ______PM, Continuous

Manual start__________, Time clock start________

Chilled water pump HP______, Volt________, Phase________

Chilled water pumps amps L_1_______, L_2______, L_3________

Operating hours ______AM to ______PM, Continuous

Manual start________, Time clock start__________

Chilled water temp. setting__________

AIR COOLED COMPRESSOR UNITS

Building Served________________________ Unit No.________

Mechanical system served_______________

MFR. Name_____________________________ Model No._______

Tons_________, Age__________

No. of Compressors__________

Compressor KW_________, Voltage_________, Phase__________

BOILER SYSTEMS

Building Served________________ Unit No.________

MFR. Name________________ Model________

Rated Capacity________

Type fuel - Gas______, Oil______, Electric________

Type boiler - Cast iron________, Fire tube________, Water Tube________

Type generation - steam______, Hot water______, Pressure______

Type burner - ATM________, Forced draft________

Forced draft data:

Motor HP________, Volts______, Phase________

Operating Conditions:

Age______, Poor______, Fair______, Good______

Seasonal Operation:

Continuous______, No. Months________

BOILER SYSTEMS

Building Served______________________ Unit No.__________

MFR. Name____________________________ Model_______________

Rated Capacity_______________

Type fuel - Gas______, Oil_______, Electric__________

Type boiler - Cast iron_________, Fire tube_________, Water Tube________

Type generation - steam_______, Hot water_______, Pressure________

Type burner - ATM_________, Forced draft__________

Forced draft data:

Motor HP________, Volts________, Phase__________

Operating Conditions:

Age________, Poor______, Fair______, Good________

Seasonal Operation:

Continuous________, No. Months__________

DOMESTIC HOT WATER SYSTEMS

Building served__________ Unit No.________

Type: Gas______, Steam________, Hot water_______

MFR.____________________, Model______________, Age______

Circulator pump HP________, Voltage_____Phase_________

Pump operation Continuous________, Week-ends______

Condition: Poor______, Fair_____, Good______

Temperature setting________ °F

Primary Service:

Rest Rooms_______

Showers_________

Kitchens_________

Laboratories______

Dormatories________

BUILDING DESCRIPTION FORM

Building Name__________________________ No.________

Building Age __________

Primary Occupancy Use______________________________

____

Floor Area__________ Number of Floors________

%Glass in Walls, N______, E______, S______, W______

Operation Schedule:

Week day______AM to ______P M, No. Per Year________

Week-end______AM to ______PM No. Per Year________

Holiday______AM to ______PM No Per Year________

Approximate No. of peiple at max. occupancy____________

Requirements For Continual Temperature Control.

a. Laboratories ______Temp. R.H.______%

b. Computers ______Temp. R.H.______%

c. Occupancy ______Temp. R.H.______%

d. Material Storage ______Temp. R.H.______%

e. Security Office ______Temp. R.H.______%

Type Air Handling Systems:

Multizone________, Single Zone________, Fan Coil Unit______

Double Duct______, Variable Volume______, Terminal Reheat________

Packaged Unitary or Roof Top__________

Type Heating-Cooling Service:

Air Cooled Dx______________ Water Cooled Dx____________

CHW - From Central Plant________________, Local Chiller_____________

Steam from Central Plant____________________

Hot Water Boiler_____________, Electric_____________, Gas_________

Electric Heating_____________, Gas Furnace_______________________

Lighting - Fluorescent____________, Incandescent___________, Other________

Available Building Information:

Plans and specifications___________________

Separate metering - Gas_________, Electricity_________, Steam__________

No. of Elevators - Escalators________________

Building Distribution Voltage_______________ Phase_____________, Wires______

Index